Edited by
David J. Constable and
Concepción Jiménez-
González

Handbook of Green Chemistry

Volume 11
Green Metrics

Related Titles

Marteel-Parrish, A., Abraham, M.A.

Green Chemistry and Engineering
A Pathway to Sustainability

2014
Print ISBN: 978-0-470-41326-5 (Also available in a variety of electronic formats)

Reniers, G.L., Sörensen, K., Vrancken, K. (eds.)

Management Principles of Sustainable Industrial Chemistry
Theories, Concepts and Industrial Examples for Achieving Sustainable Chemical Products and Processes from a Non-Technological Viewpoint

2013
Print ISBN: 978-3-527-33099-7 (Also available in a variety of electronic formats)

Greene, J.P.

Sustainable Plastics
Environmental Assessments of Biobased, Biodegradable, and Recycled Plastics

2014
Print ISBN: 978-1-118-10481-1 (Also available in a variety of electronic formats)

Rajagopal, R.

Sustainable Value Creation in the Fine and Speciality Chemicals Industry

2014
Print ISBN: 978-1-118-53967-5 (Also available in a variety of electronic formats)

Centi, G., Perathoner, S. (eds.)

Green Carbon Dioxide
Advances in CO2 Utilization

2014
Print ISBN: 978-1-118-59088-1 (Also available in a variety of electronic formats)

Edited by David J. Constable and Concepción Jiménez-González

Handbook of Green Chemistry

Volume 11: Green Metrics

Verlag GmbH & Co. KGaA

The Editor

Prof. Paul T. Anastas
Yale University
Center for Green Chemistry & Green Engineering
225 Prospect Street
New Haven, CT 06520
USA

Volume Editors

Dr. David J. Constable
American Chemical Society
Green Chemistry Institute
1155 Sixteenth St., N.W.
Washington, DC 20036
USA

Dr. Concepción Jiménez-González
GlaxoSmithKline
Corporate Environment, Health and Safety
2245 Plum Frost Dr.
Raleigh, NC 27603
USA

Cover

The cover picture contains images from Corbis Digital Stock (Dictionary) and PhotoDisc, Inc./Getty Images (Flask containing a blue liquid).

All books published by **Wiley-VCH** are carefully produced. Nevertheless, authors, editors, and publisher do not warrant the information contained in these books, including this book, to be free of errors. Readers are advised to keep in mind that statements, data, illustrations, procedural details or other items may inadvertently be inaccurate.

Library of Congress Card No.: applied for

British Library Cataloguing-in-Publication Data
A catalogue record for this book is available from the British Library.

Bibliographic information published by the Deutsche Nationalbibliothek
The Deutsche Nationalbibliothek lists this publication in the Deutsche Nationalbibliografie; detailed bibliographic data are available on the Internet at <http://dnb.d-nb.de>.

© 2018 Wiley-VCH Verlag GmbH & Co. KGaA, Boschstr. 12, 69469 Weinheim, Germany

All rights reserved (including those of translation into other languages). No part of this book may be reproduced in any form – by photoprinting, microfilm, or any other means – nor transmitted or translated into a machine language without written permission from the publishers. Registered names, trademarks, etc. used in this book, even when not specifically marked as such, are not to be considered unprotected by law.

Print ISBN: 978-3-527-32644-0
ePDF ISBN: 978-3-527-69526-3
ePub ISBN: 978-3-527-69527-0
Mobi ISBN: 978-3-527-69525-6

Cover Design Adam-Design, Weinheim, Germany
Typesetting Thomson Digital, Noida, India
Printing and Binding betz-druck GmbH, Darmstadt, Germany

Printed on acid-free paper
10 9 8 7 6 5 4 3 2 1

Contents

List of Contributors XI
Preface XIII

1 **Green Chemistry Metrics** *1*
David J.C. Constable
1.1 Introduction and General Considerations *1*
1.2 Feedstocks *5*
1.3 Chemicals *6*
1.3.1 Hazard and Risk *6*
1.4 General Chemistry Considerations and Chemistry Metrics *10*
1.5 Evolution of Green Chemistry Metrics *11*
1.6 Andraos: Tree Analysis *14*
1.7 Process Metrics *15*
1.8 Product Metrics *16*
1.9 Sustainability and Green Chemistry *17*
1.10 Making Decisions *18*
References *19*

2 **Expanding Rational Molecular Design beyond Pharma: Metrics to Guide Safer Chemical Design** *29*
Nicholas D. Anastas, John Leazer, Michael A. Gonzalez, and Stephen C. DeVito
2.1 Introduction to Safer Chemical Design *29*
2.2 Life Cycle Thinking *30*
2.2.1 Sustainability, Green Chemistry, and Green Engineering *30*
2.2.2 Life Cycle Considerations *31*
2.2.3 Life Cycle Assessment *32*
2.2.4 Chemical Process Sustainability Evaluation – Metrics *34*
2.3 Attributes of Chemicals of Good Character *36*
2.4 Tools for Characterizing the Attributes of Chemicals of Good Character *37*
2.4.1 Strive to Reduce or Eliminate the Use of Chemicals *40*
2.4.2 Maximize Biological and Use Potency and Efficacy *40*

2.4.3	Strive for Economic Efficiency 40
2.4.4	Limited Bioavailability 41
2.4.5	Limited Environmental Mobility 41
2.4.6	Design for Selective Reactivity: Toxicity 41
2.4.7	Minimize the Incorporation of Known Hazardous Functional Groups: Toxicophores and Isosteres 42
2.4.8	Minimize the Use of Toxic Solvents 42
2.4.9	Limited Persistence and Bioaccumulation 43
2.4.10	Quick Transformation to Innocuous Products 44
2.4.11	Avoid Extremes of pH 44
2.5	A Decision Framework 44
2.5.1	A Suggested Protocol for Approaching Safer Chemical Design 45
2.5.2	Alternatives and Chemical Risk Assessment 45
2.6	The Road Ahead: Training of a Twenty-First Century Chemist 46
	References 46

3	**Key Metrics to Inform Chemical Synthesis Route Design** 49
	John Andraos and Andrei Hent
3.1	Introduction 49
3.2	Material Efficiency Analysis for Synthesis Plans 50
3.3	Case Study I: Bortezomib 56
3.3.1	Millennium Pharmaceuticals' Process 59
3.3.2	Pharma-Sintez Process 62
3.3.3	Material Efficiency – Local and Express 64
3.3.4	Synthesis Strategy for Future Optimization 72
3.3.5	Summary 73
3.4	Case Study II: Aspirin 74
3.4.1	Reaction Network 74
3.4.2	Material Efficiency 76
3.4.3	Environmental and Safety–Hazard Impact 78
3.4.4	Input Energy 84
3.4.5	Case I 84
3.4.6	Case II 85
3.4.7	Case III 85
3.4.8	Case IV 85
3.4.9	Case V 86
3.4.10	Case VI 86
3.4.11	Concluding Remarks and Outlook for Improvements 88
	References 91

4	**Life Cycle Assessment** 95
	Concepción Jiménez-González
4.1	Introduction 95
4.2	The Evolution of Life Cycle Assessment 96
4.3	LCA Methodology at a Glance 97

4.3.1	Goal and Scope	*98*
4.3.2	Inventory Analysis	*98*
4.3.3	Impact Assessment	*99*
4.3.4	Interpretation	*99*
4.3.5	LCI/A Limitations	*100*
4.3.6	Critical Review	*101*
4.3.7	Streamlined Life Cycle Assessment	*102*
4.4	Measuring Greenness with LCI/A – Applications	*103*
4.4.1	Probing Case Studies	*103*
4.4.2	Chemical Route Comparison	*106*
4.4.3	Material Assessment	*109*
4.4.4	Product LCAs	*112*
4.4.5	Footprinting	*115*
4.5	Final Remarks	*117*
	References	*118*

5 Sustainable Design of Batch Processes *125*
Tânia Pinto-Varela and Ana Isabel Carvalho

5.1	Introduction	*125*
5.2	State of the Art	*126*
5.2.1	Design and Retrofit of Batch Processes	*127*
5.2.2	Sustainability Assessment	*131*
5.3	Framework for Design and Retrofitting in Batch Processes	*136*
5.3.1	Economic Assessment	*138*
5.3.2	Environmental Assessment	*139*
5.3.3	Social Assessment	*140*
5.3.4	Methodologies	*141*
5.4	Case Studies	*142*
5.4.1	Retrofit Sustainable Batch Design	*142*
5.4.2	Design of Batch Process	*147*
5.5	Conclusions	*150*
	References	*152*

6 Green Chemistry Metrics and Life Cycle Assessment for Microflow Continuous Processing *157*
Lihua Zhang, Qi Wang, and Volker Hessel

6.1	Introduction	*157*
6.1.1	Green Chemistry and Green Engineering in the Pharmaceutical Industry	*157*
6.1.2	Green Metrics and Life Cycle Assessment	*158*
6.1.3	Continuous Processing at Small Scale	*159*
6.2	Environmental Analysis through Green Chemistry Metrics and Life Cycle Assessment	*162*
6.2.1	Green Chemistry Metrics	*162*
6.2.2	Life Cycle Assessment (LCA)	*163*

6.3	Application of Green Chemistry Metrics and Life Cycle Assessment to Assess Microflow Processing	*163*
6.3.1	Use as Benchmarking Tool for Continuous versus Batch; at Lab and Production Scale	*164*
6.3.2	Use as Decision Support Tool for Single Innovation Drivers – Choice of Type of Microreactor and Type of a Catalyst (Including Use/Not Use)	*167*
6.3.2.1	Reaction Conditions of Batch Process and Continuous Microflow Process	*167*
6.3.2.2	SLCA Results	*168*
6.3.2.3	Economic Evaluation	*170*
6.3.2.4	Conclusions	*171*
6.3.3	Use as Decision Support Tool for Single Innovation Drivers – Solvent Choice and Role of Recycling	*171*
6.3.4	Use as Decision Support Tool for Bundled Innovation Drivers Such as Multifacetted Process Optimization versus Process Intensification	*174*
6.3.4.1	API Production Process at Sanofi	*174*
6.3.4.2	Process Alternatives for Optimization and Intensification	*174*
6.3.4.3	Ecological Profile Comparison of Crude Batch and Continuous Operation	*175*
6.3.4.4	Cost Analysis of Batch and Continuous Operation	*178*
6.3.4.5	Conclusions	*179*
6.3.5	Cascading Reactions Into a Microreactor Flow Network – Greenness of Multistep Reaction/Separation Integration	*179*
6.3.5.1	LCA Study for Single-Step Analyses in Batch and Flow	*181*
6.3.5.2	LCA Study for "Two-Reactor Network" Process Designs	*184*
6.3.5.3	LCA Study for "Three-Reaction Network" Process Designs	*184*
6.3.6	Use as Process-Design Guidance and Benchmarking Tool Against Conventional Processes	*186*
6.3.6.1	Process Simulation and CAPEX Cost Study	*188*
6.3.6.2	LCA for Continuous Flow Synthesis of ADA	*190*
6.3.6.3	LCA for Two-Step Conventional Synthesis of ADA	*191*
6.3.6.4	Complete LCA Picture	*191*
6.3.6.5	Green Metrics Compared for the Direct Microflow Route and Conventional Two-Step Route	*192*
6.3.6.6	Conclusions	*194*
6.4	Economic Analysis and Snapshot on Applications with Continuous Microflow Processing	*195*
6.4.1	Life Cycle Costing (LCC)	*195*
6.4.2	Snapshot on LCC Applications with Continuous Microflow Processing	*196*
6.5	Conclusions and Outlook	*199*
	References	*201*

7	**Benchmarking the Sustainability of Biocatalytic Processes** *207*	
	John M. Woodley	
7.1	Introduction *207*	
7.2	Biocatalytic Processes *207*	
7.3	Biocatalytic Process Design and Development *210*	
7.4	Sustainability of Biocatalytic Processes *210*	
7.5	Quantitative Measuring of the Sustainability of Biocatalytic Processes *212*	
7.6	Early Stage Sustainability Assessment *213*	
7.6.1	Evaluation of Route Feasibility *214*	
7.6.1.1	Atom Economy *214*	
7.6.1.2	Carbon Mass Efficiency *214*	
7.6.2	Evaluation of Biocatalyst and Reaction Development *215*	
7.6.2.1	Process Mass Intensity *215*	
7.6.2.2	Solvent Intensity *215*	
7.6.2.3	Water Intensity *216*	
7.6.2.4	E-factor *216*	
7.7	Benchmarking *216*	
7.7.1	Route Selection *216*	
7.7.2	Biocatalyst and Reaction Development *217*	
7.8	Examples *217*	
7.8.1	Biocatalytic Route to Atorvastatin *218*	
7.8.2	Biocatalytic Route to Sitagliptin *219*	
7.9	Future Perspectives *221*	
7.9.1	Process Development *221*	
7.9.2	Methodology *223*	
7.10	Concluding Remarks *224*	
	References *225*	
8	**How Chemical Hazard Assessment in Consumer Products Drives Green Chemistry** *231*	
	Lauren Heine and Margaret H. Whittaker	
8.1	Introduction *231*	
8.2	What Drives Consumer Product Companies to Look for Less Hazardous Chemical Ingredients *232*	
8.2.1	Chemical Substitution and Regrettable Substitution *233*	
8.2.2	Nonprofit Organization (NPO) Campaigns *235*	
8.2.3	Retailer Initiatives *237*	
8.2.4	State Initiatives *240*	
8.2.5	Consumer Product Sector Leaders: Setting the Example for Others *242*	
8.3	What is Chemical Hazard Assessment? *243*	
8.3.1	Globally Harmonized System of Classification and Labelling of Chemicals (GHS) *244*	
8.3.2	Comprehensive and Abbreviated Forms of CHA *247*	

8.3.2.1	GreenScreen for Safer Chemicals	248
8.3.2.2	Quick Chemical Assessment Tool (QCAT)	252
8.3.2.3	GreenScreen List Translator (GS LT)	253
8.4	How Chemical Hazard Assessment is Used	255
8.4.1	Chemical Footprint Project	255
8.4.2	Health Product Declaration Version 2.0 (HPD)	259
8.4.3	Red List – Declare Label	259
8.4.4	United States Environmental Protection Agency: Safer Choice Program	260
8.4.5	International Living Future Institute's Living Product Challenges	262
8.4.6	Cradle to Cradle Certified Product Program	262
8.4.7	Chemical Alternatives Assessment	263
8.5	Case Studies Showing How CHA Leads to Safer Consumer Products	264
8.5.1	Case Study 1. US EPA Safer Choice Product Certification	264
8.5.2	Case Study 2. Levi Strauss & Co. Screened Chemistry	267
8.5.3	Case Study 3. Development of an Alternative Food Can Liner	269
8.6	Challenges: Beyond Chemical Hazard Assessment	271
8.6.1	Transparency	271
8.6.2	Filling Data Gaps for Existing and Emerging Hazards: Predictive Toxicology and Tox21	272
8.6.3	Integrating CHA into Green Product Design	272
8.7	Conclusion	274
	References	275
9	**Tying it all together to drive Sustainability in the Chemistry Enterprise** 281	
	David J.C. Constable and Concepción Jiménez-González	
9.1	New Areas of Sustainable and Green Chemistry Metrics Research	286
	References	290
	Index 291	

List of Contributors

Nicholas D. Anastas
US Environmental Protection Agency
National Risk Management Research Laboratory
5 Post Office Square
Boston, MA 02109-3912
USA

John Andraos
CareerChem
504-1129 Don Mills Road
Toronto, ON M3B 2W4
Canada

Ana Isabel Carvalho
University of Lisbon
Instituto Superior Técnico
Centre for Management Studies (CEG-IST)
Av. Rovisco Pais
1049-001 Lisboa
Portugal

David J.C. Constable
ACS Green Chemistry Institute
1155 16th Street, N.W.
Washington, DC 20036
USA

Stephen C. DeVito
US Environmental Protection Agency
Data Quality and Analysis Branch
Toxics Release Inventory Program
Mail code 7410M
1200 Pennsylvania Avenue NW
Washington, DC 20460
USA

Michael A. Gonzalez
Emerging Chemistry and Engineering Branch
Land and Materials Management Division
Life Cycle Assessment Center of Excellence

and

US EPA Office of Research and Development
National Risk Management Research Laboratory
26 W. Martin Luther King Drive
MS 483
Cincinnati, OH 45268
USA

Lauren Heine
Northwest Green Chemistry
8108 S Krell Ridge
Spokane, WA 99223
USA

List of Contributors

Andrei Hent
University of Toronto
Department of Chemistry
80 St. George St.
Toronto, ON M5S 3H6
Canada

Volker Hessel
Eindhoven University of Technology
Micro Flow Chemistry and Process Technology
Laboratory of Chemical Reactor Engineering
P.O. Box 513
5600 MB Eindhoven
The Netherlands

Concepción Jiménez-González
GlaxoSmithKline
Global Manufacturing and Supply
5 Moore Dr
Research Triangle Park, NC 27709-3398
USA

John Leazer
US Food & Drug Administration
Northeast Food and Feed Laboratory
158-15 Liberty Avenue
Jamaica, NY 11433
USA

Tânia Pinto-Varela
University of Lisbon
Instituto Superior Técnico
Centre for Management Studies (CEG-IST)
Av. Rovisco Pais
1049-001 Lisboa
Portugal

Qi Wang
Eindhoven University of Technology
Micro Flow Chemistry and Process Technology
Laboratory of Chemical Reactor Engineering
P.O. Box 513
5600 MB Eindhoven
The Netherlands

Margaret H. Whittaker
ToxServices LLC
1367 Connecticut Avenue, NW
Washington, DC 20036
USA

John M. Woodley
Technical University of Denmark (DTU)
Department of Chemical and Biochemical Engineering
Søltofts Plads
2800 Lyngby
Denmark

Lihua Zhang
Eindhoven University of Technology
Micro Flow Chemistry and Process Technology
Laboratory of Chemical Reactor Engineering
P.O. Box 513
5600 MB Eindhoven
The Netherlands

and

Kunming University of Science and Technology
Faculty of Metallurgical and Energy Engineering
650093 Kunming
China

Preface

Although there has been much activity in sustainability, green chemistry, and engineering since we started working in these areas in the early 1990s, there is still considerably more progress that needs to be made. While it is true that there has been a significant amount of research into more sustainable chemistries and processes, there is continuing debate about how to measure sustainability.

The reason for the continuing debate in part stems from the origins of green chemistry and engineering. Green chemistry and engineering initially stemmed from general principles that were intended to prompt people to proactively think about how to develop more sustainable products and processes in a way that avoided increased regulations. However, the principles lacked a standard structure to measure success. In other words, there were no rigorous scientific frameworks to evaluate aims and progress. This is not uncommon in many new fields of science and technology, and typically these frameworks evolve over time as the field matures.

Thankfully, there has been considerable progress and there is now a deeper more precise understanding of what is meant by "green chemistry" and "green engineering." Even with the progress to date, there is still much work to be done. Sustainability, green chemistry, and green engineering are inherently complex concepts in which a single metric approach is not only insufficient but also misleading, and a multivariate approach is required. If several aspects of sustainability are not simultaneously assessed, one runs the risk of taking the wrong decision by missing trade-offs, ignoring other impacts, or miscalculating the comprehensive sustainability impacts and benefits associated with a process, product, or service.

Unfortunately, not everyone is comfortable with or trained in using a multivariate approach. Thus, some researchers still either claim that it is impossible to assess the sustainability profile of something or insist that taking a single metric approach is valid. We understand the allure of a single mythical metric that would easily, accurately, and precisely guide all decisions as in a "A is better than B" white and black outcome. It is enticing to justify new innovations focusing on a single metric (e.g., global warming potential) forgetting about other effects that may be trade-offs (e.g., toxicity, water scarcity). However, complex, interrelated systems are rarely simple, and designing processes and products is inherently complex, requiring a rigorous, systematic, and systemic approach. In addition, the potential environmental, health, and safety implications make it essential to evaluate different impacts, particularly as they are closely interrelated. To truly drive more sustainable chemical processes, it is imperative to evaluate them from a systems standpoint, which necessarily calls for a range of complementary metrics.

Given the inherent complexity, the communication of green chemistry metrics for effective use is an ongoing challenge that will require continuous attention, as one can easily either overcomplicate or oversimplify the results. All of this can be overwhelming, and could drive any organization to play catch up with developments in the external environment. With additional developments in artificial intelligence, machine learning, data analytics, and data visualization, we expect that some of the challenges of using and communicating sustainability metrics will be solved in due course. In the meantime, as the application of green chemistry and engineering metrics becomes more ingrained in standard processes, it helps to have examples of application.

Our motivation to work on this book is precisely to highlight the evolution of green chemistry and engineering metrics, including practical examples of a multivariate approach. We aim to provide a survey of the current approaches, particularly focusing on application examples of the robust use metrics. This survey could be used as a baseline for the next generation of sustainability metrics that may benefit from more developed data analytics and visualization. We hope that the examples presented here will motivate the community to continuously improve the quest for more sustainable products and processes.

Green Chemistry Institute David J. Constable
American Chemical Society
Washington, DC, USA

GlaxoSmithKline Concepción Jiménez-González
Corporate Environment, Health and Safety
Raleigh, NC, USA

1
Green Chemistry Metrics
David J.C. Constable

1.1
Introduction and General Considerations

Green chemistry has recently been recognized as having achieved its twenty-fifth anniversary [1,2]. Across this span of time there has been a steady growth in green chemistry-related research and a deepening understanding of what green chemistry is and what it is not, although it is fair to say that it is still consistently debated. Since the earliest discussions about green chemistry, it is also fair to say that the field of chemistry has not embraced green chemistry nor seen the pursuit of green chemistry for the good science and innovation opportunities that are inherent to it. Green chemistry is fundamentally how one thinks about chemistry and how one performs or practices chemistry. Interestingly and perhaps unfortunately, green chemistry has been seen by many chemists as something akin to a social movement similar to other environmental movements [3]. A consequence of this perception is that green chemistry is sometimes seen as not being worthy of serious scientific consideration, in spite of the fact that the major proponents of green chemistry have always pointed to the fact that it is intended to spur innovation and promote the very best science while seeking to avoid or prevent human health and environmental problems. Green chemistry is not about environmental chemistry; to do the best green chemistry, one must however, understand environmental chemistry. It is also not about end-of-pipe environmental improvement, although this is often an important area to pursue.

Part of the problem for this mistaken perception of green chemistry, especially when the term was beginning to be used, was a lack of precision or rigor in providing evidence when making claims about chemistry and labeling a new innovation as "environment-friendly" or "eco-friendly" or "green" or "greener." The good news is that over the past 15 or so years, there has been considerable work on the part of many in the green chemistry and engineering community to develop methodologies and approaches to systematically and rigorously assess whether or not something is green or sustainable [4–23] (Curzons, A.D., Constable, D.J.C., and Cunningham, V.L. (2002) *Bond Economy: An Alternative*

Approach to Synthetic Efficiency, Unpublished, GlaxoSmithKline.). Ironically, perhaps, despite all the work to develop metrics, there are still a large number of researchers and practitioners in chemistry publishing research in the top green chemistry and engineering journals who either ignore the use of metrics, claim that it is impossible to assess whether or not something is truly green or sustainable, or use a single metric to justify calling their chemistry innovation "green." This is truly unfortunate because much of what is published in the green chemistry and engineering literature, if viewed from a multivariate metrics perspective, is decidedly lacking in sufficient justification for a chemistry innovation to be called "green," "greener," sustainable, or more sustainable.

It is worth taking a moment to discuss some general ideas about metrics before getting into the details of green chemistry and engineering metrics. Generally speaking, it is commonly accepted that metrics must be clearly defined, simple to interpret and use, measurable, rely on objective determinations rather than subjective approaches, and should ultimately drive some kind of desired behavior or practice. The best way to use metrics is in a comparative sense; it is generally not very productive to engage in assessments from an absolute or absolutist frame of mind, particularly when focusing on sustainability. Rather, one should choose a frame or point of reference against which to apply a metric or make a comparison. In this way, one is able to say with some confidence that one particular outcome or impact is better or worse, greener or more sustainable than the alternatives. Absolutes tend to get in the way of people moving forward and making progress, and the lack of knowing something absolutely is frequently the reason given as to why nothing can or should be done. For example, "there are many possible approaches to metrics and who knows which ones are the best" is frequently offered as a reason for not applying metrics of any kind. This is a spurious argument at best.

Another thing to consider is that one's approach to metrics should be from a multivariate perspective. That is to say that the assessment of "green" or "sustainable" should be from multiple discreetly different kinds of measurements and adapted for a given context. Stated slightly differently, a single metric is insufficient to characterize a chemical, a type of chemistry, a process, or a product as being green, greener, or more sustainable. For example, just because a chemical transformation is done with a catalyst does not mean that it is as green as may be presumed. If, for example, that catalytic transformation is carried out using a platinum group element, and the catalyst is a homogeneous one (as opposed to a heterogeneous one), it is not green or sustainable from multiple perspectives. First, platinum group metals currently favored by many chemists, for example, iridium or platinum, are extremely rare, with relatively low abundance in the Earth's crust. Second, the mining and subsequent extraction of these metals from ore, followed by the refining of those metals to separate the various elements that commonly occur in the ore is a highly mass and energy intensive process and results in significant environmental impacts from waste ore, spent extraction liquors, and so on. Finally, running a catalytic reaction homogeneously where the metal is not extracted from the spent mother liquors, it is likely for that metal to end up as a trace contaminant of incinerator slag, unlikely to be recycled, and most likely disposed of as hazardous waste.

The last example about platinum group metals is also a good example of thinking about chemistry from a systems perspective, and/or using life cycle thinking to make a more realistic assessment of whether or not something is "green" or sustainable. Systems thinking is something that is more common in the biological sciences, where dependencies and connections between and among living systems are seen in the context of their ecological niche, the ecosystem an organism inhabits, and the broader environment containing that ecosystem. In other words, life is seen to exist as a complex web of interactions and interdependencies where perturbations to that system are felt in multiple parts of the system, although they may not be immediately noticeable. Systems thinking is also more common in chemical engineering, where one is forced to see a particular unit operation in the context of the overall process, or one is thinking about mass and energy integration across a plant. In the latter instance, mapping the mass and energy inputs and outputs of every unit operation and thinking about how these might be better utilized to increase the overall mass and energy efficiency of the plant is now quite common and a well-developed practice.

Among many chemists, however, research is focused on the immediate environment of two reactants in a round-bottom flask, along with some additional reagents, solvents, and catalysts. The idea that the choice of reactants, reagents, or solvents made in a laboratory for a given experiment has an impact on a broader system, the environment, or humans seems to be largely irrelevant to what may appear as the much more interesting consideration of whether or not two chemicals may react in a novel way. The previous discussion about platinum group metals is a great example of systems thinking and how few chemists are equipped to employ it. If they were, they might think twice about using an element like iridium as a homogeneous catalyst in a biomass conversion process. While its immediate benefit for catalytically converting lignin to a useful framework molecule, for example, may illustrate interesting chemistry that has not been previously done, using an extremely rare element that is dispersed as waste and effectively lost is a dubious application at best, and the approach is unlikely to be ever commercially applied.

A familiarity with life cycle thinking and the even more desirable life cycle inventory/assessment methodology would help chemists in systems thinking, the idea of boundary conditions, and human or environmental impacts trading. Life cycle inventory/assessment in the context of green chemistry is covered in great detail in Chapter 4 as well as elsewhere [24–37]. The idea of the boundary in which an assessment is carried out that is associated with life cycle impact/assessment is very important if one wants to perform a sustainability or green assessment. Where one draws the boundary for the assessment will likely make considerable differences in the outcome of the assessment. For example, just performing an environment, health and safety, or a life cycle impact category assessment (i.e., the ozone depleting potential, greenhouse gas equivalent, eutrophication potential, etc.) of materials used in a particular reaction (boundary is limited) may be a good starting point, but it neglects the cumulative impacts associated with the materials throughout their life cycle.

If one does actually do a fully burdened cradle-to-gate or cradle-to-cradle impact assessment, one will invariably be faced with many difficult questions. For example, which is more important – the cumulative greenhouse gas impact or the ozone-depleting impact, or the nonrenewable resource impact?

Another aspect of metrics that is worth keeping in mind is that metrics should not be collected just for the sake of keeping metrics. Good metrics should be systematically analyzed, promote strategic analysis of trends, outcome, and impacts, and promote continuous improvement. In some instances once metrics are established, these are not revisited on a regular basis and assessed as to whether or not they are successfully changing behavior. This is unfortunate, because metrics should change behavior in the desired direction.

Virtually every discipline of chemistry can and should be seeking to develop and apply metrics to evaluate whether or not the chemistry they are practicing is being done in a green, greener, or more sustainable fashion than existing methods. Chemists seem to understand and accept that the chemistry innovation they report in the literature needs to be superior in some fashion to an existing method or approach; indeed, part of demonstrating the novelty of an approach is in reviewing precedent and comparing with what has been done previously. This same mind set is recommended to be carried over and the green or sustainability aspects of the innovation should be equally proven. A variety of tools created from applying metrics have been created over many years that enable chemists to evaluate the chemicals, solvents, reagents, and so on that are routinely used in chemistry of all kinds [38–48],[1),2),3)] and these tools can now be used routinely.

It is worth pointing to several chemistry disciplines where metrics are being applied as success stories of the value of using green chemistry metrics. None of these approaches are perfect, but they are certainly a step in the right direction and illustrate the breadth of what is possible. The first discipline to be discussed is synthetic organic chemistry, where metrics approaches have a long history and have been reported for assessing individual reactions, synthesis routes, and design approaches [49–82]. This is one area of chemistry that now has very complete and well-articulated precedent for applying metrics. A second area that serves as a good example is analytical chemistry, where metrics have been developed for analytical instrumental techniques, such as chromatography and spectroscopy, but have also been applied to sample preparation and waste [83–94]. Finally, it is worth noting that green chemistry metrics approaches are being successfully integrated into chemistry education [95–103]. This is an extremely significant development given the importance of educating the next generation of chemists who will not only see that it is eminently

---

1) www.reagentguides.com/ (accessed January 14, 2017).
2) https://www.acs.org/content/acs/en/greenchemistry/research-innovation/tools-for-green-chemistry.html (accessed January 14, 2017).
3) http://learning.chem21.eu/methods-of-facilitating-change/ (accessed January 14, 2017).

possible to integrate green chemistry metrics into their professional careers, but that these metrics can be used to intelligently and confidently move the world toward more sustainable practices.

1.2 Feedstocks

As it currently exists, most of the global chemistry enterprise is deeply rooted in the use of petroleum and a variety of key inorganic elements and compounds to make the products society uses every day. From a sustainable and green chemistry perspective, most of what is used by chemists is currently obtained in ways that are completely unsustainable. For example, while one may debate how much petroleum there is to be extracted and how long it will last, eventually it will run out, and long before that, extracting it from the ground will come at an increasingly greater economic, social, and environmental cost. From the perspective of chemistry, this is important because the majority of the basic organic framework molecules that chemists use come from petroleum, and it has been recently shown that about 120 of these molecules are used in most organic syntheses [104]. There is over a 100-year tradition of chemistry being performed on petroleum-based molecules that are in a highly reduced state and very unreactive. Comparing molecules obtained from petroleum to those found above ground from bio-based sources, bio-based molecules are generally highly oxidized and/or functionalized and generally not yet available at comparable volumes as would be required, for example, to supply high-volume plastics manufacturing. Moreover, the types of chemistries one would use to convert bio-based framework molecules to products of interest are not very efficient at this point. However, there has been increasing interest and development of bio-based and renewable feedstocks. There are basically two different strategies here. The first is to make use of molecules as they are from Nature and selectively remove some of the functional groups or convert them to functionalized analogues of chemicals that are currently in widespread use. The second is to completely convert them to chemicals, such as benzene/toluene/xylene (BTX) or aniline, or any of the type of molecules you obtain from a petrochemical supply chain. One could propose variations of metrics around either of these approaches, or just stick with the kinds of metrics that will be described later. Regardless, it could be argued that converting highly functionalized molecules to a reduced molecule like BTX is not desirable based on the inherent environmental, health, and safety hazards associated with BTX.

Another approach to assessing feedstocks from renewable and bio-based sources is to use mass-based metrics to calculate the amount or proportion of the biomass put into a process to make a chemical or product. This approach has been discussed elsewhere, as a resource efficiency metric analogous to mass intensity but it is by no means the only approach [105–112]. Another extensive treatment of biomass utilization may be found in work reported by Iffland *et. al*, where a biomass utilization efficiency metric is described for application to bio-based chemicals, polymers, and fuels [113].

Regardless of the approach used, it is important to keep in mind the supply chain associated with a renewable, bio-based chemical or product, just as with one made from a petrochemical source. One should not lose sight of the complexity associated with sourcing a renewable and bio-based chemical or product, and this should necessarily bring one to assess renewability from a life cycle perspective. Growing, harvesting, transporting the biomass, processing the bulk biomass to remove the desired fraction of interest, purifying, and finally isolating the final product requires a significant amount of mass and energy.

One example of sourcing a renewable and bio-based chemical might be furfural, which can be used to produce furan or THF. The supply chain required to produce furan includes sulfuric acid, methanol, or carbon monoxide that are often derived from fossil feedstocks, and process-related energy that is likely to be sourced from a mix of renewable and nonrenewable primary energy sources.

If one is interested in a broader sustainability assessment, the production of furan from renewable and bio-based sources must account for land use and the consequential impacts on the environment, but also the potential for competition with the primary production of food. Metrics for renewable and bio-based chemicals is a complex area of continuing development where there is ongoing national and international debate and where there are no easy answers.

1.3
Chemicals

1.3.1
Hazard and Risk

Any discussion of green chemistry and engineering metrics needs to include a discussion of hazard and risk. Unfortunately, despite clear definitions for each, there is a great tendency for people to use these terms interchangeably. Chemical hazards are associated with an inherent physical or physicochemical property or an effect that a chemical has on a living organism. Examples of inherent hazards are things like acidity or basicity, acute or chronic toxicity, reactivity to air or water, and so on. There has been an enormous amount of study of chemical hazards to better understand the potential impact of chemicals on the environment, to ensure worker safety, and to avoid acute and chronic human health impacts. There is now a considerable body of good information available for many of the most used, high-volume commodity chemicals. These data have been developed to better understand and avoid chemical hazards and the development of hazard information has been a key aspect of chemicals legislation since the 1970s when laws like the US Toxic Substances Control Act (TSCA), recently reauthorized as the Frank Lautenberg Chemical Safety Act of 2016, were introduced, or in the EU, with the REACH (Registration, Evaluation,

Authorization and Restriction of Chemicals) legislation, and with many other countries throughout the world adopting legislation similar to REACH. This legislation requires manufacturers to determine environmental, safety, and human health hazards associated with chemicals before giving approval to chemical manufacture, and these data are important parts of any risk assessment. Table 1.1 contains an illustrative but not comprehensive list of common ecotoxicity hazard data that might be collected to assess chemicals. It should be understood that one would also collect analogous human toxicity data (acute and chronic toxicity, dermal, sensitization, etc.) and worker safety data (e.g., flammability, explosivity, flash point, etc.) as part of a routine chemical hazard or risk assessment. If one tries to gather data like these for most any chemical, they will rapidly find that there may be a lot of data, very little data, or some data with many gaps in the data set. When there is a significant amount of data, the most common observation is that chemicals have a large number of hazards associated with them, so trying to decide what to do about this presents a challenge. When there are data gaps, the tendency is to say it is okay to use the chemical, especially if there are no hazards identified in the existing data. Historically, when data is missing, many chemists assumed the chemical must be safe because there is no data available; if there was a problem, it is assumed that someone would have identified any problems. This is an unfortunate and poor assumption to make and certainly not recommended.

A variety of methods and processes have been developed to assess hazard data and on the basis of the hazard assessment, certain actions are recommended. One of the most commonly used in the last few years is GreenScreen [114–116], but other similar approaches have been developed [117]. Regardless of the approach taken, the important thing is that the assessment be transparent, that data gaps are documented, and whether or not the data used for the assessment is experimentally derived or derived from some kind of quantitative structure activity model. Another issue where the state of the art in chemical hazard assessment is not developed is in how to handle mixtures. Despite this being the case, it should not prevent one from looking at chemical hazards and taking action on the basis of the assessment. Ideally, one would choose chemicals that have comparable or superior technical performance with the least number of EHS issues associated with them.

One final way of thinking about EHS hazards of chemicals is to think about major global impacts of chemicals as is common in life cycle inventory/assessment. Table 1.2 contains a small, illustrative list of common but not in any way comprehensive impact categories that are considered to be the midpoint of the assessment. What is meant by this is that in a life cycle assessment, one can group chemicals according to broad categories such as global warming equivalents or ozone depletions equivalents, acidification equivalents, and so on, and these are considered the midpoint. An endpoint analysis means that ozone depletion can have one of many environmental or human health impacts such as increased cancer mortality, broad ecosystem impacts, and so on. Typically, it is perfectly acceptable to remain at the midpoint and assess chemicals according

Table 1.1 Illustrative ecotoxicity hazard data.

Test parameter	Results
Physical properties	
Water solubility	mg/l at 20 °C
UV/visible spectrum (photolysis)	nm (Absorption above 290 nm, photolysis may be possible)
Vapor pressure	mm Hg at 25 °C
Dissociation constant	pK1 = at 25 °C
	pK2 = at 25 °C
Partition coefficients	
Octanol/water partition coefficient (Log Kow)	
Distribution coefficient (Log Dow)	pH5 = X
	pH7 = Y
	pH9 = Z
Soil organic carbon distribution coefficient (Log Koc)	X to Y
Soil distribution coefficient (log K_D)	X to Y
Depletion	
Biodegradation	Aerobic – inherent
	Percent degradation: X%, Y days, modified Zahn-Wellens (primary, loss of parent)
	Aerobic – soil
	Percent degradation: X%, Y days
Hydrolysis rate	Half-life > X year, pH = 7, chemically stable in water (Y/N)

Ecotoxicity	Results
Activated sludge respiration	IC_{50} <, >, = X mg/l
Algal inhibition	IC_{50} = <, >, = X mg/l, 72 h, Scenedesmus subspicatus
	NOEC <, >, = X mg/l
Acute toxicity to daphnids	EC_{50} <, >, = X mg/l, 48 h, Daphnia pulex
	NOEC <, >, = X mg/l, 48 h
Chronic toxicity to daphnids	LOEC <, >, = X mg/l, 8 d, reproduction, Ceriodaphnia dubia
	NOEC <, >, = X mg/l, 8 d, reproduction
Acute toxicity to fish	EC_{50} <, >, = X mg/l, 96 h, juvenile Oncorhynchus mykiss (rainbow trout)
	NOEC <, >, = X mg/l
Earthworm toxicity	EC_{50} <, >, = X mg/kg, 28 d, Eisenia foetida (manure worm)
	NOEC <, >, = X mg/kg

Table 1.2 Brief list of LCI/A impact categories.

Greenhouse gasses	Acidification	Eutrophication
Photochemical ozone creation potential	Volatile organic compounds	Human toxicity
Heavy metals	Hazardous waste	Solids to landfill

to their potential impacts as a midpoint category. This will be explained in greater detail in Chapter 4 and the reader is referred elsewhere for a more comprehensive treatment of this subject.

Risk is defined as:

Risk = f(the inherent hazard of a material, the potential or likelihood for exposure).

Risk is often expanded to include a severity rating at given frequencies and probabilities of occurrence.

For most of the world, chemical hazard is seen as a driver for policy and regulation. For industry, there is a desire to make risk-based assessments of chemicals. You might ask why this distinction is important to a discussion of green chemistry metrics. Basically, all chemicals are hazardous in some fashion depending on the context. You can drink too much water and die, or drown in a 2″ puddle of water, if the conditions are correct, yet our lives depend on water and humans are mostly water in chemical composition. This example is perhaps overstating the issue, but most policy and regulation of chemicals is driven from the perspective that any exposure to hazardous chemicals constitutes an unacceptable degree of risk. Another related idea, the precautionary principle, states that if the chemical hazards of a chemical compound are not known, or are suspected of being harmful to people or the environment, one should not create or use that chemical unless or until there is proof that there is no risk of harm from that chemical. While regulators promote hazard-based assessments as a means for managing chemicals, industry operates using a wide variety of hazardous substances and rigorously controls exposure to decrease risk to human health and the environment and consequently, this often puts industry at odds with regulators.

For a chemist in a laboratory, risk is generally not something that is top of mind when they are performing an experiment. In the case of hazard, hazard is accepted as a normal part of chemistry and most chemists are rather cavalier about chemical hazards since the hazard is related to why chemicals react, and chemists are very interested in making chemicals react. For the second part of risk, that is, exposure, this may be easy for chemists to ignore because they work in hoods and wear personal protective equipment as a well-accepted, routine, and expected practice while working in the laboratory. The emphasis is on making something new and chemically interesting, not on protecting the environment or human health. It is also conventional wisdom that interesting chemistry cannot be done unless one is using hazardous chemicals to make a reaction proceed to completion in as quantitative and rapid a fashion as possible.

1.4
General Chemistry Considerations and Chemistry Metrics

The traditional way of thinking of efficiency in chemical synthesis is to think of it in terms of chemical selectivity, rates of reaction, and yield. It is worth having a brief, closer look at each one of these ideas.

Chemical Selectivity
One could argue that a lack of selectivity in a reaction is perhaps one of the biggest drivers of waste in most chemical reactions, so it is worth a deeper dive thinking about the ways chemists think about selectivity. In the case of chemoselectivity, a chemoselective reagent is one that reacts with one functional group (e.g., a halide, R-X), but not another (e.g., a carbonyl group, R-C=O). The problem for many reagents is that they can be somewhat promiscuous and will react with more than one portion of the molecule and that leads to the formation of impurities or undesirable by-products. This lack of selectivity also leads to the use of protecting groups as a synthetic strategy, but this is an inherently wasteful strategy. In the case of enantioselectivity, an inactive substrate (a molecule of interest) is converted selectively to only one of two enantiomers. Enantiomers are isomers (i.e., compounds with the same numbers and types of atoms but possessing different structures, properties, etc.) who differ only in the left- and right-handedness of their orientations in molecular space. Enantiomers rotate polarized light in equal but opposite directions and typically react at different rates with other chiral compounds. In drug synthesis, or in the case of certain crop protection agents, for example, a lack of enantioselectivity can lead to as much as 50% of the product being lost, although there are many synthetic strategies like dynamic kinetic resolution that are used to drive the synthetic product toward a single enantiomer. In the case of stereoselectivity, an inability to direct a synthesis toward the exclusive or predominant formation of a specific isomer results in the formation of many undesirable compounds. Finally, reactions can potentially lead to the formation of two or more structural isomers (e.g., R-O-C=N or R-N=C=O); that is, chemicals possessing the same chemical formula but very different chemical properties. Regioselective reactions are reactions that lead only to the formation of one of the structural isomers. In each of the mentioned cases, a lack of selectivity can lead to tremendous waste of the starting materials, reagents, and catalysts that were used in the synthetic process.

Kinetics
Kinetics is the study of reaction rates; that is, how quickly or slowly a reaction proceeds to completion. Reaction rates are fundamentally controlled by the difference in chemical potential that exists between the reactants and the product, but rates can be dramatically influenced by, to name a few of the most important, reactant concentrations, solvent/solubility effects, applied energy (e.g., temperature, light, etc.) mixing, or the degree of steric hindrance in either or both of

the reactants. Each of these influences can lead to a reduction in chemical selectivity and a loss of efficiency. In many synthetic organic chemistry papers, it is not uncommon to see reactions that take a very long time to go to completion. Stated differently, in synthetic organic chemistry the emphasis is usually on obtaining the desired product – not on how quickly one obtains it, or at what temperature, or for how many days the reaction proceeds – in a yield that is high enough to isolate the desired product.

Yield

The most ubiquitous measure of chemical efficiency employed by chemists and chemical engineers alike is undoubtedly percentage yield. For any given reaction

$$A + B \rightarrow C$$

Theoretical yield = ([A] moles of limiting reagent) × (stoichiometric ratio: [C/A] desired product/limiting reagent) × (FW of desired product [C])

Percentage yield = (actual yield/theoretical yield) × 100

The interesting thing about yield is that it really is a very poor measure of the overall efficiency of a process and from that perspective, it is effectively useless when assessing whether or not a particular chemical or process is green. Yield ignores the reality that you rarely have only chemicals A and B reacting in isolation, and in many cases, B is added in large stoichiometric excess to drive the reaction to make as much of C as possible. Reagents, catalysts, solvents, and so on are not included in yield and in most cases, kinetics are not considered. So, a chemist can heat a reaction to 120 °C, hold at reflux for 3 days to obtain a yield of 50%, and subsequently claim victory for an awesome reaction no one has done. It does not matter that such a reaction would never be used in a manufacturing situation, nor would it matter if one had to employ massive amounts of solvent in workup or isolation.

1.5
Evolution of Green Chemistry Metrics

For anyone who has worked in the chemical manufacturing and processing industries and who also had responsibility for managing their company's environmental performance, using the traditional chemistry metrics already described as a means for measuring and tracking company's environmental, safety and health performance, efficiency, or compliance would find measures of chemical efficiency woefully inadequate. This is why there has been a focus on metrics that looked at waste and put that in terms of mass and energy rather than in moles. A brief review of the evolution of green chemistry and engineering metrics follows.

Atom Economy

In 1991, noted synthetic organic chemist Barry Trost introduced the term atom economy to prompt synthetic organic chemists to pursue "greener chemistry" as part of their search for synthetic efficiency [118]. Briefly stated, atom economy is a calculation of how many atoms in two chemicals that react remain in the final molecule or product. Final product applies equally to a single chemical transformation, a series of chemical transformations in a single stage of a multistage synthetic route, or to all the reactants in a complete synthetic route to a final product. Unfortunately, atom economy does not include a consideration of yield, stoichiometric excesses, or any of the other things that usually go into making a reaction proceed to completion. A complete review of atom economy and its weaknesses may be found elsewhere [119,120], although it is recognized that atom economy continues to be used by chemists [121–124]. Atom economy is mentioned here for the sake of its historical place, but as a metric to drive green chemistry or for greening a reaction, it is not a particularly practical, informative, or useful metric. It should be noted that the idea of atomic or molecular efficiency; that is, how much of a molecule in a reaction is retained in the final product, is one that has been explored by a number of groups over the years. For example, carbon efficiency [8], bond economy (Curzons, A.D., Constable, D.J.C., and Cunningham, V.L. (2002) *Bond Economy: An Alternative Approach to Synthetic Efficiency*, Unpublished, GlaxoSmithKline.), and molar efficiency [76] have all been proposed and investigated for their usefulness as metrics for assessing synthetic efficiency and their ability to promote greener chemistry. However, none of these measures have proven to be as useful as hoped since they don't correlate well with mass, energy, or waste associated with chemical reactions. Unfortunately, it is a fact that chemistry lacks atomic and molecular precision, especially for complex molecules; it usually takes a relatively large amount of mass and energy to get molecules to react.

E-Factor

E-factor, proposed by Roger Sheldon in the early 1990s [4,5], is defined as follows:

$$\text{E-factor} = \frac{\text{Total waste (kg)}}{\text{kg product}}$$

In the original publication, it is not explicitly stated whether or not this metric included or excluded water, but it could be used to describe either case. E-factor is relatively simple to understand and its application has drawn attention to the waste produced for a given quantity of product. As part of his original publication, Sheldon produced a comparison of the relative wastefulness of different sectors of the chemical processing industries as diverse as petrochemicals, specialties, and pharmaceuticals and this comparison has been tremendously influential over the past nearly 25 years.

While the metric has been tremendously helpful, it may in practice be subject to a lack of clarity depending on how waste is defined by the user. As in most of green chemistry and engineering, where one draws the boundaries in any given comparison or assessment impacts the outcome. For example, is only waste from one manufacturing plant in view, and is waste from emissions treatment (e.g., acid gas scrubbing, pH adjustment in wastewater treatment plants, etc.) included? Is waste from energy production (heating or cooling reactions, abatement technology, etc.) included? Or perhaps something like waste solvent passed on to a waste handler and burned in a cement kiln is not included. Depending on how these are handled, you can arrive at very different answers.

It is also generally true, at least based on industrial experience, that drawing attention to waste does not generally capture the attention or imagination of chemists, unless, of course, they have spent most of their careers trying to commercialize a chemical product and/or process. Among chemists, there may be a tendency to discount the importance of all the other things that go into the reaction and focus on the "good science" that goes into getting two reactants to proceed quantitatively to a desired product. Chemical process wastes also do not capture most business leader's attention unless profit margins for their chemicals are very small.

Step and Pot Economy

In 1997, Paul Wender and colleagues published a paper introducing the concept of step economy in the context of an ideal synthesis for a new drug [125,126]. While it may seem obvious now, it was not appreciated that completing a synthesis in as few steps as possible is actually a desirable strategy or one that makes a synthesis greener. The connection between limiting steps and making a synthesis greener is especially true if there are a limited number of solvent switches, isolations, and recrystallizations in the overall synthesis route. A related idea to step economy reported by Clarke *et al.* is known as pot economy [127,128]. Basically, having a series or a set of cascade or multicomponent reactions in a single pot and of high efficiency means that you have fewer steps, fewer isolations, and fewer solvents, reagents, and/or catalysts. If the reaction mass efficiency of a step and pot economic synthesis is high, the overall process mass efficiency will also be quite high, a very desirable outcome from a green chemistry perspective.

Effective Mass Yield

In 1999, Hudlicky *et al.* proposed a metric known as effective mass yield [49], defined "as the percentage of the mass of desired product relative to the mass of all non-benign materials used in its synthesis." Or, stated mathematically as follows:

$$\text{Effective mass yield} = \frac{\text{Mass of products}}{\text{Mass of nonbenign reagents}} \times 100$$

This metric was arguably the first in green chemistry that focused attention on the fact that not all mass that passes through a process has an equivalent impact and in fact, there is a relatively small amount of nontoxic mass associated with many processes to produce chemicals. The metric also tried to bridge from a commonly used term in chemistry, yield, and worked to tie that to the fact that many toxic reagents are used in chemistry. Adding reagent toxicity is an extremely important consideration in any assessment of what is "green" and toxicity is certainly something that is absent from any chemists discussion about yield.

Despite the positives of this metric, the attempt by Hudlicky *et al.* to define benign as "those by-products, reagents or solvents that have no known environmental risk associated with them for example, water, low-concentration saline, dilute ethanol, autoclaved cell mass, and so on," suffers from a lack of definitional clarity. In addition to this lack of clarity, it should be understood that wastes, such as saline, ethanol, and autoclaved cell mass, have environmental impacts of one kind or another that would have to be evaluated and addressed. There is also a very practical problem that defining "nonbenign" is difficult, especially when you are working with complex reagents and reactants that have limited environmental or occupational toxicity information. Also, as noted previously, environmental risk is a function of hazard and exposure, and merely determining hazard for many materials in commerce is currently not possible. Adding to that the difficulty of assessing exposure and the challenge is even more daunting. Unless and until robust human and environmental toxicity information or credible quantitative structure activity estimates are routinely available for the wide diversity of chemicals used, trying to use this metric for most synthetic chemical operations is effectively impossible.

Reaction Mass Efficiency

In an attempt to get over some of the failings of atom economy while retaining a focus on reactants and the efficiency of a reaction, the metric reaction mass efficiency was investigated [8]. This is another mass-based metric that incorporates atom economy, yield, and stoichiometry and is the percentage of the mass of reactants in the final product.

There are two ways to calculate RME.

For a generic reaction $A + B \rightarrow C$

$$\text{Reaction mass efficiency} = \left(\frac{\text{m.w. of product C}}{\text{m.w. of A} + (\text{m.w. of B} \times \text{molar ratio B/A})} \right) \times \text{yield}$$

or more simply

$$\text{Reaction mass efficiency} = \left(\frac{\text{mass of product C}}{\text{mass of A} + \text{mass of B}} \right) \times 100$$

This was a small step forward, but it still does not include all the other materials – reagents, solvents, catalysts – that go into making a reaction proceed to completion.

1.6
Andraos: Tree Analysis

In 2005, Andraos published several papers [54,55] detailing his work to unify reaction metrics for green chemistry and performed detailed reaction analyses on a number of different synthetic routes to target molecules using routes available from the literature. This work was expanded a year later in a paper [56] that used tree analysis to quantify mass, energy, and cost throughput efficiencies of simple and complex synthesis plans and networks. Tree analysis provides a straightforward graphical illustration of key metrics for any reaction sequence, and a variety of spreadsheets have been created to facilitate the analysis. Andraos has continued to expand his work to include environmental, safety, and health metrics into his reaction analysis [68]. He and his coworkers have also developed a series of spreadsheets and educational modules for integrating green chemistry metrics into chemistry education [101–103]. More about Andraos' work may be found in Chapter 3. Given the extensive literature precedent that has been developed, and the ready availability of spreadsheets to facilitate calculations, it is a bit of a mystery as to why green chemistry metrics are not routinely reported by all chemists.

1.7 Process Metrics

A very good treatment of process metrics that is still very relevant was published in a prior book on green chemistry metrics, so the information found there will not be repeated here. Chapters 5–7 of this book also contain an extended treatment of metrics for batch, continuous, and bioprocessing that go well beyond what was covered in the earlier work. There has been, in general, an extensive amount published about process metrics in the green chemistry and engineering literature and the reader is referred elsewhere [129–154].

Process Mass Intensity
Process mass intensity was first published in 2001 [8], although it had been extensively applied within SmithKline Beecham for about 4 years prior to being described in a publication. Process mass intensity was an attempt to focus attention on the inefficiency associated with a typical pharmaceutical process, which at the time, was not the subject of much interest within the pharmaceutical industry. Process mass intensity was built on reaction mass efficiency and included all the reactants, reagents, catalysts, solvents, and any other materials used in product work-up and isolation. It is relatively straightforward to calculate

$$\text{Mass intensity (MI)} = \frac{\text{Total mass in reaction vessel (kg)}}{\text{Mass of product (kg)}}$$

It may also be useful to compare MI with E-factor where

$$\text{E-factor} = \text{MI} - 1$$

At the time the metric was introduced, the amount of water used in a process was excluded, despite the fact that this was routinely calculated. In recent years, water has been included in mass intensity calculations by most pharmaceutical companies for a variety of reasons, but at the time this metric was introduced, it was quite difficult to get chemists to think that organic solvents were important as significant process cost and environmental drivers let alone getting them to think about the total cost of producing high-purity water.

Mass intensity may also be expressed as its reciprocal and converting it to a percentage; in this form it may be compared to metrics like effective mass yield and atom economy although there is generally no correlation between these metrics. This metric is generally known as mass productivity or mass efficiency, and it was proposed as a means of making mass intensity more accessible to managers in business. Productivity and efficiency are more easily understood and valued than an intensity metric ever would be in the business world.

$$\text{Mass productivity} = \frac{1}{\text{MI}} \times 100 = \frac{\text{Mass of product}}{\text{Total mass in reaction vessel}} \times 100$$

1.8
Product Metrics

There has been increasing societal concern about chemicals in products, especially in food, water, personal care, and a growing number of other consumer products. As a result, leading consumer product companies have increased their focus and activities to ensure that chemicals in products they put on the market are safe for humans and the environment. Safer does not mean that there is no inherent hazard associated with any given chemical, but it does mean that under conditions of use, the product will not contain chemicals that will cause harm under normal conditions of use and if used as intended. Legislation like the Frank Lautenberg Chemical Safety Act of 2016 and REACH in the EU are illustrative of government responses to consumer's desires to ensure that chemicals in commercial production, and those that are ultimately in products, will not have adverse impacts on human health and the environment. In the United States, consumer concerns have also given rise to programs like the US EPAs Design for the Environment, Safer Choice program, a voluntary labeling program for companies that use chemicals in their products that are on the US EPAs Safer Chemicals List.

Societal concerns have also spurred in the United States the promulgation of legislation, like the misnamed CA Green Chemistry Act. The California Act has, among other things, legislated that selected chemicals undergo an alternatives assessment; that is, an evaluation of chemicals that provide the same function as an existing chemical but which hopefully do not have the various kinds of human health or environmental hazards associated with that chemical. There are a few concerns with the alternatives assessment approach. First, there is a presumption that one or more alternative compounds are readily available, that these alternatives perform the desired function, and that they will be cost competitive with the incumbent. A second issue is the presumption that there are a sufficient number of alternatives available that do not have the human health and environmental impacts associated with the compound under review. A third issue is that there is still not a defined or agreed standard for how alternatives assessments are to be carried out, and one last issue that is of great concern to industry is that the assessments are all based on an assessment of the inherent hazards of a chemical and not the risk associated with that chemical in use.

Despite these concerns, there are at least two opportunities for green chemistry. First, the identification of compounds to be assessed may eventually be a direct result of the development of "safer" chemicals, safer at least from a human and environmental hazard perspective, through the systematic application of green chemistry and engineering principles during chemical design and development. The second is in contributing to a broader discussion that expands green chemistry considerations beyond merely inherent hazard or "safer" chemicals, and incorporates more sustainability considerations into the development of alternative chemicals. Much of what is covered in the remaining chapters of this book will be very helpful in equipping chemists to evaluate their chemistry from an objective sustainable and green chemistry and engineering perspective. Chapter 8, in particular, will cover in greater depth how some of what was discussed already about green chemistry and its application to products is being handled by a variety of organizations today.

1.9
Sustainability and Green Chemistry

For a very long time in green chemistry circles, there has been debate about what green chemistry includes and what it does not, what sustainable chemistry is, and how the two ideas relate to each other. Despite all the publications over the past 25 years, there are actually many different opinions about what defines something as being green chemistry, and more recently, there has been revived discussion about what sustainable chemistry is and is not. Restricting the discussion for the moment to green chemistry and speaking broadly, the two biggest camps in green chemistry are composed of people that believe green chemistry is restricted to the development of safer chemicals (i.e., pay attention to toxicity and secondarily, to waste) and that all other considerations are not as important, or more under the umbrella of sustainability than belonging to green chemistry. At the opposite end of the spectrum are those that believe the principles of green chemistry and engineering, that is, all the principles that have been written and best articulated in a recent publication on design principles of sustainable and green chemistry [155], encompass most sustainability considerations that are relevant to the integration of green chemistry and engineering into the practice of chemistry. More will be discussed about sustainability and green chemistry in the final chapter of this book, but it is important to state now that principles are not metrics and that metrics need to be derived that can be mapped to principles.

1.10
Making Decisions

As stated in Section 1.1, the purpose of metrics should be to help people make better decisions about their work. In the case of green chemistry, metrics should be guiding what kinds of choices people make in their use of reactants, reagents, solvents, catalysts, reactors, how they isolate and purify their chemical products, and how they make formulations or specify chemicals that are part of complex materials or consumer products. Unfortunately, the world is not very savvy when it comes to making decisions, and we have an especially poor understanding of how to use statistically valid, risk-based decision-making [156]. Ironically, when it comes to making important decisions that require some change in behavior, even when faced with one's own mortality, fewer than 1 in 5 are successful in making changes to develop positive habits that will increase their lifespan [157].

As mentioned previously, there has been increased interest and awareness in recent years for making better decisions about choosing "safer" chemicals that go into consumer products of one kind or another. A recent report by the US National Academy of Science has done a good job proposing a framework for chemical alternatives assessment and compiling, state-of-the-art information about alternatives assessment [158]. There are a variety of online resources becoming available as well [159,160]. Regardless of which alternatives assessment methodology or approach is taken, they all rely on some kind of multicriteria or multivariate decision-making process to guide decision-making. They are also invariably hazards based, and they do not generally include life cycle or systems level thinking in the decision frameworks.

The greatest value of chemical alternatives assessment work to date is in formalizing and standardizing the decision-making methodology, and in identifying the areas that require additional research. For example, data gaps are quite common and how one approaches data gaps in making decisions is critical. In general, there are a variety of strategies employed to fill data gaps such as read-across, nearest-neighbor estimations, quantitative structure activity relationships [161–163], and expert opinion, to name a few. At this point in time, it is fair to say that each approach has its strengths and limitations, with no clear winner; except obtaining the data through standard, accepted experimental protocols. However, lack of data is no excuse for ignoring or not using a multivariate approach to assessment. It is extremely important to undertake the assessment and be transparent about missing data and how decisions are made in the face of data gaps.

References

1 Anastas, P., Han, B., Leitner, W., and Poliakoff, M. (2016) Happy silver anniversary: green chemistry at 25. *Green Chemistry*, **18** (1), 12–13.
2 Anastas, P.T. and Allen, D.T. (2016) Twenty-five years of green chemistry and green engineering: the end of the beginning. *ACS Sustainable Chemistry & Engineering*, **4** (11), 5820–5820.
3 Woodhouse, E.J. (2005) Green chemistry as social movement? *Science, Technology & Human Values*, **30** (2), 199–222.
4 Sheldon, R.A. (1994) Consider the environmental quotient. *ChemTech*, **24** (3), 38–47.
5 Sheldon, R.A. (2000) Atom utilisation, E factors and the catalytic solution. *Comptes Rendus de l'Académie des Sciences - Series IIC - Chemistry*, **3** (7), 541–551.
6 Constable, D.J.C., Curzons, A.D., Freitas dos Santos, L.M., Geen, G.R., Kitteringham, J., Smith, P., Hannah, R.E., McGuire, M.A., Webb, R.L., Yu, M., Hayler, J.D., and Richardson, J.E. (2001) Green chemistry measures for process research and development. *Green Chemistry*, **3** (1), 7–9.
7 Winterton, N. (2001) Twelve more green chemistry principles. *Green Chemistry*, **3** (6), G73–G75.
8 Constable, D.J.C., Curzons, A.D., and Cunningham, V.L. (2002) Metrics to green chemistry which are the best? *Green Chemistry*, **4** (6), 521–527.
9 Marteel, A.E., Davies, J.A., Olson, W.W., and Abraham, M.A. (2003) Green chemistry and engineering: drivers, metrics, and reduction to practice. *Annual Review of Environment and Resources*, **28** (1), 401–428.
10 Sheldon, R.A. (2007) The E factor: fifteen years on. *Green Chemistry*, **9** (12), 1273.
11 Monteiro, J.G.M.-S., de Queiroz Fernandes Araújo, O., and de Medeiros, J.L. (2008) Sustainability metrics for eco-technologies assessment, part I: preliminary screening. *Clean Technologies and Environmental Policy*, **11** (2), 209–214.
12 Sheldon, R.A. (2008) E factors, green chemistry and catalysis: an odyssey. *Chemical Communications*, (29), 3352.
13 Calvo-Flores, F.G. (2009) Sustainable chemistry metrics. *ChemSusChem*, **2** (10), 905–919.
14 Glaser, J.A. (2009) Green chemistry metrics. *Clean Technologies and Environmental Policy*, **11** (4), 371–374.
15 Henderson, R.K., Constable, D.J.C., and Jiménez-González, C. (2010) Green chemistry metrics, in *Green Chemistry in the Pharmaceutical Industry* (eds P.J. Dunn, A.S. Wells, and M.T. Williams), Wiley-VCH Verlag GmbH, Weinheim, pp. 21–48.
16 Jiménez-González, C., Constable, D.J.C., and Ponder, C.S. (2011) Evaluating the "Greenness" of chemical processes and products in the pharmaceutical industry – a green metrics primer. *Chemical Society Reviews*, **41** (4), 1485.
17 Ribeiro, M.G.T.C. and Machado, A.A.S.C. (2013) Greenness of chemical reactions – limitations of mass metrics. *Green Chemistry Letters and Reviews*, **6** (1), 1–18.
18 Dicks, A.P. and Hent, A. (2014) Green chemistry and associated metrics, in *Green Chemistry Metrics*, SpringerBriefs in Molecular Science, Springer, pp. 1–15.
19 Dicks, A.P. and Hent, A. (2014) Selected qualitative green metrics, in *Green Chemistry Metrics*, SpringerBriefs in Molecular Science, Springer, 69–79.
20 Albini, A. and Protti, S. (2015) Green metrics, an abridged glossary, in *Paradigms in Green Chemistry and Technology*, SpringerBriefs in Molecular Science, Springer, pp. 11–24.
21 McElroy, R.C., Constantinou, A., Jones, L.C., Summerton, L., and Clark, J.H. (2015) Towards a holistic approach to metrics for the 21st century pharmaceutical industry. *Green Chemistry*, **17** (5), 3111–3121.
22 Phan, T.V.T., Gallardo, C., and Mane, J. (2015) Green motion: a new and easy to use green chemistry metric from laboratories to industry. *Green Chemistry*, **17** (5), 2846–2852.

23 Roschangar, F., Sheldon, R.A., and Senanayake, C.H. (2015) Overcoming barriers to green chemistry in the pharmaceutical industry – the green aspiration Level™ concept. *Green Chemistry*, **17** (2), 752–768.

24 Domènech, X., Ayllón, J.A., Peral, J., and Rieradevall, J. (2002) How green is a chemical reaction? Application of LCA to green chemistry. *Environmental Science & Technology*, **36** (24), 5517–5520.

25 Jiménez-González, C., Curzons, A.D., Constable, D.J.C., and Cunningham, V.L. (2004) Cradle-to-gate life cycle inventory and assessment of pharmaceutical compounds. *The International Journal of Life Cycle Assessment*, **9** (2), 114–121.

26 Curzons, A.D., Jiménez-González, C., Duncan, A.L., Constable, D.J.C., and Cunningham, V.L. (2007) Fast life cycle assessment of synthetic chemistry (FLASC™) tool. *The International Journal of Life Cycle Assessment*, **12** (4), 272–280.

27 Kümmerer, K. (2007) Sustainable from the very beginning: rational design of molecules by life cycle engineering as an important approach for green pharmacy and green chemistry. *Green Chemistry*, **9** (8), 899.

28 Huebschmann, S., Kralisch, D., Hessel, V., Krtschil, U., and Kompter, C. (2009) Environmentally benign microreaction process design by accompanying (simplified) life cycle assessment. *Chemical Engineering & Technology*, **32** (11), 1757.

29 Monteiro, J.G.M.-S., de Queiroz Fernandes Araújo, O., and de Medeiros, J.L. (2009) Sustainability metrics for eco-technologies assessment, part II. Life cycle analysis. *Clean Technologies and Environmental Policy*, **11** (4), 459–472.

30 Tabone, M.D., Cregg, J.J., Beckman, E.J., and Landis, A.E. (2010) Sustainability metrics: life cycle assessment and green design in polymers. *Environmental Science & Technology*, **44** (21), 8264–8269.

31 Wernet, G., Conradt, S., Isenring, H.P., Jiménez-González, C., and Hungerbühler, K. (2010) Life cycle assessment of fine chemical production: a case study of pharmaceutical synthesis. *The International Journal of Life Cycle Assessment*, **15** (3), 294–303.

32 Tufvesson, L.M., Tufvesson, P., Woodley, J.M., and Börjesson, P. (2012) Life cycle assessment in green chemistry: overview of key parameters and methodological concerns. *The International Journal of Life Cycle Assessment*, **18** (2), 431–444.

33 Kressirer, S., Kralisch, D., Stark, A., Krtschil, U., and Hessel, V. (2013) Agile green process design for the intensified Kolbe–Schmitt synthesis by accompanying (simplified) life cycle assessment. *Environmental Science & Technology*, **47** (10), 5362–5371.

34 Yang, S., Kraslawski, A., and Qian, Y. (2013) Revision and extension of Eco-LCA metrics for sustainability assessment of the energy and chemical processes. *Environmental Science & Technology*, **47** (24), 14450–14458.

35 Jiménez-González, C. and Overcash, M.R. (2014) The evolution of life cycle assessment in pharmaceutical and chemical applications – a perspective. *Green Chemistry*, **16** (7), 3392.

36 Cespi, D., Beach, E.S., Swarr, T.E., Passarini, F., Vassura, I., Dunn, P.J., and Anastas, P.T. (2015) Life cycle inventory improvement in the pharmaceutical sector: assessment of the sustainability combining PMI and LCA tools. *Green Chemistry*, **17** (6), 3390–3400.

37 Eckelman, M.J. (2016) Life cycle inherent toxicity: a novel LCA-based algorithm for evaluating chemical synthesis pathways. *Green Chemistry*, **18** (11), 3257–3264.

38 Thurston, D.L. and Srinivasan, S. (2003) Constrained optimization for green engineering decision-making. *Environmental Science & Technology*, **37** (23), 5389–5397.

39 Jimenez-Gonzalez, C., Curzons, A.D., Constable, D.J.C., and Cunningham, V.L. (2004) Expanding GSK'S solvent selection guide: application of life cycle assessment to enhance solvent selections. *Clean Technologies and Environmental Policy*, **7** (1), 42–50.

40 Alfonsi, K., Colberg, J., Dunn, P.J., Fevig, T., Jennings, S., Johnson, T.A., Kleine,

H.P., Knight, C., Nagy, M.A., Perry, D.A., and Stefaniak, M. (2008) Green chemistry tools to influence a medicinal chemistry and research chemistry based organisation. *Green Chemistry*, **10** (1), 31–36.
41 Henderson, R.K., Jiménez-González, C., Constable, D.J.C., Alston, S.R., Inglis, G.G.A., Fisher, G., Sherwood, J., Binks, S.P., and Curzons, A.D. (2011) Expanding GSK's solvent selection guide – embedding sustainability into solvent selection starting at medicinal chemistry. *Green Chemistry*, **13** (4), 854.
42 Slater, C.S. and Savelski, M. (2007) A method to characterize the greenness of solvents used in pharmaceutical manufacture. *Journal of Environmental Science and Health, Part A*, **42** (11), 1595–1605.
43 Jiménez-González, C., Ollech, C., Pyrz, W., Hughes, D., Broxterman, Q.B., and Bhathela, N. (2013) Expanding the boundaries: developing a streamlined tool for eco-footprinting of pharmaceuticals. *Organic Process Research & Development*, **17** (2), 239–246.
44 Kopach, M.E. and Reiff, E.A. (2012) Use of the electronic laboratory notebook to facilitate green chemistry within the pharmaceutical industry. *Future Medicinal Chemistry*, **4** (11), 1395–1398.
45 Adams, J.P., Alder, C.M., Andrews, I., Bullion, A.M., Campbell-Crawford, M., Darcy, M.G., Hayler, J.D., Henderson, R.K., Oare, C.A., Pendrak, I., Redman, A.M., Shuster, L.E., Sneddon, H.F., Walker, M.D. (2013) Development of GSK's reagent guides – embedding sustainability into reagent selection. *Green Chemistry*, **15** (6), 1542–1549.
46 Koster, K. and Cohen, M. (2013) Practical approaches to sustainability: iSUSTAIN (tool for green chemistry case study, in *Treatise on Sustainability Science and Engineering*, Springer Science + Business Media, pp. 81–108.
47 Martínez-Gallegos, J.F., Burgos-Cara, A., Caparrós-Salvador, F., Luzón-González, G., and Fernández-Serrano, M. (2015) Dihydroxyacetone crystallization: process, environmental, health and safety criteria application for solvent selection. *Chemical Engineering Science*, **134**, 36–43.
48 Tobiszewski, M., Tsakovski, S., Simeonov, V., Namieśnik, J., and Pena-Pereira, F. (2015) A solvent selection guide based on chemometrics and multicriteria decision analysis. *Green Chemistry*, **17** (10), 4773–4785.
49 Hudlicky, T., Frey, D.A., Koroniak, L., Claeboe, C.D., and Brammer, L.E. Jr. (1999) Toward a "reagent-free" synthesis. *Green Chemistry*, **1** (2), 57–59.
50 Eissen, M. and Metzger, J.O. (2002) Environmental performance metrics for daily use in synthetic chemistry. *Chemistry – A European Journal*, **8** (16), 3580.
51 Dunn, P.J., Galvin, S., and Hettenbach, K. (2003) The development of an environmentally benign synthesis of sildenafil citrate (Viagra™) and its assessment by green chemistry metrics. *Green Chemistry*, **6** (1), 43.
52 Eissen, M. and Hungerbuhler, K. (2003) Mass efficiency as metric for the effectiveness of catalysts. *Green Chemistry*, **5** (2), G25–G27.
53 Gronnow, M.J., White, R.J., Clark, J.H., and Macquarrie, D.J. (2005) Energy efficiency in chemical reactions: a comparative study of different reaction techniques. *Organic Process Research & Development*, **9** (4), 516–518.
54 Andraos, J. (2005) Unification of reaction metrics for green chemistry: applications to reaction analysis. *Organic Process Research & Development*, **9** (2), 149–163.
55 Andraos, J. (2005) Unification of reaction metrics for green chemistry II: evaluation of named organic reactions and application to reaction discovery. *Organic Process Research & Development*, **9** (4), 404–431.
56 Andraos, J. (2006) On using tree analysis to quantify the material, input energy, and cost throughput efficiencies of simple and complex synthesis plans and networks: towards a blueprint for quantitative total synthesis and green chemistry. *Organic Process Research & Development*, **10** (2), 212–240.
57 Van Aken, K., Strekowski, L., and Patiny, L. (2006) EcoScale, a semi-quantitative

tool to select an organic preparation based on economical and ecological parameters. *Beilstein Journal of Organic Chemistry*, **2** (1), 3.
58 Kinen, C.O., Rossi, L.I., and de Rossi, R.H. (2009) The development of an environmentally benign sulfideoxidation procedure and its assessment by green chemistry metrics. *Green Chemistry*, **11** (2), 223–228.
59 Andraos, J. (2009) Global green chemistry metrics analysis algorithm and spreadsheets: evaluation of the material efficiency performances of synthesis plans for oseltamivir phosphate (tamiflu) as a test case. *Organic Process Research & Development*, **13** (2), 161–185.
60 Fringuelli, F., Lanari, D., Pizzo, F., and Vaccaro, L. (2010) An E-factor minimized protocol for the preparation of methyl β-hydroxy esters. *Green Chemistry*, **12** (7), 1301.
61 Andraos, J. (2010) Parameterization and tracking of optimization of synthesis strategy using computer spreadsheet algorithms. *Green Catalysis* (ed. R.H. Crabtree), Wiley-Blackwell, New Jersey.
62 Andraos, J. (2010) *The Algebra of Organic Synthesis: Green Metrics, Design Strategy, Route Selection, and Optimization*, CRC Press, Boca Raton.
63 Andraos, J. (2011) A database tool for process chemists and chemical engineers to gauge the material and synthetic efficiencies of synthesis plans to industrially important targets. *Pure and Applied Chemistry*, **83** (7). doi: 10.1351/pac-con-10-10-07
64 Marcinkowska, M., Rasała, D., Puchała, A., and Gałuszka, A. (2011) A comparison of green chemistry metrics for two methods of bromination and nitration of bis-pyrazolo[3,4-b;4′,3′-e]pyridines. *Heterocyclic Communications*, **17** (5–6). doi: 10.1515/hc.2011.038
65 Augé, J. and Scherrmann, M.-C. (2012) Determination of the global material economy (GME) of synthesis sequences – a green chemistry metric to evaluate the greenness of products. *New Journal of Chemistry*, **36** (4), 1091.
66 Andraos, J. (2012) Green chemistry metrics: material efficiency and strategic synthesis design, in *Encyclopedia of Sustainability SciencCe and Technology*, Springer, pp. 4616–4642. doi: 10.1007/978-1-4419-0851-3_224
67 Andraos, J. (2012) Green chemistry metrics: material efficiency and strategic synthesis design, in *Innovations in Green Chemistry and Green Engineering*, Springer, pp. 81–113. doi: 10.1007/978-1-4614-5817-3_4
68 Andraos, J. (2012) Inclusion of environmental impact parameters in radial pentagon material efficiency metrics analysis: using benign indices as a step towards a complete assessment of "Greenness" for chemical reactions and synthesis plans. *Organic Process Research & Development*, **16** (9), 1482–1506.
69 Assaf, G., Checksfield, G., Critcher, D., Dunn, P.J., Field, S., Harris, L.J., Howard, R.M., Scotney, G., Scott, A., Mathew, S., Walker, G.M.H., and Wilder, A. (2012) The use of environmental metrics to evaluate green chemistry improvements to the synthesis of (S, S)-reboxetine succinate. *Green Chemistry*, **14** (1), 123–129.
70 Bonollo, S., Lanari, D., Longo, J.M., and Vaccaro, L. (2012) E-factor minimized protocols for the polystyryl-BEMP catalyzed conjugate additions of various nucleophiles to α,β-unsaturated carbonyl compounds. *Green Chemistry*, **14** (1), 164–169.
71 Leng, R.B., Emonds, M.V.M., Hamilton, C.T., and Ringer, J.W. (2012) Holistic route selection. *Organic Process Research & Development*, **16** (3), 415–424.
72 Andraos, J. (2013) On the probability that ring-forming multicomponent reactions are intrinsically green: setting thresholds for intrinsic greenness based on design strategy and experimental reaction performance. *ACS Sustainable Chemistry & Engineering*, **1** (5), 496–512.
73 Andraos, J. (2013) Safety/hazard indices: completion of a unified suite of metrics for the assessment of "Greenness" for chemical reactions and synthesis plans.

Organic Process Research & Development, **17** (2), 175–192.

74 Machado, A.A.S.C. (2014) Battery of metrics for evaluating the material greenness of synthesis reactions. *Química Nova*, **37** (6), 1094–1109.

75 White, S.D. (2013) Applying the principles of green chemistry to selected traditional organic chemistry reactions. Available at: http://research.library.mun.ca/6379/ (accessed September 20, 2015).

76 McGonagle, F.I., Sneddon, H.F., Jamieson, C., and Watson, A.J.B. (2014) Molar efficiency: a useful metric to gauge relative reaction efficiency in discovery medicinal chemistry. *ACS Sustainable Chemistry & Engineering*, **2** (3), 523–532.

77 Toniolo, S., Aricò, F., and Tundo, P. (2014) A comparative environmental assessment for the synthesis of 1,3-oxazin-2-one by metrics: greenness evaluation and blind spots. *ACS Sustainable Chemical Engineering*, **2** (4), 1056–1062.

78 Duarte, R.C.C., Ribeiro, M.G.T.C., and Machado, A.A.S.C. (2015) Using green star metrics to optimize the greenness of literature protocols for Syntheses. *Journal of Chemical Education*, **92** (6), 1024–1034.

79 Andraos, J. (2016a) Complete green metrics evaluation of various routes to methyl methacrylate according to material and energy consumptions and environmental and safety impacts: test case from the chemical industry. *ACS Sustainable Chemistry & Engineering*, **4** (1), 312–323.

80 Andraos, J. (2016b) Critical evaluation of published algorithms for determining material efficiency green metrics of chemical reactions and synthesis plans. *ACS Sustainable Chemistry & Engineering*, **4** (4), 1917–1933.

81 Andraos, J., Mastronardi, M.L., Hoch, L.B., and Hent, A. (2016) Critical evaluation of published algorithms for determining environmental and hazard impact green metrics of chemical reactions and synthesis plans. *ACS Sustainable Chemistry & Engineering*, **4** (4), 1934–1945.

82 Subramaniam, B., Helling, R.K., and Bode, C.J. (2016) Quantitative sustainability analysis: a powerful tool to develop resource-efficient catalytic technologies. *ACS Sustainable Chemistry & Engineering*, **4** (11), 5859–5865.

83 He, Y., Tang, L., Wu, X., Hou, X., and Lee, Y.I. (2007) Spectroscopy: the best way toward green analytical chemistry? *Applied Spectroscopy Reviews*, **42** (2), 119–138.

84 Armenta, S. and de la Guardia, M. (2009) Green spectroscopy: a scientometric picture. *Spectroscopy Letters*, **42** (6–7), 277–283.

85 Gaber, Y., Törnvall, U., Kumar, M.A., Ali Amin, M., and Hatti-Kaul, R. (2011) HPLC-EAT (environmental assessment tool): a tool for profiling safety, health and environmental impacts of liquid chromatography methods. *Green Chemistry*, **13** (8), 2021.

86 Hartman, R., Helmy, R., Al-Sayah, M., and Welch, C.J. (2011) Analytical method volume intensity (AMVI): a green chemistry metric for HPLC methodology in the pharmaceutical industry. *Green Chemistry*, **13** (4), 934.

87 Moseley, J.D. and Kappe, C.O. (2011) A critical assessment of the greenness and energy efficiency of microwave-assisted organic synthesis. *Green Chemistry*, **13** (4), 794.

88 Gałuszka, A., Migaszewski, Z.M., Konieczka, P., and Namieśnik, J. (2012) Analytical eco-scale for assessing the greenness of analytical procedures. *TrAC Trends in Analytical Chemistry*, **37**, 61–72.

89 Turner, C. (2013) Sustainable analytical chemistry – more than just being green. *Pure and Applied Chemistry*, **85** (12). doi: 10.1351/pac-con-13-02-05

90 Soto, R., Fité, C., Ramírez, E., Bringué, R., and Iborra, M. (2014) Green metrics analysis applied to the simultaneous liquid-phase etherification of isobutene and isoamylenes with ethanol over Amberlyst™ 35. *Green Processing and Synthesis*, **3** (5), 321–333.

91 Tobiszewski, M., Tsakovski, S., Simeonov, V., and Namieśnik, J. (2013)

Application of multivariate statistics in assessment of green analytical chemistry parameters of analytical methodologies. *Green Chemistry*, **15** (6), 1615.

92 Turner, C. (2013) Sustainable analytical chemistry – more than just being green. *Pure and Applied Chemistry*, **85** (12). doi: 10.1351/pac-con-13-02-05

93 Tobiszewski, M., Marć, M., Gałuszka, A., and Namieśnik, J. (2015) Green chemistry metrics with special reference to green analytical chemistry. *Molecules*, **20** (6), 10928–10946.

94 Tobiszewski, M. (2016) Metrics for green analytical chemistry. *Analytical Methods*, **8** (15), 2993–2999.

95 Andraos, J. and Sayed, M. (2007) On the use of "green" metrics in the undergraduate organic chemistry lecture and lab to assess the mass efficiency of organic reactions. *Journal of Chemical Education*, **84** (6), 1004.

96 Capello, C., Fischer, U., and Hungerbühler, K. (2007) What is a green solvent? A comprehensive framework for the environmental assessment of solvents. *Green Chemistry*, **9** (9), 927.

97 Ribeiro, M.G.T.C., Costa, D.A., and Machado, A.A.S.C. (2010) "Green Star": a holistic green chemistry metric for evaluation of teaching laboratory experiments. *Green Chemistry Letters and Reviews*, **3** (2), 149–159.

98 Mercer, S.M., Andraos, J., and Jessop, P.G. (2012) Choosing the greenest synthesis: a multivariate metric green chemistry exercise. *Journal of Chemical Education*, **89** (2), 215–220.

99 Ribeiro, M.G.T.C. and Machado, A.A.S.C. (2013) Holistic metrics for assessment of the greenness of chemical reactions in the context of chemical education. *Journal of Chemical Education*, **90** (4), 432–439.

100 Ribeiro, M.G.T.C., Yunes, S.F., and Machado, A.A.S.C. (2014) Assessing the greenness of chemical reactions in the laboratory using updated holistic graphic metrics based on the globally harmonized system of classification and labeling of chemicals. *Journal of Chemical Education*, **91** (11), 1901–1908.

101 Andraos, J. (2016) Using balancing chemical equations as a key starting point to create green chemistry exercises based on inorganic syntheses examples. *Journal of Chemical Education*, **93** (7), 1330–1334.

102 Andraos, J. and Hent, A. (2015) Simplified application of material efficiency green metrics to synthesis plans: pedagogical case studies selected from organic syntheses. *Journal of Chemical Education*, **92** (11), 1820–1830.

103 Andraos, J. and Hent, A. (2015) Useful material efficiency green metrics problem set exercises for lecture and laboratory. *Journal of Chemical Education*, **92** (11), 1831–1839.

104 Bishop, K.J.M., Klajn, R., and Grzybowski, B.A. (2006) The core and most useful molecules in organic chemistry. *Angewandte Chemie, International Edition*, **45** (32), 5348–5354.

105 Sheldon, R.A. and Sanders, J.P.M. (2015) Toward concise metrics for the production of chemicals from renewable biomass. *Catalysis Today*, **239**, 3–6.

106 Sheldon, R.A. (2011) Reaction efficiencies and green chemistry metrics of biotransformations. *Biocatalysis for Green Chemistry and Chemical Process Development*, John Wiley & Sons, Inc., Hoboken, NJ, 67–88.

107 Sheldon, RogerA. (2011) Utilisation of biomass for sustainable fuels and chemicals: molecules, methods and metrics. *Catalysis Today*, **167** (1), 3–13.

108 Patel, A.D., Meesters, K., den Uil, H., de Jong, E., Worrell, E., and Patel, M.K. (2013) Early-stage comparative sustainability assessment of new bio-based processes. *ChemSusChem*, **6** (9), 1724–1736.

109 Lima-Ramos, J., Tufvesson, P., and Woodley, J.M. (2014) Application of environmental and economic metrics to guide the development of biocatalytic processes. *Green Processing and Synthesis*, **3** (3). doi: 10.1515/gps-2013-0094

110 Cséfalvay, E., Akien, G.R., Qi, L., and Horváth, I.T. (2015) Definition and application of ethanol equivalent: sustainability performance metrics for

biomass conversion to carbon-based fuels and chemicals. *Catalysis Today*, **239**, 50–55.
111 Juodeikiene, G., Vidmantiene, D., Basinskiene, L., Cernauskas, D., Bartkiene, E., and Cizeikiene, D. (2015) Green metrics for sustainability of bio-based lactic acid from starchy biomass vs chemical synthesis. *Catalysis Today*, **239**, 11–16.
112 Pinazo, J.M., Domine, M.E., Parvulescu, V., and Petru, F. (2015) Sustainability metrics for succinic acid production: a comparison between biomass-based and petrochemical routes. *Catalysis Today*, **239**, 17–24.
113 Iffland, K., Sherwood, J., Carus, M., Raschka, A., Farmer, T., Clark, J., Baltus, W., Busch, R., deBie, F., Diels, L., van Haveren, J., Patel, M.K., Potthast, A., Waegeman, H., and Willems, P. (2015) Definition, calculation and comparison of the "Biomass Utilization Efficiencies (BUE)" of various bio-based chemicals, polymers and fuels. Nova Institute for Ecology and Innovation. Available at http://bio-based.eu/nova-papers/ (accessed January 15, 2017).
114 GreenScreen for Safer Chemicals (2017) www.greenscreenchemicals.org/ (accessed January 15, 2017).
115 Lavoie, E.T., Heine, L.G., Holder, H., Rossi, M.S., Lee, R.E., Connor, E.A., Vrabel, M.A., DiFiore, D.M., and Davies, C.L. (2010) Chemical alternatives assessment: enabling substitution to safer chemicals. *Environmental Science & Technology*, **44** (24), 9244–9249.
116 Lavoie, E.T., Heine, L.G., Holder, H., Rossi, M.S., Lee II, R.E., Connor, E.A., Vrabel, M.A., DiFiore, D.M., and Davies, C.L. (2011) Chemical alternatives assessment: enabling substitution to safer chemicals. *Environmental Science & Technology*, **45** (4), 1747–1747.
117 GreenSuite (2017) www.chemply.com/greensuite (accessed January 15, 2017).
118 Trost, B. (1991) The atom economy – a search for synthetic efficiency. *Science*, **254** (5037), 1471–1477.
119 Jiménez-González, C. and Constable, D.J.C. (2011) *Green Chemistry and Engineering: A Practical Design Approach*, John Wiley & Sons, Inc., New Jersey.
120 Jiménez-González, C., Constable, D.J.C., and Ponder, C.S. (2012) Evaluating the "Greenness" of chemical processes and products in the pharmaceutical industry – a green metrics primer. *Chemical Society Reviews*, **14**, 1485–1498.
121 Trost, B.M. (2002) On inventing reactions for atom economy. *Accounts of Chemical Research*, **35** (9), 695–705.
122 Eissen, M., Mazur, R., Quebbemann, H.-G., and Pennemann, K.-H. (2004) Atom Economy and yield of Synthesis Sequences. *Helvetica Chimica Acta*, **87** (2), 524–535.
123 Wang, W., Lü, J., Zhang, L., and Li, Z. (2011) Real atom economy and its application for evaluation the green degree of a process. *Frontiers of Chemical Science and Engineering*, **5** (3), 349–354.
124 Dicks, A.P. and Hent, A. (2014) Atom economy and reaction mass efficiency. *SpringerBriefs in Molecular Science*, Springer Science + Business Media, pp. 17–44.
125 Wender, P. A.; Miller, B. L. (1993) Toward the ideal synthesis: connectivity analysis and multibond-forming processes, in Organic Synthesis: Theory and Applications, vol. 2 (ed. T. Hudlicky), JAI Press, Greenwich, CT, pp 27–66.
126 Wender, P.A. (2013) Toward the ideal synthesis and transformative therapies: the roles of step economy and function oriented synthesis. *Tetrahedron*, **69** (36), 7529–7550.
127 Clarke, P.A., Martin, W.H.C., Hargreaves, J.M., Wilson, C., and Blake, A.J. (2005) The one-pot, multi-component construction of highly substituted tetrahydropyran-4-ones using the Maitland–Japp reaction. *Organic & Biomolecular Chemistry*, **3** (19), 3551.
128 Clarke, P.A., Santos, S., and Martin, W.H.C. (2007) Combining pot, atom and step economy (PASE) in organic synthesis. Synthesis of tetrahydropyran-4-ones. *Green Chemistry*, **9** (5), 438.
129 Heinzle, E., Weirich, D., Brogli, F., Hoffmann, V.H., Koller, G., Verduyn, M.A., and Hungerbühler, K. (1998)

Ecological and economic objective functions for screening in integrated development of fine chemical processes. 1. Flexible and expandable framework using indices. *Industrial & Engineering Chemistry Research*, **37** (8), 3395–3407.

130 Koller, G., Fischer, U., and Hungerbühler, K. (2000) Assessing safety, health, and environmental impact early during process development. *Industrial & Engineering Chemistry Research*, **39** (4), 960–972.

131 Curzons, A.D., Mortimer, D.N., Constable, D.J.C., and Cunningham, V.L. (2001) So you think your process is green, how do you know? – using principles of sustainability to determine what is green – a corporate perspective. *Green Chemistry*, **3** (1), 1–6.

132 Gonzalez, M.A. and Smith, R.L. (2003) A methodology to evaluate process sustainability. *Environmental Progress*, **22** (4), 269–276.

133 Rahman, M., Heikkilä, A.-M., and Hurme, M. (2005) Comparison of inherent safety indices in process concept evaluation. *Journal of Loss Prevention in the Process Industries*, **18** (4–6), 327–334.

134 Rosini, G., Borzatta, V., Paolucci, C., and Righi, P. (2008) Comparative assessment of an alternative route to (5-benzylfuran-3-yl)methanol (Elliott's alcohol), a key intermediate for the industrial production of resmethrins. *Green Chemistry*, **10** (11), 1146.

135 Zhang, X., Li, C., Fu, C., and Zhang, S. (2008) Environmental impact assessment of chemical process using the green degree method. *Industrial & Engineering Chemistry Research*, **47** (4), 1085–1094

136 Righi, P., Rosini, G., and Borzatta, V. (2009) Introducing green metrics early in process development. Comparative assessment of alternative industrial routes to Elliott's alcohol, A key intermediate in the production of Resmethrins, in *Sustainable Industrial Chemistry*, Wiley-VCH Verlag GmbH, Weinheim, pp. 551–562.

137 Wernet, G., Papadokonstantakis, S., Hellweg, S., and Hungerbühler, K. (2009) Bridging data gaps in environmental assessments: modeling impacts of fine and basic chemical production. *Green Chemistry*, **11** (11), 1826.

138 Wernet, G., Mutel, C., Hellweg, S., and Hungerbühler, K. (2010) The environmental importance of energy use in chemical production. *Journal of Industrial Ecology*, **15** (1), 96–107.

139 Jimenez-Gonzalez, C., Ponder, C.S., Broxterman, Q.B., and Manley, J.B. (2011) Using the right green yardstick: why process mass intensity is used in the pharmaceutical industry to drive more sustainable processes. *Organic Process Research & Development*, **15** (4), 912–917.

140 Hessel, V., Cortese, B., and de Croon, M.H.J.M. (2011) Novel process windows – concept, proposition and evaluation methodology, and intensified superheated processing. *Chemical Engineering Science*, **66** (7), 1426–1448.

141 Tugnoli, A., Santarelli, F., and Cozzani, V. (2011) Implementation of sustainability drivers in the design of industrial chemical processes. *AIChE Journal*, **57** (11), 3063–3084.

142 Van der Vorst, G., Dewulf, J., Aelterman, W., De Witte, B., and Van Langenhove, H. (2011) A systematic evaluation of the resource consumption of active pharmaceutical ingredient production at three different levels. *Environmental Science & Technology*, **45** (7), 3040–3046.

143 Dach, R., Song, J.J., Roschangar, F., Samstag, W., and Senanayake, C.H. (2012) The eight criteria defining a good chemical manufacturing process. *Organic Process Research & Development*, **16** (11), 1697–1706.

144 Ruiz-Mercado, G.J., Smith, R.L., and Gonzalez, M.A. (2012) Sustainability indicators for chemical processes: I. Taxonomy. *Industrial & Engineering Chemistry Research*, **51** (5), 2309–2328.

145 Tian, J., Shi, H., Li, X., Yin, Y., and Chen, L. (2012) Coupling mass balance analysis and multi-criteria ranking to assess the commercial-scale synthetic alternatives: a case study on glyphosate. *Green Chemistry*, **14** (7), 1990.

146 Kjell, D.P., Watson, I.A., Wolfe, C.N., and Spitler, J.T. (2013) Complexity-based

metric for process mass intensity in the pharmaceutical industry. *Organic Process Research & Development*, **17** (2), 169–174.

147 Li, T. and Li, X. (2014) Comprehensive mass analysis for chemical processes, a case study on L-Dopa manufacture. *Green Chemistry*, **16** (9), 4241

148 Garba, M.D. and Galadima, A. (2015) Efficiencies of green chemistry metrics in the activities of petroleum refinery process. *International Science and Investigation Journal*, **4** (2), 65–87.

149 Gómez-Biagi, R.F. and Dicks, A.P. (2015) Assessing process mass intensity and waste via an aza -Baylis–Hillman reaction. *Journal of Chemical Education*, **92** (11), 1938–1942.

150 Morais, A.R.C., Dworakowska, S., Reis, A., Gouveia, L., Matos, C.T., Bogdał, D., and Bogel-Łukasik, R. (2015) Chemical and biological-based isoprene production: green metrics. *Catalysis Today*, **239**, 38–43.

151 Sheldon, R.A. (2012) Fundamentals of green chemistry: efficiency in reaction design. *Chemical Society Reviews*, **41** (4), 1437–1451.

152 Shokrian, M., High, K.A., and Sheffert, Z. (2014) Screening of process alternatives based on sustainability metrics: comparison of two decision-making approaches. *International Journal of Sustainable Engineering*, **8** (1), 26–39.

153 Vaccaro, L., Lanari, D., Marrocchi, A., and Strappaveccia, G. (2014) Flow approaches towards sustainability. *Green Chemistry*, **16** (8), 3680.

154 Li, J., Simmons, E.M., and Eastgate, M.D. (2017) A data-driven strategy for predicting greenness scores, rationally comparing synthetic routes and benchmarking PMI outcomes for the synthesis of molecules in the pharmaceutical industry. *Green Chemistry*, **19** (1), 127–139.

155 ACS, Green Chemistry Institute, Design Principles for Sustainable and Green Chemistry and Engineering. Available at https://www.acs.org/content/acs/en/greenchemistry/what-is-green-chemistry/principles/design-principles-for-green-chemistry-and-engineering.html (accessed January 16, 2017)

156 Kahneman, D. (2011) *Thinking, Fast and Slow*. Farrar, Straus and Giroux, New York.

157 Kegan, R. and Lahey, L.L. (2009) *Immunity to Change: How to Overcome It and Unlock the Potential in Yourself and Your Organization (Center of Public Leadership)*. Harvard Business Review Press, Boston, MA.

158 Board on Chemical Sciences and Technology and Board on Environmental Studies and Toxicology (2014) *A Framework to Guide Selection of Chemical Alternatives*. National Academies Press, Washington, DC, U S.

159 OECD (2017) OECD Substitution and Alternatives Assessment Toolbox, www.oecdsaatoolbox.org/ (accessed January 16, 2017).

160 Interstate Chemicals Clearinghouse (2017) www.theic2.org/ (accessed January 16, 2017).

161 OECD (2017) OECD QSAR Toolbox. Available at www.qsartoolbox.org/ (accessed January 16, 2017).

162 ToxPredict (2017) apps.ideaconsult.net/ToxPredict (accessed January 16, 2017).

163 EPA (2017) EPI Suite – Estimation Program Interface. Available at https://www.epa.gov/tsca-screening-tools/epi-suitetm-estimation-program-interface (accessed January 16, 2017).

2
Expanding Rational Molecular Design beyond Pharma: Metrics to Guide Safer Chemical Design

Nicholas D. Anastas, John Leazer, Michael A. Gonzalez, and Stephen C. DeVito

2.1
Introduction to Safer Chemical Design

Chemicals, both natural and synthetic, are ubiquitous in modern society and have provided benefits for medicine, consumer products, building materials, and other chemical resources. However, production and use of some of these chemicals have adversely affected public health and the environment due in part to the intrinsic hazard of the chemical or of a material.

An ever-increasing demand for new products made from renewable, sustainable sources, that pose minimal or no toxicity, requires a scientific and systematic approach to move from legacy chemicals that have not been tested for toxicity to a safer chemistry paradigm. A major challenge to the practice of designing safer chemicals is making a chemical product or process that is simultaneously effective and safe [1]. In other words, use efficacy and potency of use must be maximized, while toxicity is concurrently minimized – a formidable challenge. Designing chemicals and products for reduced hazard is not among the goals of industrial chemicals. Changing the structure of a molecule results in a different chemical. Often times, this fundamental concept is either assumed when discussing the process of developing design guidelines or is not considered at the time of design.

There is a long history of innovation by the pharmaceutical industry in designing chemicals aimed at therapeutic applications that meet efficacy and potency needs while minimizing the potential for adverse patient outcomes (i.e., side effects), as much as feasible for a medicine to reach the market. Implementation of green chemistry practices in the pharmaceutical sector has reduced the quantities of hazardous chemicals used and released to the environment, thereby reducing risks to public health and the environment [2].

Can this knowledge coupled with methods and examples put forth by green chemistry and green engineering continue to advance society toward the goal of designing and producing chemicals in a greener and sustainable manner? We believe demonstrably yes.

Handbook of Green Chemistry Volume 11: Green Metrics, First Edition. Edited by David J. Constable and Concepción Jiménez-González.
© 2018 Wiley-VCH Verlag GmbH & Co. KGaA. Published 2018 by Wiley-VCH Verlag GmbH & Co. KGaA.

2.2
Life Cycle Thinking

2.2.1
Sustainability, Green Chemistry, and Green Engineering

Just as in the business, financial, and education sectors, the concepts of sustainability are also infiltrating the scientific disciplines with a particular emphasis on chemistry and engineering. Moreover, sustainability and its implementation are becoming increasingly important in the marketplace and throughout everyday activities. However, chemists and engineers frequently approach sustainability from a very narrow perspective, without consideration for the life cycle implications of the product, service, or process being developed, or implemented, or any changes made in the hopes of improvement to its efficacy, efficiency, or overall sustainability.

Over the past 20+ years there have been tremendous efforts put forth on solidifying chemical and engineering concepts that focus on greening the approaches used to design, develop, and produce a product, good, or service. In the chemical sciences, significant discoveries have been contributed regarding chemical design, property manipulation, activity, and toxicity. Building upon these chemical and biological advances in the toxicological sciences has successfully generated computational and mechanistic data that elucidate how chemicals, by-products, and degradation products exert adverse effects upon biological organisms and the ecosystem. While chemical scientists have been effective in using the 12 principles of green chemistry [3] and the 12 principles of green engineering [4], they frequently fail to look beyond the chemical product or process to evaluate the entire life cycle impact to the environment, the economics, or the social structures.

Sustainability transcends traditional disciplinary boundaries, and there is a greater need to evaluate the life cycle impact of a chemical product or process on the environment, the economics, or the social structures. With thousands of chemicals in use today and new chemicals being introduced annually, a more holistic approach is needed to evaluate their impacts. There are opportunities for scientists and engineers to become more exposed to new approaches that incorporate concepts from other disciplines that contribute to the holistic approach to molecular and process design. Therefore, it is necessary to provide the tools, methods, and examples that green chemists, toxicologists, and engineers will need to properly assess the sustainability and environmental impact of a chemical or chemical process throughout its life cycle.

Along with the successes and gains resulting from green chemistry and green engineering approaches, it is now time for these new concepts, as well as those from other disciplines, to be introduced to chemists and engineers that will allow them to realize the magnitude of influence they have "beyond the bench." By focusing in this direction, novel and innovative research opportunities for interdisciplinary concepts will contribute to more sustainable chemical design, synthesis, process, and use, thereby making scientists and engineers better equipped to utilize their knowledge, as molecular and process architects, to influence and accelerate the development of sustainable chemical design, synthesis, manufacture, and use. The end goal is the creation and design of truly environmentally benign or sustainably designed chemicals. Such approaches, in addition to those previously mentioned, to meet this goal include the following:

- *Sustainable molecular design*: Designing molecules to have the desired purpose and efficacy while minimizing adverse impacts to human health and the environment, including the use and application of new computational methods to achieve these properties.
- *Design for the environment and fit-for-purpose in the twenty-first century*: The focus on reduction of materials use, reuse of materials, energy use, water consumption, recycling, and the design of processes that facilitate these opportunities.
- *Life cycle considerations*: Consideration to the design, manufacture, and use of a product across its entire life cycle, from raw material extraction and conversion, to its manufacture and distribution, through to use/reuse/recycling, and ultimate disposal.
- *Life cycle assessment (LCA)*: This methodology and attendant quantitative approaches provide in-depth understanding on the impacts associated with a product, from the perspective of its chemical structure through to its process manufacture within the framework of traditional cradle to grave considerations to more advanced cradle-to-cradle paradigms.
- *Sustainability metrics*: Developing the means to quantify product and process sustainability.

2.2.2
Life Cycle Considerations

In order to achieve sustainability with respect to the life cycle of chemical and a process, researchers must have the ability not only to minimize or eliminate this risk across the life cycle but also be able to access and quantify any remaining risk and ensure the applied activities taken are in a more sustainable direction. As the life cycle of a chemical is mapped out, it is evident there are many areas of opportunity that exist for improvements to the current state, as well as regions for new research. It is now the goal of research to demonstrate that a chemical must not only be improved at the synthesis stage but also can be manipulated and improved at any stage of the life cycle with direct and indirect benefits and consequences. Also, when introducing a new chemical or attempting to improve an existing chemical, the importance of applying a holistic, multidisciplinary approach needs to be demonstrated.

Technology plays a significant role in virtually all aspects of a society. However, in the area of environmental protection, its role has largely been limited to the remediation of environmental problems and the implementation of strategies to achieve goals has been highly regulated. It is imperative that we as a society become as proactive to resource needs and environmental challenges as we are reactive to handling environmental damages. By applying a proactive and holistic approach, we can begin to minimize or eliminate these impacts across the entire chemical life cycle and increase protection of the environment and human health.

A transformation in the way we design, manufacture, and manage materials, compounds, and chemicals is needed to meet the challenge of developing safe and effective materials, compounds, and chemicals. The principles of green chemistry emphasize consideration of environmental factors [5]. These principles are used as a pathway to minimize potential public health and environmental risks.

The mapping out of the life cycle of a chemical or a technology can lead to identifying "hotspots" and opportunities for improving the current entity or technologies, as well as research areas for the development of innovative chemicals or processes [6]. Such opportunities offer a proactive approach that, coupled with a holistic view, provides the best path for increasing the sustainability of a system. This means minimizing impacts at one stage of the life cycle, or in one geographic region, or in a particular impact category, all while avoiding unrecognized increased impacts elsewhere. Taking a life cycle perspective requires a policy developer, environmental manager, or product designer to look beyond their own system, knowledge, or in-house operations.

Upon determining the need for designing a new chemical, and that the functionality for it has been satisfactorily addressed, the next stage is synthesis. In this stage, approaches such as green chemistry and engineering are introduced. While these areas do not constitute new disciplines, they are new approaches to performing chemistry and engineering in an environmentally conscientious manner. While it is ultimately important to design a chemical to have its desired efficacy and minimal toxicity, it is equally important that the design process is also as sustainable as possible. As previously discussed, the principles of sustainability are now being applied in the chemical sector with the emphasis of reducing negative impacts upon the environment and human health. To achieve sustainability across the life cycle of a chemical, researchers must be able to assess and quantify risks though a comprehensive framework and valuable set of metrics.

2.2.3
Life Cycle Assessment

Life cycle assessment is a multimedia environmental management approach that is used to look holistically and comprehensively at a wide range of impacts for products, processes, and activities. LCA methods have been standardized as part of the International Standards Organization (ISO) environmental management standards in ISO 14 040:2006 and 14 044:2006. LCA is intended to be a quantitative approach; however, qualitative aspects can, and should, be taken into

Figure 2.1 Key stages of a life cycle.

account when quantitative data are not obtainable in order to form a complete picture of the product system and its stages [7]. Figure 2.1 provides a graphical representation of the life cycle environmental impacts that are involved in the life cycle of a chemical, product, or service.

The major components of LCA are the goal and scope definition, developing the life cycle inventory (LCI), conducting a life cycle impact assessment (LCIA), and interpreting the results.

- *Goal and scope definition*: Define and describe the product, process, or activity. Establish the context in which the assessment is to be made and identify the boundaries and environmental effects to be reviewed for the assessment.
- *Life cycle inventory*: Identify and quantify energy, water, and material usage and the environmental releases (e.g., air emissions, solid waste disposal, waste water discharges).
- *Life cycle impact assessment*: Assess the potential human and ecological effects of energy, water, and material usage and the environmental releases identified in the inventory analysis.
- *Interpretation*: Evaluate the results of the inventory analysis and impact assessment to select the preferred product, process, or service with a clear understanding of the uncertainty and the assumptions used to generate the results.

The expansion of LCA and life cycle considerations to incorporate economic and societal factors has also been an area of intense research focus. This expansion allows for the formal LCA methods with their initial emphasis on the characterization of environmental impact(s) to be further elucidated to include the environmental aspect of sustainability. To elaborate, sustainability includes consideration of all three pillars – environment, economy, and society – across the life cycle stages.

Upon completing an LCA, the resulting life cycle inventory quantitatively represents the system under evaluation [8]. To extend the assessment, these inputs and outputs within the LCI are evaluated for their potential environmental and human health impact attributed to the environmental resources and releases identified by the LCI. Impact assessment methodology addresses the ecological and human health effects, as well as resource depletion (e.g., water, land, minerals). An LCIA attempts to establish a linkage between the product or process and its potential environmental impacts.

Comprehensive impact assessments such as sustainability metrics, LCA, and industrial ecology involve the quantification of a large number of impacts. The tie-in with LCA and LCIA principles and methodologies, that is, having scientists and engineers look across systems holistically and consider life cycle impacts using green chemistry and green engineering approaches, should be readily apparent [9]. A chemist only considering or optimizing a single reaction or a reaction sequence that is part of a process will easily miss the use of a very hazardous chemical in another part of the value chain. By looking across the entire value chain, it is possible to optimize the use of chemicals and energy.

2.2.4
Chemical Process Sustainability Evaluation – Metrics

The finite availability and accelerated depletion of ecological goods and services that sustain societal needs are compelling the chemical industry to take a more sustainable path. A main limitation to applying this approach to process development is the assessment of its sustainability. This assessment is required in order to provide guidance, methodology, and tools to designers for the creation of new chemical processes and/or modification of existing designs, all the while proceeding along the path to sustainability. The quantification of process sustainability can be attained through indicators capable of translating chemical process performance and operating conditions into metrics that have significance on a sustainability measurement scale [10].

GREENSCOPE (Gauging Reaction Effectiveness for the Environmental Sustainability of Chemistries with a multi-Objective Process Evaluator) is a methodology and tool that can be used for process design, integration, optimization, and sustainability assessment [11,12]. This methodology and tool can provide an evaluation of a process that can be interpreted to determine how to minimize raw material consumption and energy loads while minimizing/eliminating releases and maintaining the economic feasibility of the process. Furthermore, GREENSCOPE provides an approach that encompasses process design from the problem definition to the final design, through all stages of product and process design such as the application of molecular design, incorporation of the 12 principles of green chemistry, carried over to the manufacturing phase to include the use of the 12 green engineering principles, the incorporation of sustainable by design for the use phase, and application of reverse engineering concepts for recycle and disposal phases. By assessing at the early process design stages, minimal implementation costs to perform changes are realized and a higher potential to influence the sustainability behavior of the process and operation is experienced.

Several tools, such as process simulators, pure component property databases, thermodynamic models, equations of state, and experimental measurements, can provide the required data for computing GREENSCOPE sustainability indicators [13]. These tools complete the data directly obtained from the analyzed process. Different free online databases are available to provide physicochemical property data and the results of standard toxicology tests (US Environmental Protection Agency; US National Library of Medicine). Governmental organizations such as EPA, NIST (US National Institute of Standards and Technology), OECD (Organization for Economic Cooperation and Development), and the European Chemicals Agency provide online databases for risk and hazard assessment, quantitative structure–activity relationships (QSARs) models, physicochemical property databases, prediction and correlation of properties, and environmental, health, and safety (EHS) impact assessment tools. Several of EPA's computational tools and database products such as the Aggregated Computational Toxicology Resource (ACToR) database, the Integrated Risk Information System (IRIS) database, the Toxicity Estimation Software Tool (TEST), the EPI Suite™ toolbox, the Waste Reduction Algorithm (WAR), and the Tool for the Reduction and Assessment of Chemical and other environmental Impacts (TRACI) will be integrated to GREENSCOPE sustainability assessment tool.

If the process sustainability assessment decreases at some level, there are opportunities for process improvements to make topological (arrangement of equipment) and parametric (operating condition) changes without having to obtain a new detailed design [14].

Using GREENSCOPE for a sustainability assessment can provide information to the LCA and a prediction of the real data needs for process design at early stages, where data availability for conducting an LCA is more difficult to obtain [15]. The simultaneous implementation of this methodology with LCA facilitates the design of sustainable chemical processes and results in a positive global (i.e., value chain) impact. For most process sustainability improvements, a large component of the sustainability performance is attributable to a small fraction of the contributing process design and operational factors. A detailed assessment will help to identify those contributing factors that account for the bulk of the total sustainability performance. Based on a sensitivity analysis of the effects on the sustainability performance, key decision aspects will be identified and chosen. Manifesting an improved sustainability performance requires knowledge of the effect of process design and operational factors on each indicator.

2.3
Attributes of Chemicals of Good Character

Living a long, healthy, and productive life requires individuals to adopt and practice good habits and reject practices and behaviors that are considered detrimental to our longevity and health. In other words, we should do certain beneficial things and avoid other undesired things. A similar approach can be articulated for designing safer, healthier, sustainable chemicals by expressing outcomes in advance and design *for* beneficial attributes and design *against* unwanted features. DeVito has described these outcomes in terms of characteristics of a safer chemical [16]. Here we offer our approach for guiding the design of safer chemicals (Figure 2.2).

Methods of characterizing the potential unsavory attributes of a chemical are required to develop robust profiles of toxic chemicals and materials as tools to rank preferred safer alternatives and to ultimately design less hazardous products. An emerging area of research is computational toxicology that represents the next step forward in the quest to understand the mechanisms and modes of toxicity more completely. Elucidating critical molecular interactions that lead to

*Includes biological and use endpoints

Figure 2.2 Seven virtues and four turpitudes of safer chemical design.

perturbations in biochemical pathways that result in adverse health effects provided invaluable information to characterize existing products and design safer chemicals. The Toxicity Forecaster or ToxCast is a software tool developed by the USEPA that provides public access to high-throughput screening (HTS) data for hundreds of biological indicators and endpoints [17]. Investigators evaluate and integrate the information gleaned from these assays to determine potential indicators of toxic effects. Among the toxic effects of concern are cancer, developmental and reproductive toxicity, and endocrine disruption. Other available tools, such as EPISuite™, ACToR, and GREENSCOPE, to evaluate the attributes of preferred chemicals that are not primarily toxicology focused were described earlier in this chapter.

2.4
Tools for Characterizing the Attributes of Chemicals of Good Character

The foundation of any endeavor to design safer, sustainable products, process, or materials is best guided by a comprehensive understanding of the structure–hazard relationship of a molecule and its target(s). Several framework options have been proposed as strategies for organizing holistic, comprehensive approaches to designing safer chemicals [16,18–22]. There is not one correct way to go about designing a safer chemical when an individual decides, for whatever reason, to design a new chemical or to redesign an existing chemical.

The art and science of toxicology has profoundly evolved over the past two decades, from a fundamentally descriptive discipline to one that takes advantage of a number of advancements in molecular biology, computer science, and elucidating mechanisms of toxic action to describe toxicological pathways from molecular initiating events (MIE) to apical outcomes [23]. Molecular initiating events are those initial interactions between a molecule and a biochemical or biological target that can be linked to the advent of an adverse outcome or toxicity. Traditional methods used for toxicological investigation heavily relied on animal testing to determine the hazard profile of a potential toxicant. Over the years these methods were refined and improved as data quality and quantity advanced, thereby providing greater confidence in a toxicity profile.

A potentially very powerful set of emerging tools to inform toxicology through computational methods (*in silico*) includes a suite of *in vitro* HTS assays, developed by government, academic, and private researchers. Computational chemistry applied to the practice of safer chemical design uses computational toxicology, which is the application of mathematical and computer models to help assess hazards and risks to humans and the environment [24]. Understanding mechanisms of hazard, to infer and incorporate molecular adjustments to reduce toxicity and other hazard endpoints, is the ultimate goal of designing safer chemicals.

Chemical toxicity is associated with a myriad of harmful biological effects. As such, there is an urgent need to quickly and reliably evaluate existing and proposed chemicals for possible toxicological effects. Experimental methodologies for estimating toxicity values are well known and can be both cost-prohibitive and time-consuming [25]. With these facts in mind, computational methodologies for the reliable toxicity estimation of endpoints may offer significant value to both research (industrial, academic, and government) and regulatory organizations. EPA's Office of Research and Development (ORD) developed another useful tool known as Toxicity Estimation Software Tool (*aka* TEST) for this purpose. TEST is a software tool that was developed in response to a need for estimating the toxicity of compounds for estimating toxicological and physical properties of chemicals, so they can be better evaluated in life cycle considerations.

TEST uses a series of QSAR methodologies based on hierarchical clustering, group contribution, and multilinear regression. The results are typically presented using a consensus method that averages the results from several methods. At present, TEST has the following endpoints: fathead minnow LC_{50}, daphnia magna LC_{50}, *tetrahymena pyriformis* IGC_{50}, oral rat LD_{50}, bioconcentration factor (BCF), developmental toxicity, and Ames mutagenicity assay. In addition, the software program can calculate the following chemical properties: boiling point, flash point, surface tension, melting point, viscosity, vapor pressure, thermal conductivity, density, and water solubility. Examples of endpoints that are relevant to human health that TEST can predict include mutagenicity and acute mammalian toxicity. Quantitative predictions of toxicity via QSAR methodology has significantly improved over the last several years and has finally reached the point where such predictions are routinely used to fill in missing comparison data [26].

Using molecular descriptor data, hierarchical clustering divides toxicity data training sets into a series of clusters. Each cluster is developed into a QSAR model, thereby allowing for facile toxicity estimation of chemicals that do not neatly fit into a single chemical class. The confidence in the prediction increases as the structure of the chemical more closely matches structures in a given cluster [27].

The chemical structure to be analyzed by TEST can be readily drawn utilizing the program's own chemical sketcher window. Alternatively, the chemical structure can be imported using common chemical drawing programs, including ChemDraw™. Moreover, structures can be entered via a text file using SMILES. Once the structure has been imported, the toxicity values can be estimated using the models in TEST. If experimental values are available, they are presented as well.

The proposed attributes of chemicals of good character (Table 2.1) are meant to be used as guidelines when developing comprehensive molecular design strategies to minimize risk through hazard or exposure mitigation. The anthropogenic analogies are included for those individuals that may be unfamiliar with the more technical aspects associated with evaluating chemical attributes, so hopefully this will provide a bridge for a wider audience. Specific reference limits for individual measurement metrics were not included reflecting the dynamic

Table 2.1 Attributes and measurement metrics of chemicals of good character.

Attribute	Colloquial equivalent	Hazard mitigation function	Measurement metric
Strive to reduce or eliminate the use of chemicals	Why are you here?	Ensure essentiality for possible removal	Careful inspection of the role of each component of a reaction or process
Maximize biological and use potency and efficacy	A little goes a long way	Minimizes the amount of chemical needed to achieve function (use potency). The chemical does whatever it is supposed to do quite well (use efficacy)	Receptor binding affinity Dose–response; application rates; quantities needed; frequency of use Longevity in application (efficacy)
Strive for economic efficiency	Can we afford this?	Benefit greater than cost	Cost–benefit/LCA
Limited environmental Bioavailability	Can't get in the house	Mitigates access to site of action	Log$P$ Aqueous solubility
Limited environmental mobility	Stay close to home	Reduces exposure to distant receptors	Water solubility Soil binding constants Half-lives
Design for selective reactivity	Not everyone is a suspect	Reduces probability of reacting with nontarget molecules	HOMO/LUMO gap (ΔE) Globularity
Minimal incorporation of known hazardous functional groups	If looks could kill	Reduces the potential for	Comparison to repository of known toxicophores; structural alerts; metabolism to reactive substituents Professional judgment
Minimal use of toxic solvents	Use only what you need	Limits exposure to workers; minimizes potential environmental pollution	Atom economy; process efficiency; mass intensity
Limited persistence and bioaccumulation	Knows when to leave the party	Limits exposure time and potential to toxicants	Log$P$ Ready biodegradability
Quick transformation to innocuous products (Characterize potential toxic metabolites)	A change for the better	Reduces exposure to potentially toxic substances	Monitoring parent and daughter products
Avoid extremes of pH	Average is OK sometimes	Reduces corrosivity; limits large deviations from physiological pH	Hydrogen ion concentration

and use-specific nature of the process of hazard assessment. Each of the attributes is described below in more detail.

2.4.1
Strive to Reduce or Eliminate the Use of Chemicals

People do not necessarily purchase or consume chemicals or products *per se*; they seek function regardless of the chemistry or the chemical itself. Scientists can achieve increased sustainability by providing these services through limiting or ideally eliminating the use or generation of hazardous chemicals. A reasonable first question that a molecular designer should ask is as follows: "What is the function of each chemical in my product or material?" If the compound is function-critical, then an examination of potentially less hazardous substitutes should be performed. This method of continuous improvement encourages innovation to reduce risk, improve product safety, and protect public health.

2.4.2
Maximize Biological and Use Potency and Efficacy

The more potent a chemical is in doing whatever it is supposed to do, such as a drug ingested to lower blood pressure, a pesticide applied to a field of wheat to protect the crop from a parasite, or a commercial dye or pigment intended to impart color to a material, the more ideal it is to use lesser quantities of that chemical to fulfill its purpose (i.e., function). With the reduced quantities of that chemical needed, there needs to be a reduction in its manufacture, transport, handling, and use. And since less is manufactured, transported, handled, and used in application, the potential for exposure to that chemical is likely to be reduced. Therewith, less risk is provided than that of a more potent chemical and is not correspondingly more toxic than a structurally analogous alternative chemical that is less toxic.

The more efficient a chemical is at achieving its function, the less likely is the need for frequent use or consuming greater amounts of the product. For example, in the case of dyes used to color fabrics, use efficacy is often expressed in terms of color fastness, or the ability of the dye to resist fading or running. A dye that imparts the desired color to a fabric that also does not readily fade or run after washing of the dyed fabric is said to be more efficacious than a dye that imparts the same desired color but fades or runs more readily. With the latter dye, the dyed material would have to be replaced more frequently, which in turn requires manufacture of additional dye and fabric.

2.4.3
Strive for Economic Efficiency

Chemists have relied on, and continue to use, reactive molecules as the primary tool for synthetic chemistry. Reactive molecules are more likely to react with biological pathways, cellular components, and tissues resulting in disruption and damage [28]. The reactivity of a molecule generally involves electrophilic, nucleophilic, or attack by radical species [29,30]. Characterizing and quantitating reactivity relies on molecular descriptors such as energy difference (ΔE), globularity, and hydrophobicity (log$P$), to cite but a few. The practice of relating a physicochemical attribute to a biological property or activity has been used for many years and includes many endpoints [31]. This approach has been used in the pharmaceutical industry and by EPA to characterize risk for chemicals with no or limited toxicity data and for pesticide fate and transport potential. However, the application of these tools for designing industrial chemicals is not as robust because it generally relied on quantitative rather than qualitative data for a limited number of endpoints.

2.4.4
Limited Bioavailability

Reducing bioavailabilty through molecular manipulation provides a relatively simple approach to reducing exposure. This is a risk reduction strategy achieved by limiting exposure to the ultimate receptor. Lipinski *et al.* developed a set of guidelines commonly referred to as "Lipinski's Rule of Five" that estimate the likelihood of a compound administered by the oral route of exposure given a chemical's physicochemical characteristics, thereby providing an estimate of its bioavailability [32]. These characteristics provide guidance to drug discovery scientists and medicinal chemists for selecting physicochemical features that are likely to be associated with good absorption through the gastrointestinal tract (GI). If industrial chemists can use these rules in reverse to limit oral absorption, so-called "reverse Lipinski" approach, toxicity is likely to be limited by the oral route of exposure.

2.4.5
Limited Environmental Mobility

Compounds that are water soluble tend to move more readily through environmental compartments and may contaminate air, water, soil, and sediments at sites quite a distance form their origin, thereby exposing humans and other organisms to potentially toxic substances. Incorporating fundamental physicochemical principles to chemical assessment and design can guide decisions for selecting materials that are minimally mobile in the environment.

2.4.6
Design for Selective Reactivity: Toxicity

Selective toxicity is a term first described by Albert [33] that is defined as injurious to one kind of living matter without being injurious to another. The principles of selective toxicity can be applied to the design of pesticides and other chemicals designed intentionally to be toxic by exploiting the knowledge of biochemical processes, physiological pathways, and anatomical difference among organisms. Reactivity is a function of the structure and energetics of both the reactant and target molecules, following the fundamental principles of chemistry and quantum mechanics.

2.4.7
Minimize the Incorporation of Known Hazardous Functional Groups: Toxicophores and Isosteres

Toxicophores, or structural alerts, are well-defined structural entities that have been associated with adverse outcome pathways [18,20,34]. Examining the 2D and 3D structures of molecules provided trained scientists with a level of insight into the behavior of a molecule in living systems that provides a starting point for making design choices. This is a task that demands much more effort than molecular voyeurism to confront the business end of a compound that may be implicated in an adverse outcome.

2.4.8
Minimize the Use of Toxic Solvents

Solvents are currently used throughout the chemical industry for a variety of reasons. In many industries, they are used as a median to dissolve solutes to allow for chemical reactions to take place, or to allow for more facile processing of materials. Solvents are also used to extract materials from mixtures of materials. The more commonly used solvents are often volatile; hence once they have performed their intended function, they can be evaporated, oftentimes to the detriment of the environment and human health. Solvents account for 75–80% of waste in the pharmaceutical industry [35]. In this one industry alone, any small localized decrease in the amount of solvent used or decrease in toxicity of a solvent used in a process could potentially result in a pronounced positive global impact in terms of human exposure and environmental consequences.

The environmental and human health concerns are obviously the more important features in solvent selection and usage; however, there is also the ever-present financial component. Hazardous (e.g., toxic, flammable) solvents require specialized safety equipment, permits, and so on, which may add additional cost to the user. Thus, from a sustainability framework, the use of a more benign solvent is preferred over the use of a toxic or flammable solvent, under the assumption that the alternative solvent functions with the same or greater productivity of the original solvent. With these factors in mind, EPA's ORD developed the Program for Assisting the Replacement of Industrial Solvents (PARIS III) that is a solvent substitution software tool that was developed in an effort to suggest alternatives to solvents currently used in chemical processing, especially in industry, to a more benign solvent or solvent mixture (i.e., decreased human and ecological impact). The program generally suggests a drop-in solvent or solvent system that can usually be used within the existing processing facilities, thereby eliminating costly upgrades.

Armed with the user-friendly graphical user interface of PARIS III, a technician inputs the chemical composition of the original solvent mixture. This allows specific properties of the solvent mixture to be determined (e.g., solubility, dielectric constant, freezing point, boiling point, surface tension, vapor pressure, thermal conductivity, molecular mass, viscosity, density, etc.). PARIS III then searches its database of over 5000 solvents to find alternative solvents with property values close to those of the original solvent. All alternative solvents suggested by PARIS III carry a decreased environmental and human health risk as compared with the original solvent. If single-solvent alternatives cannot be found, the user can move on to examine mixed-solvent systems that approximate the behavior of the original solvent. As before, properties of the solvent mixtures are calculated, and the alternative solvents or solvent mixtures are ranked by minimizing the physical and chemical differences as compared with the original solvent. In addition, the human health and environmental impacts are minimized by choosing only those solvents or solvent mixtures that have a total reduced environmental impact [36].

2.4.9
Limited Persistence and Bioaccumulation

While persistence, bioaccumulation, and environmental mobility are not inherently adverse chemical properties, reducing these properties minimizes temporal and spatial components of exposure to potentially toxic compounds. Compounds that are stable in the environment and in organisms have a greater potential to accumulate in each compartment. If a compound is also toxic in addition to these two properties, the likelihood of adverse outcomes increases. Therefore, designing products and materials to limit these attributes would reduce risks. Persistence is the tendency to remain in the environment and can be characterized by comparing half-lives of a target chemical in a medium of concern to appropriate benchmark values. Products should be designed for rapid breakdown to innocuous products quickly after use.

Bioaccumulation is referred to as a process of increasing concentrations of chemicals in living or environmental media at a rate greater than the degradation of the substance. This phenomenon is often observed for lipophilic compounds that tend to distribute to adipose tissue and lipid rich tissues like nerves as well as to soils and sediments. This behavior can be measured or modeled using a partition coefficient, P, as an indicator of bioaccumulation. Values for the logarithm of the partition coefficient ($\log P$), are available for many compounds [37]. Increasing values to an optimum value indicate a greater potential for bioaccumulation. Designing chemicals with lower $\log P$ values can reduce or eliminate the extent of bioavailability.

2.4.10
Quick Transformation to Innocuous Products

Chemicals can be transformed in the environment through biological enzymatic pathways as well as through nonenzymatic process such as photolysis and hydrolysis. These potential biotransformation pathways can be characterized by examining existing experimental data on fate and transport in the environment or by using predictive modeling tools.

2.4.11
Avoid Extremes of pH

The concentration and availability of hydrogen ions can influence a number of toxicity pathways, including necrosis, absorption potential, receptor, and enzyme binding, to cite but a few. Advancements in understanding pharmacokinetics (what the body does to the drug or molecule) and pharmacodynamics (what the drug or molecule does to the organism) can be extended to elucidating and understanding toxic mechanisms and to the design of safer chemicals. Absorption, distribution, metabolism, and excretion, represented by the acronym ADME, determine the fate and transport molecules in humans and other organisms. For a molecule to manifest toxicity, it must reach its site of action, whether that site is a macromolecular receptor, cell membrane, DNA, or other location. Understanding the physicochemical factors affecting absorption provides guidance on how to apply SAR to molecular design for hazard reduction.

Bioactivation is a physiological process that changes an unreactive moiety to an active toxicant. Designing molecules that minimize or prevent bioactivation will often reduce toxicity but not always. A major goal of sustainable design is to redirect metabolism so that chemicals can exist harmoniously with Nature. Increasing volatility may enhance the rate of elimination through exhalation; however, this amplified volatility may increase exposure through inhalation pathways. Hydrophilicity will enhance urinary excretion.

2.5
A Decision Framework

At the end of the evaluation process, there must be a mechanism to evaluate overall hazard or the relative "greenness" of a product or process. This assignment most often involves weighing the evidence of a process that includes making choices based on a combination of objective scientific data and subjective less quantitative measures. Value judgments are often a significant part of the decision process involving trade-offs among several value propositions, a common one being toxicity–benefit profiles of pharmaceuticals. The process works much more efficiently if planning is done early and described as completely as possible, is transparent and well documented.

2.5.1
A Suggested Protocol for Approaching Safer Chemical Design

- Determine whether the target molecule or any likely metabolic products is on an authoritative list of banned or restricted substances or meets risk assessment criteria of concern.
- Inventory strategic, standard molecular properties including melting and boiling points, molecular weight, water solubility, log$P$, partitioning behavior, vapor pressure, and so on
- Determine whether known toxicity data are available for the individual chemical or chemical class of concern. If there are data, characterize toxicity as completely as possible and determine whether to abandon the use of the molecule completely or continue with development.
- Identify known toxicophores and other attributes of concern. Identify opportunities to remove or replace the structural defect with an isosteric replacement.
- Make molecular design changes to reduce bioavailability.
- Attempt to predict potential toxic biotransformation products through empirical or modeled data.
- Develop predictive structure–toxicity relationships through a weight of evidence approach.
- Determine what structural modification, if any, will effectively reduce toxicity.
- Make the structural changes.
- Test new chemical entity in appropriate systems (e.g., *in vivo*).

2.5.2
Alternatives and Chemical Risk Assessment

Characterization of the potential risks associated with chemical manufacturing and use requires a fundamental understanding of the structure–toxicity relationship. Numerous methods, protocols, and frameworks have been put forward as guides to an orderly examination of hazard and risk. Many frameworks and approaches are available, and the use of the results of these approaches depends on the number and ontology of the evaluation endpoints selected as well as the ultimate use of application [38]. Alternatives assessment (AA) is one method for identifying, comparing, and selecting safer substitutes for chemicals of concern that may be hazardous.

Risk assessment is a quantitative process that characterizes the probability of manifesting adverse biological outcomes (i.e., risk), by evaluating intrinsic hazard and the extent of exposure [39]. Risk assessment guidance provides flexibility to adapt to many different exposure scenarios, thereby providing a tool flexible for inclusion in LCA or to use as an objective tool.

2.6
The Road Ahead: Training of a Twenty-First Century Chemist

Successful design of chemicals and materials for hazard reduction will require that a twenty-first century chemist to possess an expanded science skill set that is lacking in most practicing and academic chemists. To achieve this goal, the education of chemists, and all scientists, must include training in the principles of toxicology and risk assessment. This education is mandatory at the college and university level, but it should also be appropriately framed for a K-12 audience.

Although there are many scientists well trained in their individual fields, a new breed of scientist, astutely and holistically trained in chemistry, toxicology, biochemistry, and environmental science, must emerge from our new pathway forward. A modern chemist, especially those involved in synthetic chemistry, must be broadly trained in several complementary scientific areas, including fundamental and organic chemistry, medicinal chemistry, pharmacology/toxicology, environmental science, and biostatistics. This does not mean that every chemist need training equivalent to a practicing toxicologist. However, it is beneficial to personal and public health that all chemists appreciate the fundamental relationship between molecular structure and biological response and that they apply these principles to as many aspects of their work as possible, particularly in safer chemical design. Practicing chemists, professors, and students of chemistry are afforded myriad opportunities to educate themselves in the field of toxicology through formal coursework, attending professional meetings and participating in continuing education training.

Molecular design for hazard reduction challenges the skills of chemists, toxicologists, and affiliated scientists to examine the fundamental nature of the structure–hazard relationship, link chemical structure and properties to adverse effects, and make molecular design changes that render the substance intrinsically less capable of causing harm. It must be remembered, however, that predicting adverse outcomes from efforts based on *in silico* and *in vitro* approaches is very difficult given the profound complexity of physiological systems to respond to direct and bioactivated toxicants.

References

1 Garrett, R.L. (1996) Pollution prevention, green chemistry and the design of safer chemicals, in *Designing Safer Chemicals, Green Chemistry for Pollution Prevention*, American Chemical Society Symposium Series 640 (eds S.C. DeVito and R.L. Garrett), ACS, Washington, DC, pp. 2–15.

2 DeVito, S.C., Keenan, C., and Lazarus, D. (2015) Can pollutant release and transfer registers (PPTRs) be used to assess implementation and effectiveness of green chemistry practices? A case study involving the toxics Release Inventory (TRI) and pharmaceutical manufacturers. *Green Chemistry*, **17**, 2679–2692.

3 Anastas, P.T. and Warner, J.C. (1998) *Green Chemistry: Theory and Practice*, Oxford University Press, Oxford.

4 Anastas, P.T. and Zimmerman, J.B. (2003) Design through the 12 principles of green engineering. *Environmental Science & Technology*, **37**, 94A–101A.
5 Anastas, N.D. and Warner, J.C. (2005) The incorporation of hazard reduction as a chemical design criterion in green chemistry. *Chemical Health and Safety*, **12** (2), 9–13.
6 Anderson, L.A. and Gonzalez, M.A. (2012) Designing sustainable chemical synthesis: the influence of chemistry on process design, in *Green Chemistry for Environmental Remediation*, Scrivner Publishing, pp. 79–106.
7 Curran, M.A. (2009) Bio-based materials. Kirk-Othmer Encyclopedia of Chemical Technology, KOE-09-0006.R1, 23 pp.
8 Bare, J.C. (2010) Life cycle impact assessment research developments and needs. *Clean Technology and Environmental Policy*, **12** (4), 344–351.
9 Meyer, D.E., Curran, M.A., and Gonzalez, M.A. (2009) Industrial manufacture and use of nanocomponents and their role in the life cycle impact of nanoproducts. *Environmental Science & Technology*, **43** (5), 1256–1263.
10 Gonzalez, M.A. and Smith, R.L. (2003) A methodology for the evaluation of process sustainability. *Environmental Progress*, **22** (4), 269–276.
11 Ruiz-Mercado, G.J., Smith, R.L., and Gonzalez, M.A. (2012) Sustainability indicators for chemical processes: I. Taxonomy. *Industrial and Engineering Chemistry Research*, **51**, 2309–2328.
12 Ruiz-Mercado, G.J., Smith, R.L., and Gonzalez, M.A. (2012) Sustainability indicators for chemical processes: II. Data needs. *Industrial and Engineering Chemistry Research*, **51**, 2329–2353.
13 Ruiz-Mercado, G.J., Gonzalez, M.A., and Smith, R.L. (2013) Sustainability indicators for chemical processes: III. Biodiesel case study. *Industrial and Engineering Chemistry Research*, **52**, 6747–6760.
14 Smith, R.L. and Ruiz-Mercado, G.J. (2014) A method for decision making using sustainability metrics. *Clean Technologies and Environmental Policy*, **16**, 749–755.
15 Ruiz-Mercado, G.J., Gonzalez, M.A., and Smith, R.L. (2014) Expanding GREENSCOPE beyond the gate: a green chemistry and life cycle perspective. *Clean Technologies and Environmental Policy*, **16**, 703–717.
16 DeVito, S.C. (2012) The design of safer chemicals: past, present and future perspectives, in *Green Processes. Volume 9: Designing Safer Chemicals* (eds R. Boethling and A. Voutchkova), Wiley-VCH Verlag GmbH, Weinheim, Germany, pp. 1–19.
17 USEPA (2016) ACToR database: http://epa.gov/ncct/toxcast (accessed 24 February, 2016).
18 DeVito, S.C. (1996) General principles for the design of safer chemicals: toxicological considerations for chemists, in *Designing Safer Chemicals, Green Chemistry for Pollution Prevention*, American Chemical Society Symposium Series 640 (eds S.C. DeVito and R.L. Garrett), ACS, Washington, DC, pp. 16–59.
19 DeVito, S.C. (2012) Structural and toxic mechanism-based approaches to design safer chemicals, in *Green Processes. Volume 9: Designing Safer Chemicals* (eds R. Boethling and A. Voutchkova), Wiley-VCH Verlag GmbH, Weinheim, Germany, pp. 77–106.
20 Anastas, N.D. (2009) Incentives for using green chemistry and the presentation of an approach for green chemical design, in *Green Chemistry Metrics: Measuring and Monitoring Sustainable Processes* (ed. D. Constable), Blackwell Publishing, 344 pp.
21 Anastas, P.T. (2009) The transformative innovations needed by green chemistry for sustainability. *ChemSusChem*, **2** (5), 391–392.
22 Voutchkova, A.M., Osmitz, T.G., and Anastas, P.T. (2012) Toward a comprehensive molecular design framework for reduced hazard. *Chemical Reviews*, **110**, 5845–5882.
23 Ankley, G., Bennett, R.S., Erickson, R.J. et al. (2010) Adverse outcome pathways: a conceptual framework to support ecotoxicology research and risk assessment. *Environmental Science & Technology*, **29** (3), 730–741.
24 Kavlock, R. and Dix, D. (2010) Computational toxicology as implemented by the U.S. EPA: providing high

throughput decision support tools for screening and assessing chemical exposure, hazard and risk. *Journal of Toxicology and Environmental Health, Part B*, **13**, 197–217.

25 Zhu, H., Martin, T.M., Ye, L., Sedykh, A., Young, D.M., and Tropsha, A. (2009) Use of cell viability assay data improves the prediction accuracy of conventional quantitative structure-activity relationship models of animal carcinogenicity. *Chemical Research in Toxicology*, **22**, 1913–1921.

26 Ruiz, P.G., Begluitti, T., Tincheer, J., Wheeler, J., and Mumtaz, M. (2012) Prediction of acute mammalian toxicity using QSAR methods: a case of sulfur mustard and its breakdown products. *Molecules*, **17**, 8982–9001.

27 Martin, T.M., Harten, P., Venkatapathy, R., Das, S., and Young, D.M. (2008) A hierarchical clustering methodology for the estimation of toxicity. *Toxicology Mechanisms and Methods*, **18**, 251–266.

28 Guengerich, F.P. and MacDonald, J.S. (2007) Applying mechanisms of chemical toxicity to predict drug safety. *Chemical Research in Toxicology*, **20**, 344–369.

29 Guengerich, F.P. (2003) Cytochrome P450 oxidations in the generation of reactive electrophiles: epoxidation and related reactions. *Archives of Biochemistry and Biophysics*, **409**, 59–71.

30 Mandal, S., Mougdil, M., and Mandal, S.K. (2009) Rational drug design. *European Journal of Pharmaceutical*, **625**, 909–100.

31 Rusyn, I., Sedykh, A., Low, Y., Guyton, K.Z., and Tropsha, A. (2012) Predictive modeling of chemical hazard by integrating numerical descriptors of chemical structures and short-term toxicity assay data. *Toxicological Sciences*, **127** (1), 1–9.

32 Lipinski, C.A., Lombardo, F., Dominy, B.W., and Feeney, P.J. (1999) Experimental and computational approaches to estimate solubility and permeability in drug discovery and development settings. *Advanced Drug Delivery Reviews*, **23**, 3–25.

33 Albert, A. (1985) *Selective Toxicity: The Physico-Chemical Basis of Therapy*, Springer, 768 pp.

34 Blagg, J. (2010) Structural alerts for toxicity, in *Burger's Medicinal Chemistry, Drug Discovery and Development*, 7th edn (eds D.J. Abraham and D.P. Rotella), John Wiley & Sons, Inc, New York, 840 pp.

35 Constable, D.J.D., Dunn, P.J., Hayler, J.D. et al. (2007) Key green chemistry research areas-a perspective from the pharmaceutical industry. *Green Chemistry*, **9**, 411–420.

36 Harten, P.F. and Salama, G. (2004) PARIS III, the search for cleaner solvent replacements for RCRA. *Clean Technology*, **4** (11), 20–26.

37 Benfenati, E., Gini, G., Piclin, N., Roncaglioni, A., and Vari, M.R. (2003) Predicting log P of pesticides using different software. *Chemoshpere*, **53**, 1155–1164.

38 NAS (National Academy of Science) (2014) *A Framework to Guide the Selection of Chemical Alternatives*, NRC, NAS Press, Washington, DC.

39 USEPA (2014) Framework for human health risk assessment in decision-making, in *Risk Assessment Forum*, EPA/100/R14/001, Office of the Science Advisor.

3
Key Metrics to Inform Chemical Synthesis Route Design

John Andraos and Andrei Hent

3.1
Introduction

Since 2009 when the first edition of this work was released, the field of green chemistry metrics has undergone considerable maturation. To understand this progression one can begin by examining published works within two categories: (1) the development of new tools and methodology with regard to green metrics, and (2) efforts to generalize the material and demonstrate its applicability in multiple fields of chemistry at the pedagogical, academic, and professional level. Within the first category, the last 9 years saw an increased use of database algorithms [1–3] and synthesis tree diagrams [4] to facilitate the fast calculation of material efficiency metrics such as atom economy [5], overall yield, and kernel reaction mass efficiency (RME_{kernel}) [6] for an entire synthesis plan [7,8]. With this shift toward the evaluation of multistep chemical processes came new topics and problems related to decision-making in synthesis. To address these problems, chemists devised solutions that include the proper integration of green metrics analysis with synthesis and optimization strategies [9], the quantitative evaluation of environmental impact [10] coupled with health and safety information [11], the incorporation of life cycle analysis [12], and the role of uncertainty as related to missing physical or chemical data needed for a thorough analysis [13]. Furthermore, at the pedagogical level, recent publications whose intended audience are undergraduate chemistry students and faculty highlight the increased academic interest in the field of green chemistry and associated metrics including their accessibility to students [14,15]. In the academic arena, established journals like *Organic Process Research and Development* and *Green Chemistry* have begun changing their submission guidelines requiring that submitted manuscripts contain a detailed green metrics analysis as part of the supplementary material [16]. On the industrial side, well-regarded authors of articles on green chemistry [17–19] and associated metrics [20,21] have, of late, advocated for the integration of multiple areas of green chemistry, metrics, and perspectives in the general direction of a more unified total that can be used to

Handbook of Green Chemistry Volume 11: Green Metrics, First Edition. Edited by David J. Constable and Concepción Jiménez-González.
© 2018 Wiley-VCH Verlag GmbH & Co. KGaA. Published 2018 by Wiley-VCH Verlag GmbH & Co. KGaA.

assess and promote reaction greenness with a higher degree of confidence. In this respect, we have always maintained that integration of metrics is not only possible but also desirable [22]. This is because we hold that objective measurement and comparison through quantitative assessment of criteria that are relevant to a chemical synthesis forms the best means of understanding and evaluating the greenness of chemical reactions and processes. Although some authors have expressed the view that green metrics should remain as a separated field of disconnected concepts [23], a view that has often lead to false conclusions about the general metrics expressions for a multistep synthesis, we contend that integration of material, environmental, health and safety, and energy efficiency metrics provides the most powerful evaluation to answer the question of which process is the greenest among multiple candidates. In addition, we believe that applying this method also illuminates the many opportunities for improvement that may exist for a particular process, especially with regard to reaction conditions, operational and bottom-line costs, and environmental and energy impact reduction. As such, we welcome the various shades of green perspective as we contend that one can judge and select the best shade using a thorough quantitative analysis.

To showcase the practical applications of this approach, we focus the present chapter on the factors that are involved in carrying out a thorough quantitative analysis of process greenness. We thus divide the chapter into two case studies: the synthesis of bortezomib (case 1) and the synthesis of aspirin (case 2), respectively. In the first case study, we compare the material efficiency of a linear and a convergent strategy for the synthesis of bortezomib in order to highlight the benefits of using database algorithms, synthesis tree diagrams, and radial pentagons as useful tools in the evaluation of green metrics for whole processes. In particular, we highlight the integrative nature of reaction mass efficiency and how this metric can be used in conjunction with simpler metrics to better understand process optimization opportunities. In the second case study, we further demonstrate the power of integration, this time on a more general level, by comparing seven different approaches to the synthesis of phenol, a key intermediate in the manufacture of aspirin. This synthesis is analyzed based on material efficiency, environmental impact, health and safety impact, and energy efficiency, marking the first time that energy metrics are evaluated for a chemical process. Using these three categories, the various aspirin processes are ranked from greenest to least green with concluding comments reserved for possible future improvements.

3.2
Material Efficiency Analysis for Synthesis Plans

Over the last 9 years, as our expanding database of green named organic reactions [3] began to approach its current limit, the field of metrics turned to the analysis of whole synthesis plans [1,9,24]. The algorithm approach we introduced several years ago [6] was a good starting point, but it nevertheless required certain modifications in order to work for an entire process. One such modification pertains to the question of connecting individual reaction steps into a logical sequence where the moles (masses) of all input and auxiliary materials in a plan are matched with a chosen basis scale for the final target product. To maintain simplicity, we configured the REACTION spreadsheets so that a target basis scale of 1 kg was designated as the desired mass of the final product. Next, for each individual step in the synthesis, the REACTION spreadsheets calculate the appropriate scaling factor required for making the component masses and metrics analysis of that step compatible with those of the entire synthesis. The calculation is therefore computed in a backward direction starting from the final step and moving sequentially to the beginning of the plan. For example, supposing that we wanted to make 1 kg of the final target product, we can calculate the scaling factor of the final step in the synthesis according to Eq. (3.1):

$$\text{Scaling Factor} = \frac{1000}{X} \quad (3.1)$$

where X denotes the experimental yield (in grams) of the product for the final step in the synthesis. Using this value, the scaled mass of the limiting reagent that leads to that product is determined by multiplying the experimental mass of the limiting reagent and the scaling factor. This gives the scaled mass of limiting reagent for the final step of the synthesis. One can then substitute this value for the mass of the target product of the previous step (i.e., the penultimate step of the synthesis), thus continuing the sequence for the remainder of the plan. On the basis of this approach, a linear plan is integrated according to the scale of all the materials involved in each of its individual reaction steps that are required to produce a certain amount of the final target product. A linear plan of N steps, for instance, will have N scaling factors, one for each synthetic step. We note, however, that this method rests fundamentally on the assumption that the magnitude of the reaction yield (RY) is independent of the scale of the reaction. The idea of working backward to determine appropriate scaling factors becomes even more helpful in the context of a convergent synthesis. Here, the calculation follows the same sequence listed above, working backward from the final synthetic step, until one reaches the first convergent step where the synthesis splits into two separate branches. At this point, one has to determine the appropriate scaling factor for each of the branches to be able to calculate the scaled mass of the target products for those branches. After this is done, the sequence is continued as for an ordinary linear synthesis plan until the next convergent step is encountered. Altogether, this modification enables the REACTION spreadsheets to standardize and evaluate green metrics for a chemical process consisting of N steps.

Next, we note that the material efficiency metrics that are evaluated for individual reactions follow Eq. (3.2):

$$\text{RME} = \frac{1}{\text{PMI}} = \frac{1}{E+1} = (\text{AE}) \times (\varepsilon) \times \left(\frac{1}{\text{SF}}\right) \times (\text{MRP})$$
$$= \text{RME}_{\text{Kernel}} \times \left(\frac{1}{\text{SF}}\right) \times (\text{MRP}) \quad (3.2)$$

where RME is the generalized reaction mass efficiency [6], PMI is process mass intensity [20], E is the E-factor [25], AE is atom economy [5], ε is the reaction yield, SF is the stoichiometric factor [22], and MRP is the material recovery parameter [6,22].

Furthermore, in a recent pedagogical work concerning green metrics [15], a distinction was made between material efficiency at an intrinsic level, a perspective that considers the inherent nature of the chosen chemistry, and a global perspective that encompasses the amounts of all auxiliary materials used such as reaction solvents, isolation and purification solvents, catalysts, and other consumed materials. Based on this distinction, green metrics were classified according to which type of analysis they could shed light on. For example, the intrinsic efficiency category included the metrics of reaction yield, AE and RME_{kernel}, while at the global level, the analysis focused on RME, PMI, E, SF, and MRP. With this distinction in mind, researchers may proceed by selecting one or several of these metrics in order to validate claims of achieved reaction or process "greenness." Nevertheless, more often than not, the metrics analysis included within typical research articles is not presented in an integrated fashion allowing the reader to understand both the benefits and the possible shortcomings of a particular approach. And as long as the metrics are considered to be fundamentally disconnected from one another, the likelihood of omitting data or making errors when generalizing equations to entire synthesis plans becomes inevitably more pronounced. For this reason, we advocate the calculation of all relevant metrics and their presentation using appropriate visualization tools that reflect the essential connections between the metrics and their contribution to an overall perspective on efficiency.

To better understand the significance of conceptual integration with regard to material efficiency metrics, we include Figure 3.1 that depicts a bull's-eye type of diagram. At the center of this diagram, we include the fundamental intrinsic efficiency metrics of atom economy and reaction yield. We consider these metrics fundamental because they are inherent to the nature of the chosen chemistry and therefore changeable, most often, only through changes of chemistry. For example, changing reaction conditions such as choice of catalyst and/or solvent, reaction temperature, and reaction time will directly affect reaction yield

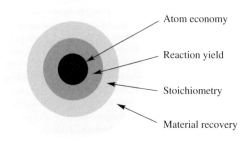

Figure 3.1 Representation of material efficiency metrics inherent to reaction mass efficiency.

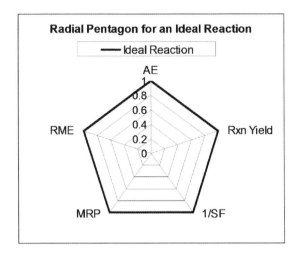

Figure 3.2 Radial pentagon depiction of ideal material efficiency conditions where each of the five parameters has a value of 1.

outcomes, whereas changing the chemical structures of reagents that will assemble to give the structure of the desired product in a chemical reaction will directly affect both atom economy and reaction yield outcomes. Nevertheless, a more global perspective is possible where the stoichiometry and amounts of auxiliary materials are included. Since these parameters do not require fundamental changes in chemistry for improvements in "greenness," we consider them as part of the global view. In previous works [22], we have shown that although each of these four metrics provide separate information about the material efficiency of a reaction or plan, their combination in the form of a mathematical product of values expressed in absolute form (as values between 0 and 1) can be used to determine the generalized reaction mass efficiency, which is also a value between 0 and 1 [6]. Finally, calculating these metrics for individual reaction steps and using the scaling factors to enable their calculation for a whole process allows one to create radial polygons to depict the results for all metrics in a clear and readily understandable diagram (Figure 3.2). Using this visualization tool, one can immediately judge both intrinsic and global efficiency performance for reactions and plans as well as areas that may require further attention and optimization.

The last tool we will use in the remainder of the chapter is the synthesis tree diagram [4]. Although this tool has been previously explained, we include it in the present discussion because of its usefulness in quickly determining the intrinsic material efficiency of processes. A synthesis tree diagram can be constructed for linear (Scheme 3.1, Figure 3.3) and convergent plans (Scheme 3.2, Figure 3.4) by omitting chemical structures and listing input reagents and process intermediates along with their molecular weights in a grid-like fashion. Reaction steps are read vertically and identified by observing their connections

Step 1: A + 2 B + C ⟶ I_1 + Q_1 Yield = ε_1

Step 2: I_1 + D ⟶ I_2 + Q_2 Yield = ε_2

Step 3: I_2 ⟶ I_3 Yield = ε_3

Step 4: I_3 + 1/2 E ⟶ P + Q_3 Yield = ε_4

Scheme 3.1 Generic linear process consisting of four steps where I_1–I_3 are intermediates and Q_1–Q_3 are by-products.

and respective yields listed on the bottom horizontal axis of the grid. In addition, connections are made so that the product for each step lies at the centroid point of all reagents and intermediates that lead directly to it. It is important to note, as always, that reagents and molecular weights are also checked for overall chemical reaction balancing and appropriate stoichiometric coefficients are added where necessary. In the case of convergent plans, one can use synthesis tree diagrams to quickly identify where the points of convergence occur. This can also enable an experimenter to perform accurate reaction scheduling simply by replacing the reaction yields at the bottom of the diagram with reaction times. But perhaps the most useful feature of synthesis tree diagrams is their ability to condense the most essential information about a process, as far as the intrinsic material efficiency metrics are concerned, namely, the molecular weights and reaction yields, into one simple diagram. Once the diagram is constructed, the intrinsic efficiency for any step as well as the entire process is simply read off of

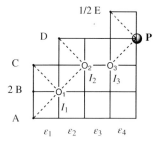

Figure 3.3 Synthesis tree diagram for the linear process in Scheme 3.1.

$$\text{Atom Economy}_{\text{Overall}} = \frac{p}{a + 2b + c + d + e/2}$$

where a, b, c, d, and e are the molecular weights of A, B, C, D, and E, respectively.

Overall Yield = $\varepsilon_1 \varepsilon_2 \varepsilon_3 \varepsilon_4$

$$\text{RME}_{\text{Kernel Overall}} = \frac{p}{\dfrac{e}{2\varepsilon_4} + \dfrac{d}{\varepsilon_2 \varepsilon_3 \varepsilon_4} + \dfrac{a + 2b + c}{\varepsilon_1 \varepsilon_2 \varepsilon_3 \varepsilon_4}}; \quad \text{RME}_{\text{Kernel Step 2}} = \frac{i_2}{\dfrac{i_1 + d}{\varepsilon_2 \varepsilon_3 \varepsilon_4}}$$

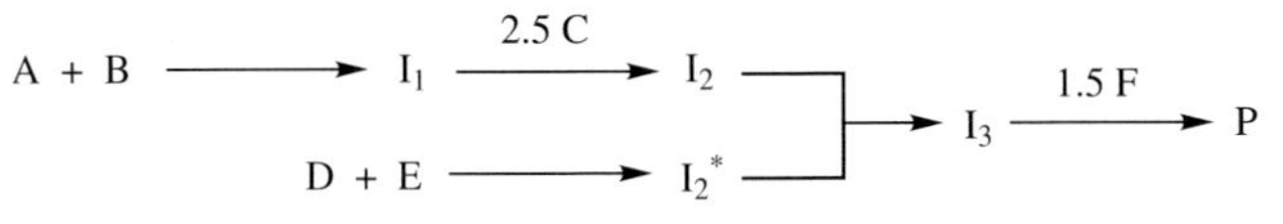

3.2 Material Efficiency Analysis for Synthesis Plans

$$A + B \longrightarrow I_1 \xrightarrow{2.5\, C} I_2$$
$$D + E \longrightarrow I_2^*$$
$$I_2, I_2^* \longrightarrow I_3 \xrightarrow{1.5\, F} P$$

step 1 $A + B \longrightarrow I_1 + Q_1$ Yield = ε_1

step 2 $I_1 + 2.5\, C \longrightarrow I_2 + Q_2$ Yield = ε_2

step 2* $D + E \longrightarrow I_2^* + Q_3$ Yield = ε_2^*

step 3 $I_2 + I_2^* \longrightarrow I_3 + Q_4$ Yield = ε_3

step 4 $1.5\, F + I_3 \longrightarrow P + Q_5$ Yield = ε_4

Scheme 3.2 Generic convergent process consisting of five steps, including one convergent step, where I_1–I_3, I_2^* are intermediates and Q_1–Q_5 are by-products.

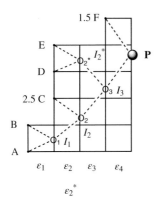

Figure 3.4 Synthesis tree diagram for the convergent process in Scheme 3.2.

$$\text{Atom Economy}_{\text{Overall}} = \frac{p}{a + b + 2.5c + d + e + 1.5f}$$

Overall Yield$_{\text{Main Branch}} = \varepsilon_1 \varepsilon_2 \varepsilon_3 \varepsilon_4$;   Overall Yield$_{\text{Convergent Branch}} = \varepsilon_2^* \varepsilon_3 \varepsilon_4$

$$\text{RME}_{\text{Kernel Overall}} = \frac{p}{\dfrac{(1.5)f}{\varepsilon_4} + \dfrac{d + e}{\varepsilon_2^* \varepsilon_3 \varepsilon_4} + \dfrac{(2.5)c}{\varepsilon_2 \varepsilon_3 \varepsilon_4} + \dfrac{a + b}{\varepsilon_1 \varepsilon_2 \varepsilon_3 \varepsilon_4}}; \quad \text{RME}_{\text{Kernel Step 2}^*} = \frac{i_2^*}{\dfrac{e + d}{\varepsilon_2^* \varepsilon_3 \varepsilon_4}}$$

the diagram. The metrics for the example syntheses in Schemes 3.1 and 3.2 are illustrated below each figure.

We conclude this section by noting two additional features. The REACTION spreadsheets have also been modified to check for reaction balancing in addition to providing further breakdowns in the E-factor according to the amounts of waste contributed from auxiliary, kernel, solvent, and other components. The last thing to note is that the material efficiency metrics presented above are *not* calculated for an overall balanced chemical process either through simple addition or through multiplication of corresponding metrics for individual reactions [23], and this is because of the requirement for the scaling factors for every step in the synthesis. For example, the overall E-factor (PMI) for a synthesis plan is *not* equal to the sum of the E-factors (PMIs) of each contributing reaction. The overall atom economy for a synthesis plan is *not* the product of the atom economies of each contributing reaction, except for the unique case of plans consisting entirely of sequential rearrangement and/or elimination reactions. More complete mathematical proofs of these statements are available in an article published in the Journal of Chemical Education [26]. With this material firmly established, we are fully equipped to apply this knowledge to two case studies. In the first case study, we analyze two syntheses of bortezomib, one linear and one convergent, and compare them for overall greenness according to material efficiency. In the second case study, we examine seven plans to make phenol that is a key intermediate for the synthesis of aspirin. For that study, since the structures of starting materials, intermediates, and final product are simple and their chemical and toxicological properties are well documented, we showcase the full spectrum of green metrics covering material efficiency, environmental impact, safety–hazard impact, and energy input consumption.

3.3
Case Study I: Bortezomib

Bortezomib, formerly PS-341, is a first-in-class proteasome inhibitor currently marketed by Millennium Pharmaceuticals under the trade name Velcade® for the treatment of multiple myeloma [26]. Multiple myeloma is one of the most common hematological diseases in North America affecting nearly 3 in 100 000 where the median survival time is 3 years [27]. More specifically, the disease is characterized by neoplastic proliferation of plasma cells in the bone marrow. Although several drugs work to prolong the survival time, bortezomib is one of the few shown to have the potential to reign in and possibly reverse the growth of malignant tumors [27]. This activity relates to the compound's selectivity for the chymotrypsin-like activity of the 20S proteasome, a protein complex the inhibition of which can result in the accumulation of proapoptotic proteins that can ultimately promote cell death. This particular type of cell cycle stage is crucial for the control and reversal of cancerous tumors in the body. Furthermore,

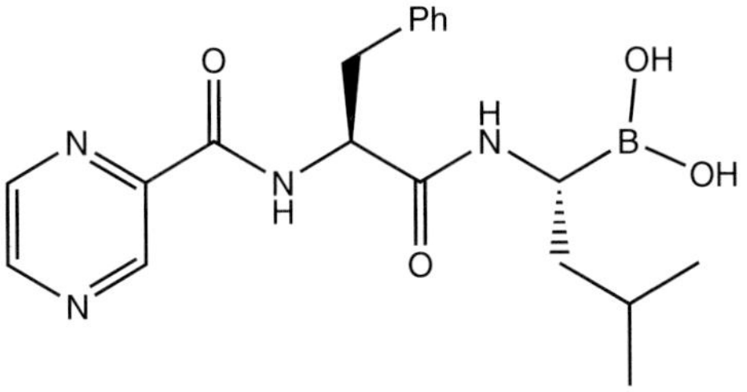

Figure 3.5 Structure of bortezomib.

in clinical research trials, patients that were given bortezomib experienced a high response to the treatment, a result that has paved the way for bortezomib to be tested as a potential medication for the treatment of other rare blood diseases, solid tumors, hematological malignancies, and cases where a combinational therapeutic approach is required. In 2007, bortezomib represents 45% of Millennium Pharmaceuticals' total cash flow with future projections expected to rise [28].

It is therefore not surprising that synthetic organic chemists have paid much attention to bortezomib over the years, and especially recently. In fact, since 1998 when Millennium patented the discovery synthesis of bortezomib [29], multiple synthetic approaches have appeared in both the academic [30–34] and patent [35–44] literature. The commercial process for bortezomib was itself patented in 2005 [45]. Although there are many synthetic approaches to bortezomib, there is also much overlap within the various strategies employed over the years. This is because bortezomib, as shown by its molecular structure in Figure 3.5, consists of four essential building blocks: a boronic acid functionality at one terminus, an *N*-acyl pyrazine group at the other terminus, and a phenylalanine plus leucine dipeptide-like analog in the middle of the structure. All synthesis strategies surveyed form the final product either through linear or through convergent approaches by first establishing these four essential building blocks. In the case of the boronic acid, for instance, all strategies include the formation of intermediate **A** (Scheme 3.3) using one of two boronic ester protection groups. Despite its larger size and complex structure, the majority of approaches surveyed utilize the pinanediol boronic ester protection because of its chiral structure and ability to direct incoming groups in order to establish the stereochemistry of the leucine side chain later on in the synthesis.

In the case of convergent approaches [32–34,36], the synthesis of bortezomib consists of coupling **A** with the pyrazine phenylalanine subunit, itself synthesized from simple starting materials (Scheme 3.4). Nevertheless, despite the choice of strategy, all approaches surveyed achieve the synthesis of the same set of target bonds in the final structure of bortezomib (Scheme 3.5).

Since the various strategies to synthesize bortezomib are very similar in the context of key intermediates and target bonds formed, we decided to

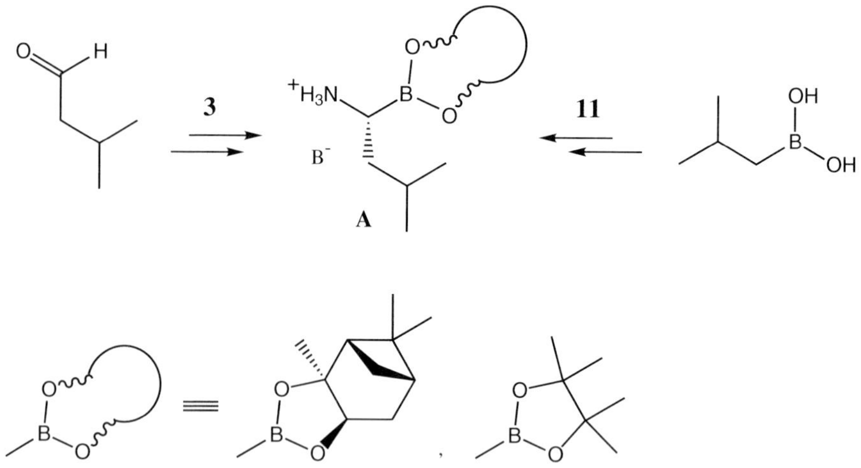

Scheme 3.3 Reaction network showing the number and variety of strategies employed to synthesize intermediate **A**.

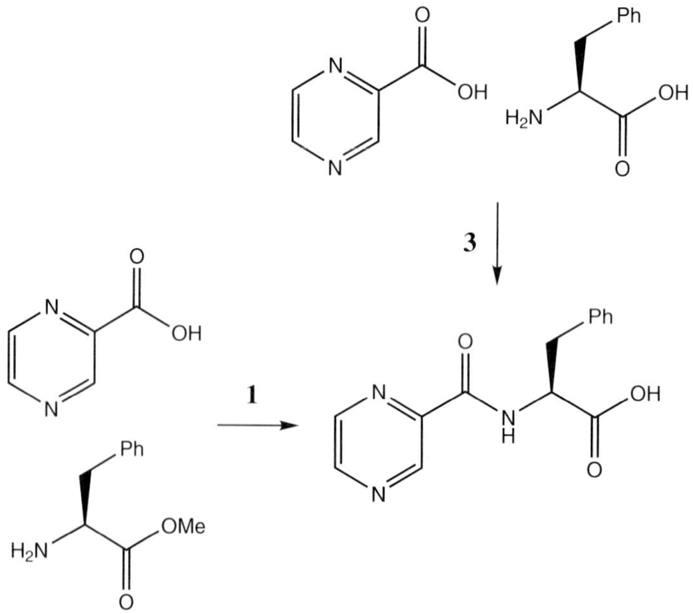

Scheme 3.4 Reaction network showing the number and variety of strategies employed to synthesize the pyrazine phenylalanine subunit of bortezomib.

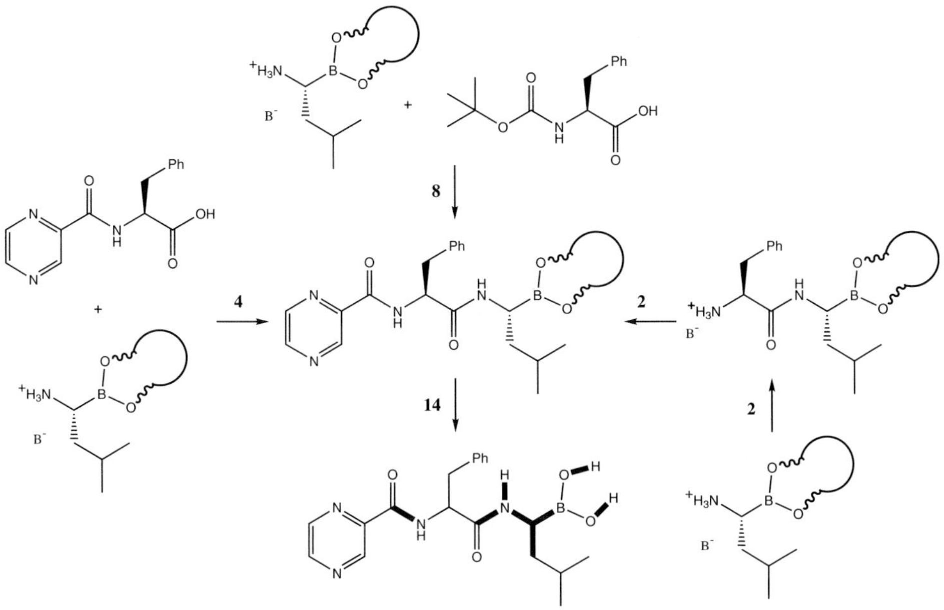

Scheme 3.5 Reaction network showing the number and variety of strategies employed to synthesize bortezomib.

compare a 2009 convergent route advertised for its material efficiency with the linear 2005 Millennium Pharmaceuticals' process to investigate the widely held belief that convergent synthesis strategies are more efficient than their linear counterparts. In the following sections, we present the synthesis steps for both routes and evaluate them using material efficiency metrics from both the intrinsic and global perspectives. It is important to note that a fuller metrics evaluation for all synthetic approaches to bortezomib referenced above is beyond the scope of this study.

3.3.1
Millennium Pharmaceuticals' Process

The Millennium Pharmaceuticals' bortezomib process [45] is a multistep linear synthesis that begins with very simple compounds, including alkyl boronic acid, glycol, Boc-protected phenylalanine, and pyrazine-2-carboxylic acid. The full synthesis is described in several patent and literature references that are not entirely relevant to the present study, in which we will focus only on the last four synthetic steps. We have chosen this sequence in order to make the comparison to the convergent synthesis simpler and more direct so as to be able to reliably verify the claims made in Ref. [31].

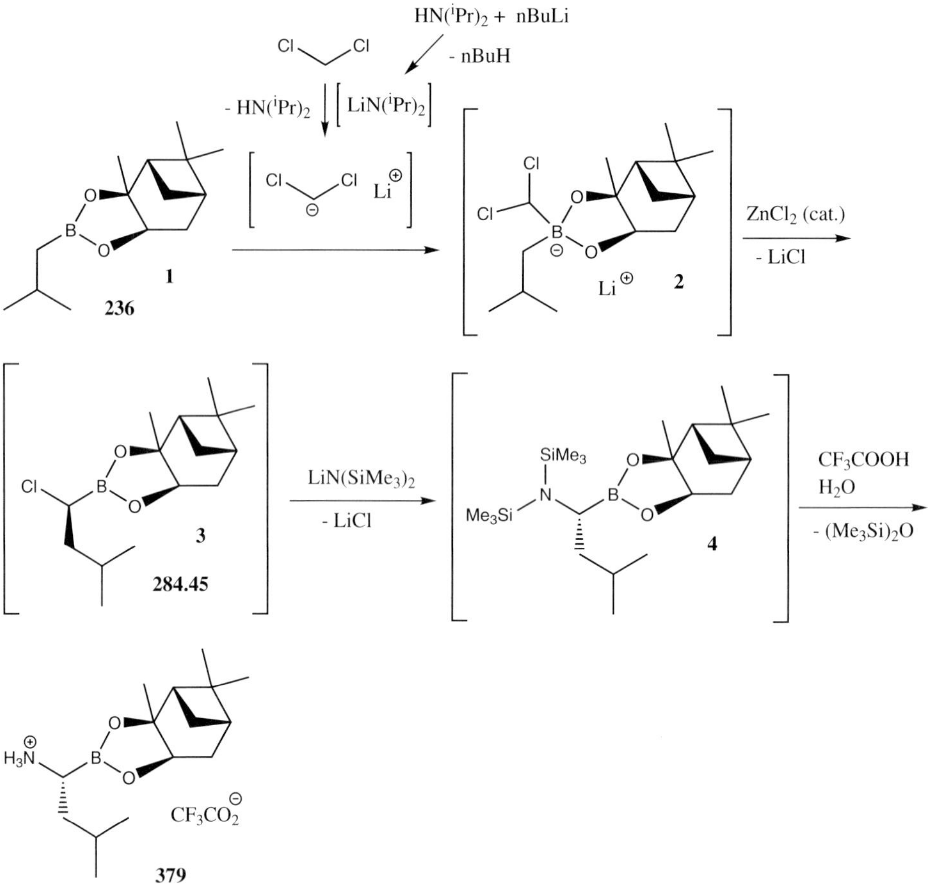

Scheme 3.6 Synthesis of intermediate **A** and the first of last four steps in the Millennium Pharmaceuticals' bortezomib process. Numbers below structures refer to molecular weights.

The first step in this sequence consists of the synthesis of intermediate **A** (Scheme 3.6). Here, Millennium researchers begin with a pinanediol boronic ester **1** that undergoes a homologation reaction in the presence of dichloromethane and lithium diisopropylamide (LDA) to synthesize a chiral glycol pinanediol intermediate **2**. The mechanism of this first reaction occurs by the attack of dichloromethyllithium (generated *in situ*) at the alpha position of **1** to form **2** that then rearranges by a Lewis acid catalyzed migration of the alkyl group to give the chiral glycol (+)-pinanediol **3**. Chirality during this step is established on account of the chiral directing activity of the boronic ester subunit. Zinc chloride is also employed as a Lewis acid in order to decrease the rate of epimerization possible when free chlorine ions are present in solution alongside **3**. After the completion of this step, **3** participates in a nucleophilic substitution reaction that utilizes lithium hexamethyldisilazane (LiHMDS) to introduce a silylated amino group with inversion of stereochemistry thus forming **4**. The final sequence requires **4** to be reacted with trifluoroacetic acid (TFA) to complete a desilylation reaction that gives the TFA salt intermediate **A** in 62% yield.

This key intermediate is then coupled with *N*-Boc-L-phenylalanine using 2-(*H*-benzotriazole-1-yl)-1,1,3,3-tetramethyluronium tetrafluoroborate (TBTU) as a coupling agent in the presence of two equivalents of a tertiary amine to give the Boc-protected dipeptide **5** (Scheme 3.7). This species is then reacted with hydrochloric acid to cleave the Boc protection group and produce the amino salt **6** in 79% yield.

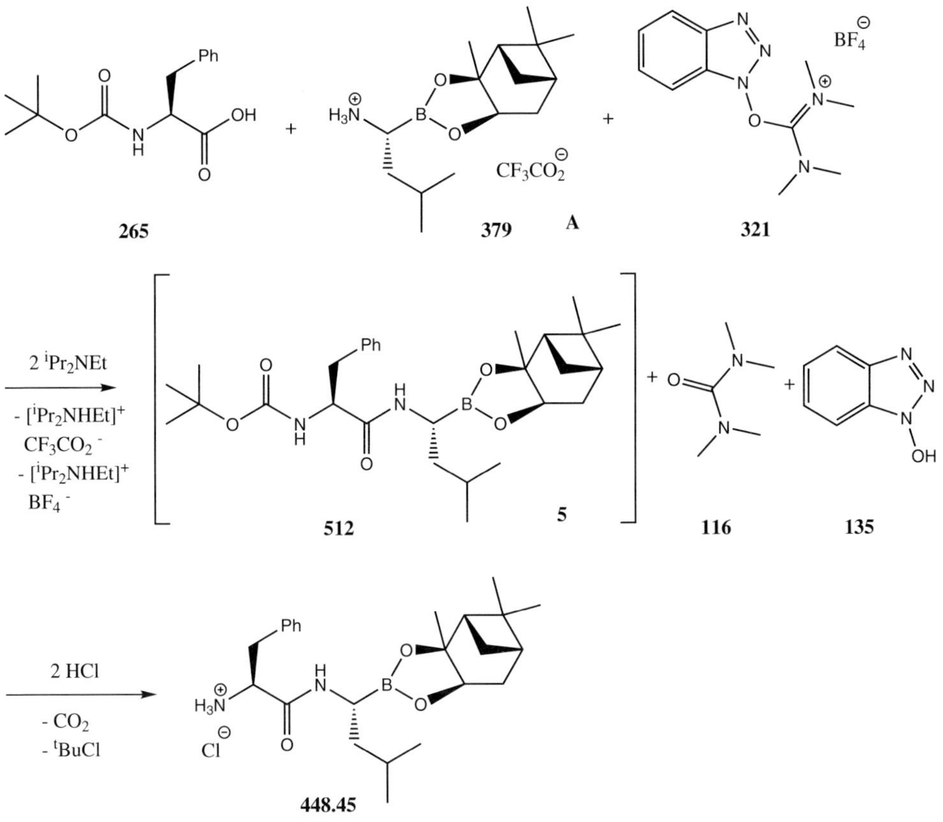

Intermediate 6 in 79% yield

Scheme 3.7 Synthesis of intermediate **6** and the second of last four steps in the Millennium Pharmaceuticals' bortezomib process. Numbers below structures refer to molecular weights.

The final two steps of the process consist of the coupling of **6** with 2-pyrazine carboxylic acid in the presence of TBTU and two equivalents of tertiary amine to give the tripeptide intermediate **7**. The boronic ester subunit of this compound is then cleaved under acidic conditions using isobutyl boronic acid, a process that also regenerates pinanediol boronic ester which can be reused in another batch run.

3.3.2
Pharma-Sintez Process

In 2009, Ivanov *et al.* proposed a convergent route to bortezomib that they stated was more efficient on account of step reduction and elimination of wasteful protection groups [32]. In this approach, two synthesis branches are used to form intermediate **7** by coupling intermediates **A** and **11** during the convergent step. The synthesis of intermediate **A** follows the same linear procedure previously described (Scheme 3.8). The synthesis of intermediate

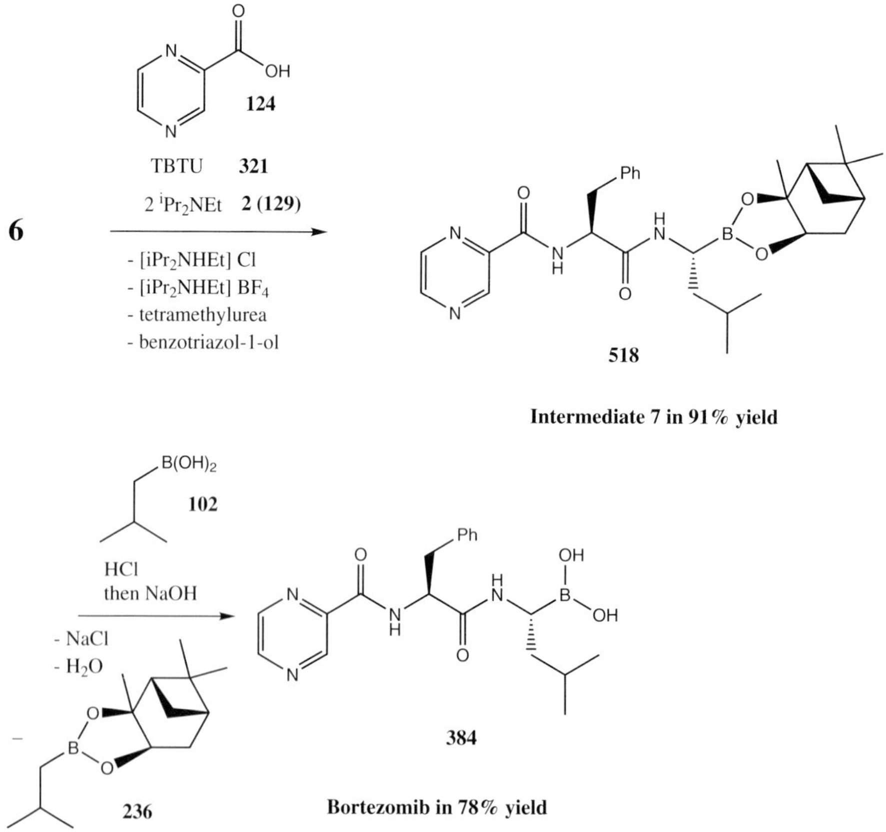

Scheme 3.8 Final two steps in Millennium Pharmaceuticals' bortezomib process. Numbers below structures refer to molecular weights.

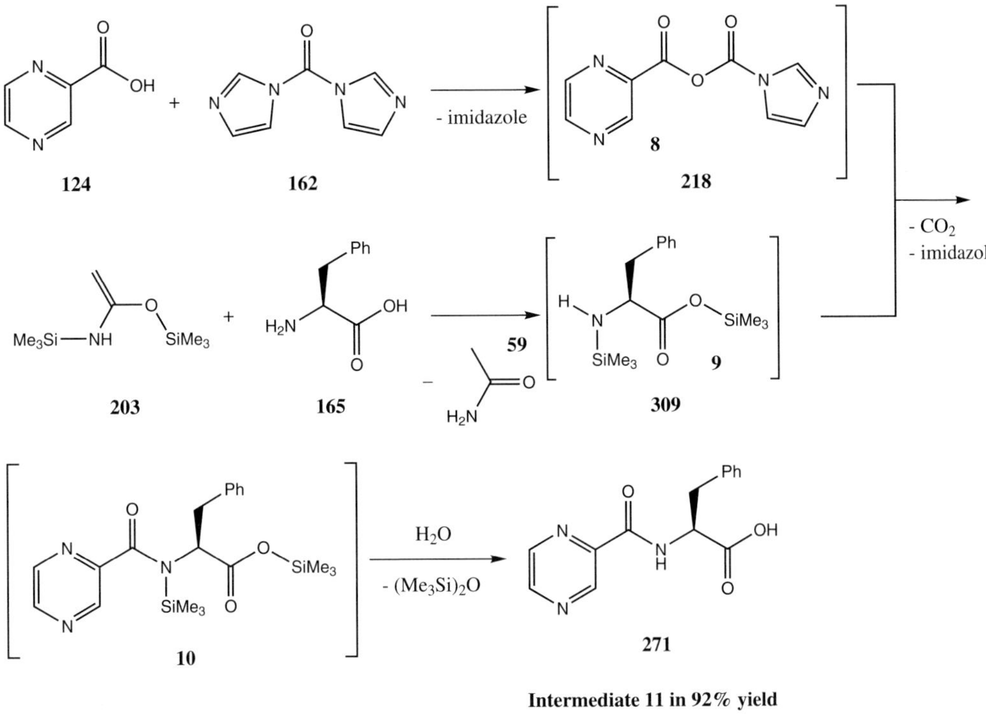

Scheme 3.9 Synthesis of key intermediate **11** in the Pharma-Sintez bortezomib process. Numbers below structures refer to molecular weights.

11 follows a more convergent approach (Scheme 3.9). Here, the authors start with an *N,O*-bis-silylation of L-phenylalanine with *N,O*-bis(trimethylsilyl)-acetamide (BSA) to produce intermediate **9**. In the same pot, a pyrazinecarboxylic acid imidazolide intermediate **8** is prepared by reacting 2-pyrazine carboxylic acid with *N,N'*-carbonyldiimidazole (CDI). With a retained reactivity at the silylated amino group, the partially protected phenylalanine analog **9** can now react with the newly formed intermediate **8**, itself an acylating agent, to form dipeptide **10**, which upon simple aqueous workup is desilylated giving **11** in 92% yield.

The newly formed dipeptide-like **11** is then coupled with **A** in the presence of TBTU and an excess amount of a tertiary amine to form the key intermediate **7** in 84% yield (Scheme 3.10). Finally, as in the Millennium process, **7** is treated with isobutyl boronic acid to form bortezomib while also releasing the chiral pinanediol boronic ester that could be reused in the synthesis of **A**. In the next section, we will consider the intrinsic and global material efficiency performance of these two processes and suggest alternative synthesis approaches to improve the performance. Afterward, we will evaluate whether

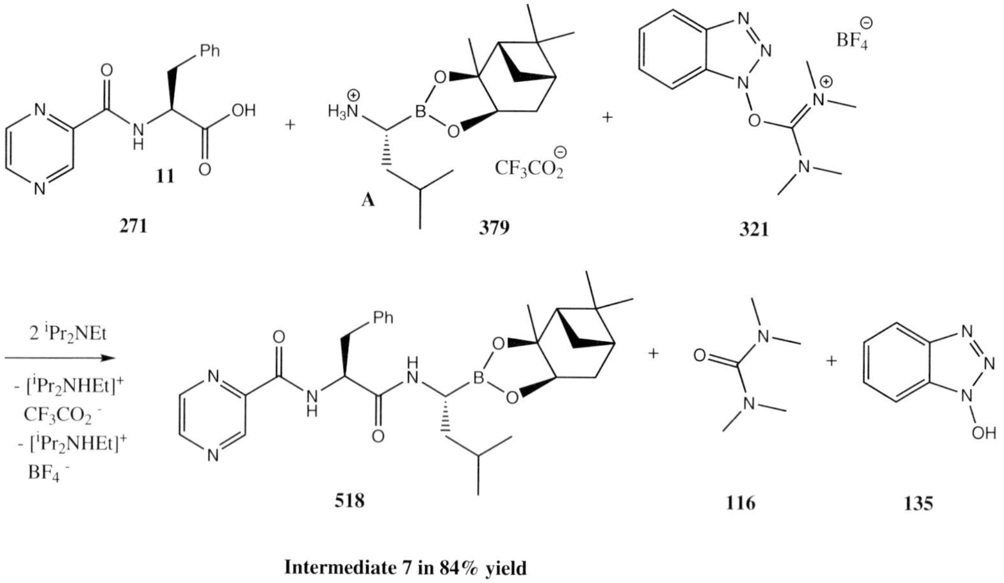

Intermediate 7 in 84% yield

Scheme 3.10 Synthesis of key intermediate **7** in the Pharma-Sintez bortezomib process. Numbers below structures refer to molecular weights.

the convergent strategy is in fact more material efficient than the commercial linear method.

3.3.3
Material Efficiency – Local and Express

We begin the metrics analysis by investigating the intrinsic material efficiency of each process. To do this quickly and effectively, we construct synthesis tree diagrams [4] for each process based on the balanced chemical equations and schemes presented in the previous section. We also reemphasize that a correct metrics analysis cannot begin without proper balancing of all materials and reactions appearing in the entire synthesis plan. With this in mind we shall begin with the synthesis tree diagram for the Millennium Pharmaceuticals' process (Figure 3.6).

From Figure 3.6, we can immediately say that the Millennium process involves four synthetic steps occurring in a linear sequence and comprising 17 input materials. The largest molecular weight material is the coupling agent TBTU that appears in steps 2 and 3, while the best yield occurs for step 3. The overall yield and atom economy of the Millennium process are calculated according to Eqs. (3.3) and (3.4), and the overall kernel RME is determined in Eq. (3.5).

$$\text{Overall Yield} = 0.62 \times 0.79 \times 0.91 \times 0.78 = 0.35 \tag{3.3}$$

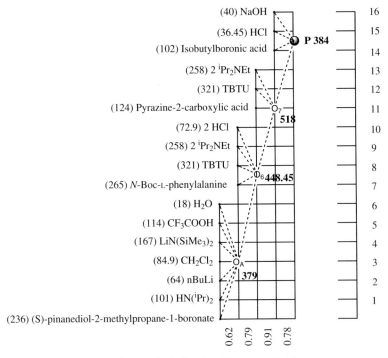

Figure 3.6 Synthesis tree diagram for Millennium Pharmaceuticals' process.

$$\text{Atom Economy}_{\text{Overall}} = \frac{384}{236 + 101 + 64 + 84.9 + 167 + 114 + 18 + 265 + 321+}$$

$$258 + 72.9 + 124 + 321 + 258 + 102 + 36.45 + 40$$
$$= 0.149 \tag{3.4}$$

$$\text{RME}_{\text{Kernel Overall}} = \frac{384}{\frac{102 + 36.45 + 40}{0.78} + \frac{258 + 321 + 124}{0.91 \times 0.78} + \frac{265 + 321 + 258 + 72.9}{0.79 \times 0.91 \times 0.78} +}$$

$$\frac{236 + 101 + 64 + 84.9 + 167 + 114 + 18}{0.62 \times 0.79 \times 0.91 \times 0.78}$$
$$= 0.075 \tag{3.5}$$

Similarly, for the Pharma-Sintez convergent bortezomib route, we can construct a synthesis tree diagram (Figure 3.7) and calculate the same intrinsic efficiency metrics (Eqs. (3.6)–(3.9)). Based on Figure 3.7, we can see that the Pharma-Sintez process consists of four synthetic steps, one of which is convergent. Furthermore, it has two synthesis branches and it necessitates 15 input materials.

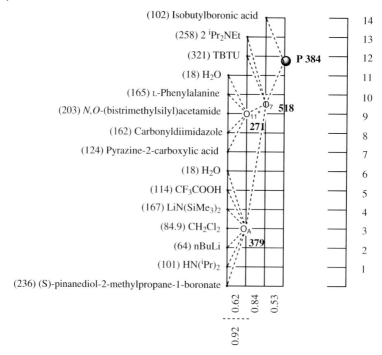

Figure 3.7 Synthesis tree diagram for Pharma-Sintez bortezomib process.

The best yield occurs for the synthesis of intermediate **11** and the largest material based on molecular weight is TBTU that only appears in the convergent step.

$$\text{Overall Yield}_{\text{Branch A}} = 0.62 \times 0.84 \times 0.53 = 0.28 \tag{3.6}$$

$$\text{Overall Yield}_{\text{Branch 11}^*} = 0.92 \times 0.84 \times 0.53 = 0.41 \tag{3.7}$$

$$\text{Atom Economy}_{\text{Overall}} = \frac{384}{236 + 101 + 64 + 84.9 + 167 + 114 + 18 + 124 + 162 + 203 + 165 + 18 + 321 + 258 + 102}$$
$$= 0.180 \tag{3.8}$$

$$\text{RME}_{\text{Kernel Overall}} = \frac{384}{\dfrac{102}{0.53} + \dfrac{321 + 258}{0.84 \times 0.53} + \dfrac{124 + 162 + 203 + 165 + 18}{0.92 \times 0.84 \times 0.53} + \dfrac{236 + 101 + 64 + 84.9 + 167 + 114 + 18}{0.62 \times 0.84 \times 0.53}} = 0.064 \tag{3.9}$$

Following the same approach, we can calculate the atom economy and kernel RME for every synthetic step in each process. This data is presented in

Table 3.1 Intrinsic efficiency metrics for Millennium Pharmaceuticals' bortezomib process.

Step	Yield	AE	RME$_{kernel}$
1	0.62	0.48	0.30
2	0.79	0.35	0.27
3	0.91	0.45	0.41
4	0.78	0.55	0.43
Overall	0.35	0.15	0.08

Table 3.2 Intrinsic efficiency metrics for Pharma-Sintez bortezomib process.

Step	Yield	AE	RME$_{kernel}$
1	0.62	0.48	0.30
1*	0.92	0.40	0.37
2	0.84	0.42	0.35
3	0.53	0.62	0.33
Overall	0.27[a]	0.18	0.06

a) This value is based on the lowest yield chosen along the longest branch.

Tables 3.1 and 3.2, respectively. Based on this information we can verify one of the claims made in the 2009 paper by Ivanov *et al.*, namely, that the convergent approach has a higher overall atom economy. This happens to be the case because of the plan's elimination of protection groups and some auxiliary materials, and thus the overall mass of input materials is reduced. Nevertheless, when reaction yield is considered, we notice that the convergent approach suffers from a lower overall yield along its main branch. This difference arises mainly from the poor yield of the last synthetic step, the step that is similar to that of the Millennium process, namely, the boronic ester cleavage. It is unfortunate that this is the case since this low yield also affects the kernel RME for the entire convergent process, including each synthetic step along the way, making it less material efficient than the Millennium synthesis. Since metrics analysis progressing from intrinsic to global efficiency can only become worse as more data are considered, due to the multiplicative effect of the metrics (Eq. (3.2)), we can say with certainty at this stage that the convergent synthesis is less material efficient on account of its lower kernel RME.

To determine the extent of the inefficiency, we shall broaden our perspective to a more global view that includes stoichiometry and material recovery. To facilitate this task we employ the REACTION spreadsheets described before to calculate these additional parameters. The results for each synthetic step of either process are illustrated in Figures 3.8 and 3.9. The overall metrics

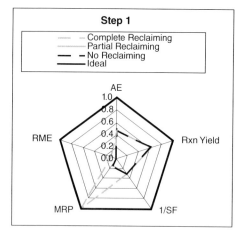
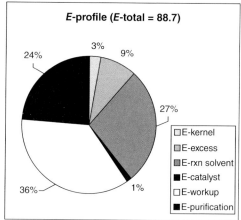
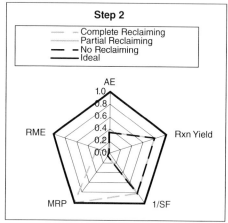
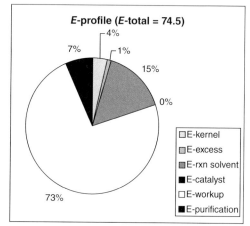
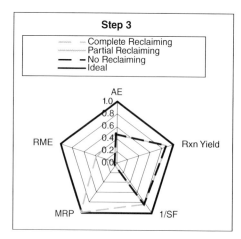
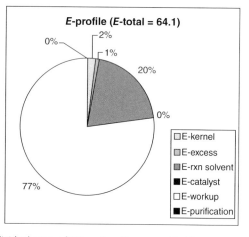

Figure 3.8 Global metrics analysis for individual steps of Millennium Pharmaceuticals' process.

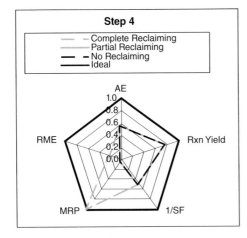 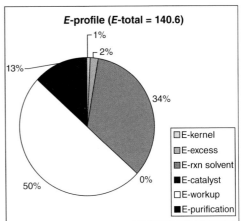

Figure 3.8 (Continued)

performances are shown in Figure 3.10, and the concrete numbers for each process are given in Tables 3.3 and 3.4, respectively. Given these data we again observe that the Millennium Pharmaceuticals' linear process is more material efficient than the convergent Pharma-Sintez route. It is interesting to note, however, that the convergent step in the Pharma-Sintez process produced much less waste in the category of unrecovered auxiliary materials, which, when coupled

Table 3.3 Global efficiency metrics for Millennium Pharmaceuticals' bortezomib process.

Step	Yield	AE	E_{kernel}	E_{excess}	E_{aux}	E_{total}	RME
1	0.62	0.48	2.34	8.03	78.36	88.7	0.0111
2	0.79	0.35	2.66	0.86	70.94	74.5	0.0132
3	0.91	0.45	1.44	0.51	62.12	64.1	0.0154
4	0.78	0.55	1.32	2.48	136.8	140.6	0.0071
Overall	0.35	0.15	11.82	16.25	494.7	522.7	0.0019

Table 3.4 Global efficiency metrics for Pharma-Sintez bortezomib process.

Step	Yield	AE	E_{kernel}	E_{excess}	E_{aux}	E_{total}	RME
1	0.62	0.48	2.34	8.03	78.36	88.7	0.0111
1*	0.92	0.40	1.69	14.14	36.99	52.8	0.0186
2	0.84	0.42	1.83	0.35	110.8	113.0	0.0088
3	0.53	0.62	2.07	0.40	>64.3	>66.8	<0.0147
Overall	0.27[a]	0.18	14.83	42.20	>585.9	>642.9	<0.0016

a) This is the lowest yield chosen along the longest branch.

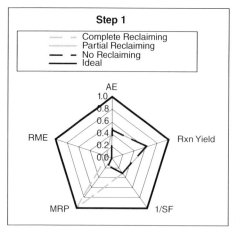
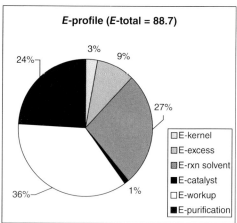
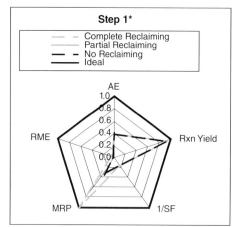
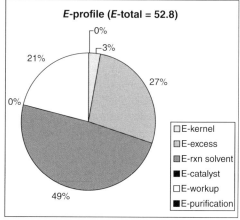
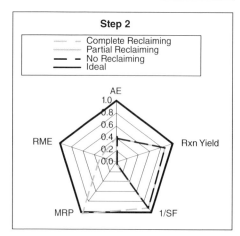
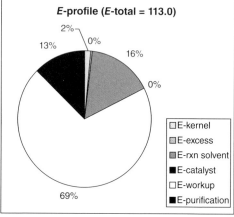

Figure 3.9 Global metrics analysis for individual steps of Pharma-Sintez process.

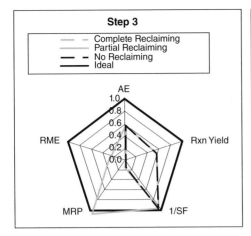
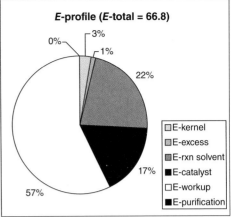

Figure 3.9 (*Continued*)

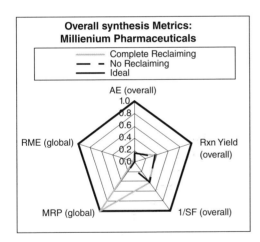
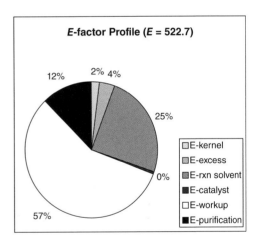

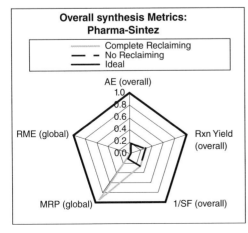
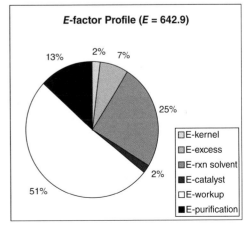

Figure 3.10 Global metrics analysis for the Millennium Pharmaceuticals and Pharma-Sintez processes.

with its higher waste due to excess materials, gave lower overall waste production than every other synthetic step in either plan. Nevertheless, we also note that this particular step also had the highest reaction yield of 92% among all the other synthesis steps. When we factor in the poor yield of the last step of the convergent route, as discussed before, the performance of the entire synthesis suffers thus lowering the overall RME significantly. Still, based on the numbers we can see that the global RME for the Millennium process is only 19% better than that of the Pharma-Sintez process as compared with a 25% difference in intrinsic efficiency according to RME_{kernel} values. This shows that although the Millennium process is both intrinsically and globally more material efficient, the convergent approach has certain strengths that make up for some of the difference in the context of overall waste reduction. Despite all of these features, both processes use significant amounts of auxiliary materials that are not recovered in any way thus contributing to global RME values that are 38–42 times lower than kernel RME values. When we consider that the Pharma-Sintez route has not been tested on a commercial scale, we can only expect this difference to be even more pronounced. In the next section, we will briefly discuss synthetic strategies that could be employed to improve the material performances of these two plans.

3.3.4
Synthesis Strategy for Future Optimization

One of the first priorities after a metrics analysis is to identify the outstanding values, both positive and negative, and connect them to the original syntheses to learn how best to make improvements, either by reduction, enhancement, or fundamental change in chemistry, if necessary. In the context of the current study, one of the most noticeable elements relates to the final step of the convergent process. Here, Ivanov *et al.* used isobutylboronic acid to cleave the boronic ester group of **7** to make bortezomib in 53% yield [32]. When we compare this step with its equivalent in the Millennium process, we can see that the same reaction is carried out under harsher conditions using hydrochloric acid and sodium hydroxide as additional reactants. We note that the use of additional components implies a lower atom economy, and indeed the Millennium process is less atom-economical for this step (7% difference). Nevertheless, these reaction conditions help improve the yield by 25% as compared with the same step in the convergent synthesis. In terms of global efficiency, it is more challenging to directly compare these two steps because there are insufficient details provided about material usage, particularly auxiliary materials used for isolation and purification, in regard to the Pharma-Sintez process. With additional information it would be possible to properly assess the pros and cons of either approach for this transformation more objectively. However, based on intrinsic efficiency alone, we can say that a 7% improvement in atom economy does not justify a 25% reduction in reaction yield, particularly in light of the fact that we are discussing the last step of the synthesis, the yield for which affects every other step that came before. This is most clearly visible in Eq. (3.9) for the calculation of overall kernel RME for the convergent route.

Another noticeable aspect relates to the use of high molecular weight materials such as TBTU and ^{i}Pr$_2$NEt to effect peptide-like coupling reactions for both processes. This sequence appears twice in the Millennium synthesis producing relatively high reaction yields of 79 and 91%, respectively. The Pharma-Sintez process also utilizes TBTU and ^{i}Pr$_2$NEt during the convergent coupling step giving a good yield of 84%. Earlier in the process, carbonyldiimidazole (CDI) is used to activate the coupling of L-phenylalanine with pyrazine-2-carboxylic acid producing the highest yield recorded for any synthetic step in either plan, namely 92%. Although the authors have discussed problems related to the use of various coupling reagents [32], for example the issue of reducing the rate of racemization in the product, which TBTU can suppress due to its 1-oxybenzotriazole unit, it is nevertheless important to ask whether there exist alternative peptide coupling reagents that are more material efficient to be used for these reactions. Interestingly, ever since amide bond formation had been selected as an area for advancing green chemistry principles [46], there has been significant attention devoted to the design of efficient coupling reagents [47] to mitigate some of the problems associated with their use [48]. A recent discovery that may have relevance here pertains to the design of a recyclable coupling reagent that can carry out racemization-free synthesis of peptide and amide functionalities [49]. The test of greenness for this technology, however, would be the maximum reaction yields it could achieve while utilizing the minimum amounts of auxiliary materials overall in both the synthetic and recycling steps. Few technologies reported since 2007 have passed this test. In terms of auxiliary material use, a final note is that every synthetic step for both processes produces at least 37 kg of waste due to auxiliary materials per kg of product synthesized, with one step being as high as 137 kg auxiliary material waste per kg of product formed. It is therefore crucial that future research should focus on developing more efficient isolation and purification protocols, perhaps using more streamlined methods such as a continuous flow reactor [50].

Finally, although we did not have an opportunity to evaluate some of the different approaches for the synthesis of bortezomib in this limited case study, such as those that utilize a simpler boronic ester protection while relying on imine chemistry to establish chiral centers [31,42,43], we do acknowledge them for future consideration for developing a green material efficient synthesis of bortezomib. We also acknowledge that a broader metrics evaluation will certainly add more shades of green to the present study and we welcome this development provided that sufficient and reliable data on the nature of the chemicals used in these two processes can be collected to enable such analysis.

3.3.5
Summary

In the present case study we have demonstrated the use of intrinsic and global material efficiency green metrics in the evaluation of the performances of two syntheses of an important pharmaceutical compound. Using smart spreadsheets, synthesis tree diagrams and radial pentagons, we have shown that a convergent synthetic approach was both intrinsically and globally less efficient than a commercial linear process that used similar chemistry. We noted that although the convergent approach demonstrated waste reduction potential in the global sense, its lack of good reaction yields for the later stages of the process significantly affected both the kernel and the global reaction mass efficiency of this approach. We ended this case study by considering strategies to achieve optimization for both the convergent and the linear syntheses of bortezomib.

3.4
Case Study II: Aspirin

In this second case study, we apply green metrics analysis under three levels of investigation to an abbreviated reaction network representing the synthesis of aspirin. For the sake of brevity, we will focus our attention only on various routes to making phenol, which is a key hub intermediate in the reaction network. The aim of this example is to illustrate to the reader the implementation of thorough metrics analysis and how such an analysis bears on decision-making within the framework of seeking a green synthesis of phenol given established industrial methods of its manufacture. The first or primary analysis involves material efficiency that covers basic metrics such as RY, AE, E, RME, and PMI. The second level of analysis takes into account the environmental impact and safety–hazard impact, which are covered by the metrics benign index (BI) and safety–hazard index (SHI), respectively. The final layer of analysis involves energy metrics. Specifically, we examine enthalpic contributions to energy consumption applied to all input materials involved in a given reaction as well as the heat of reaction for the chemical transformation. For each level of analysis, we present a brief methodology followed by the results for phenol syntheses. The merits of the best method to make phenol under each of the three levels of metrics analysis are discussed. A truly green synthesis of any target molecule is one that scores best in each of the three levels of investigation.

3.4.1
Reaction Network

Aspirin was chosen since it is a ubiquitous pharmaceutical with a long history and it has a simple chemical structure that is easily amenable to detailed metrics analysis. The main advantage is that all necessary input data needed to carry this out are well known and readily available. Such a scenario leads to a thorough investigation and meaningful ranking discussions with a minimum of uncertainty, thereby making the green decision more credible. Scheme 3.11 shows a reaction network for the synthesis of aspirin. Key reaction intermediates are shown, and the numbers above the arrows represent the number of possible routes

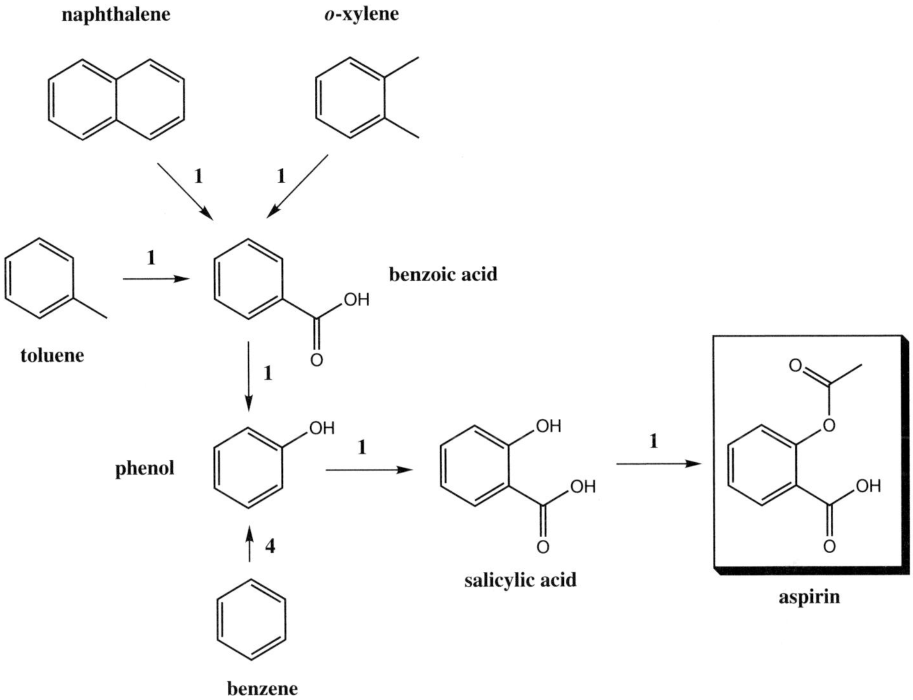

Scheme 3.11 Reaction network for aspirin synthesis

connecting pairs of intermediates. Since there are seven possible routes to phenol and only one route from it to the final target product, an optimum synthesis of aspirin will hinge on finding an optimum synthesis of phenol. Hence, the task is to find the greenest synthesis of phenol under material efficiency, environmental and safety–hazard impact, and energy efficiency considerations. Scheme 3.12 shows explicitly the seven industrial synthesis plans to phenol. Each chemical equation is balanced, reaction conditions of temperature and pressure are specified, and reaction yields and heats of reaction are given. In Figure 3.11 the plans are represented as synthesis trees. Plans 1–4 begin from benzene, plan 5 begins from toluene, plan 6 begins from o-xylene, and plan 7 begins from naphthalene. Plan 1 [51] involves sulfonation followed by fusion with sodium hydroxide. Plan 2 [51] involves chlorination with chlorine gas followed by substitution with sodium hydroxide followed by acidification. Plan 3 [51] involves chlorination with gaseous hydrochloric acid followed by acid catalyzed hydration. Plan 4 [51] is the Hock rearrangement process involving alkylation to cumene, then oxidation to cumene hydroperoxide, followed by rearrangement. Plan 5 [51] involves oxidation of toluene to benzoic acid followed by thermal decarboxylation. Plans 6 [52] and 7 [52] involve oxidation of o-xylene and naphthalene, respectively, to benzoic acid that is then thermally decarboxylated.

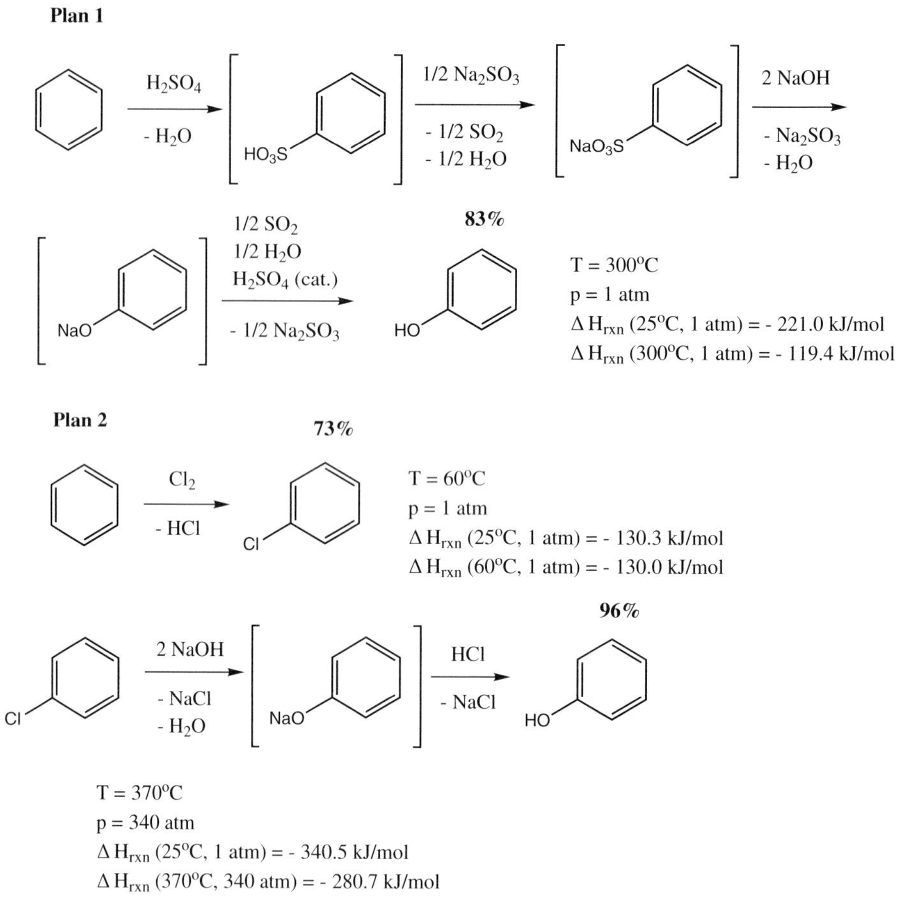

Plan 1

[Reaction scheme showing benzene → benzenesulfonic acid (HO₃S-C₆H₅) via H₂SO₄, -H₂O; then → sodium benzenesulfonate (NaO₃S-C₆H₅) via 1/2 Na₂SO₃, -1/2 SO₂, -1/2 H₂O; then → sodium phenoxide (NaO-C₆H₅) via 2 NaOH, -Na₂SO₃, -H₂O; then → phenol (HO-C₆H₅) via 1/2 SO₂, 1/2 H₂O, H₂SO₄ (cat.), -1/2 Na₂SO₃, 83%]

T = 300°C
p = 1 atm
ΔH_{rxn} (25°C, 1 atm) = - 221.0 kJ/mol
ΔH_{rxn} (300°C, 1 atm) = - 119.4 kJ/mol

Plan 2

[Benzene → chlorobenzene via Cl₂, -HCl, 73%]

T = 60°C
p = 1 atm
ΔH_{rxn} (25°C, 1 atm) = - 130.3 kJ/mol
ΔH_{rxn} (60°C, 1 atm) = - 130.0 kJ/mol

[Chlorobenzene → sodium phenoxide via 2 NaOH, -NaCl, -H₂O; → phenol via HCl, -NaCl, 96%]

T = 370°C
p = 340 atm
ΔH_{rxn} (25°C, 1 atm) = - 340.5 kJ/mol
ΔH_{rxn} (370°C, 340 atm) = - 280.7 kJ/mol

Scheme 3.12 Seven synthesis plans to make phenol.

3.4.2
Material Efficiency

Global values for the parameters RY, AE, E, RME, and PMI were determined for the seven plans to phenol according to the previously described Andraos algorithm [3,53]. Three of these parameters are connected by the simple relationship given in Eq. (3.10).

$$\text{PMI} = \frac{1}{\text{RME}} = E + 1 \qquad (3.10)$$

Results are summarized in Table 3.5. We observe that plan 5, toluene oxidation to benzoic acid followed by thermal decarboxylation, is overall the most material efficient with the lowest PMI of 2.3. This is followed very closely by the

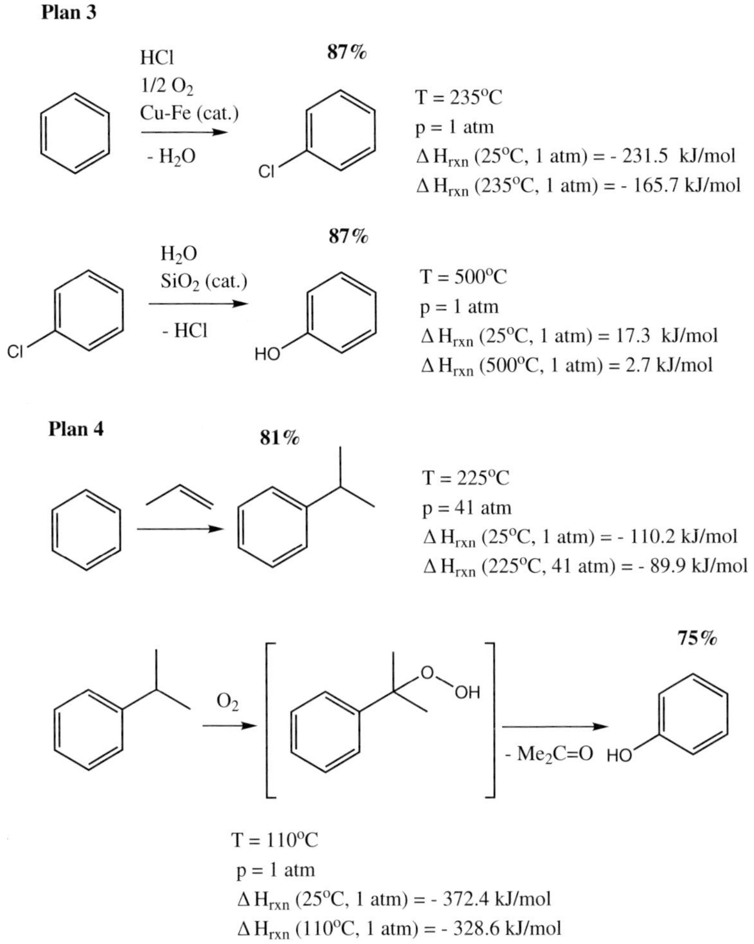

Plan 3

Benzene + HCl, 1/2 O$_2$, Cu-Fe (cat.), – H$_2$O → Chlorobenzene (87%)

T = 235°C
p = 1 atm
ΔH_{rxn} (25°C, 1 atm) = – 231.5 kJ/mol
ΔH_{rxn} (235°C, 1 atm) = – 165.7 kJ/mol

Chlorobenzene + H$_2$O, SiO$_2$ (cat.), – HCl → Phenol (87%)

T = 500°C
p = 1 atm
ΔH_{rxn} (25°C, 1 atm) = 17.3 kJ/mol
ΔH_{rxn} (500°C, 1 atm) = 2.7 kJ/mol

Plan 4

Benzene + propylene → Cumene (81%)

T = 225°C
p = 41 atm
ΔH_{rxn} (25°C, 1 atm) = – 110.2 kJ/mol
ΔH_{rxn} (225°C, 41 atm) = – 89.9 kJ/mol

Cumene + O$_2$ → [cumene hydroperoxide] → Phenol + Me$_2$C=O (75%)

T = 110°C
p = 1 atm
ΔH_{rxn} (25°C, 1 atm) = – 372.4 kJ/mol
ΔH_{rxn} (110°C, 1 atm) = – 328.6 kJ/mol

Scheme 3.12 *Continued*

Hock oxidation-rearrangement sequence given in plan 4 with a PMI of 2.6. However, plan 3, gaseous hydrochloric acid chlorination of benzene followed by hydration, has the highest atom economy at 84% and plan 1, sulfonation of benzene followed by substitution with sodium hydroxide, has the highest overall yield at 83%. The best yield performance for plan 1 is offset by its overall poor atom economy performance. In order to understand why plan 5 prevails over the others despite modest values of 73 and 60%, respectively, for RY and AE, we need to probe further the composition of its total *E*-factor from by-products, excess reagents, catalysts, reaction solvents, work-up, and purification materials. The complete breakdown of *E*-factors for all seven plans is summarized in Table 3.6. None of the plans have waste contributions from reaction solvents or auxiliary materials. Plans 3, 5, 6, and 7 utilize catalysts at a loading of 0.1 mol%. We observe

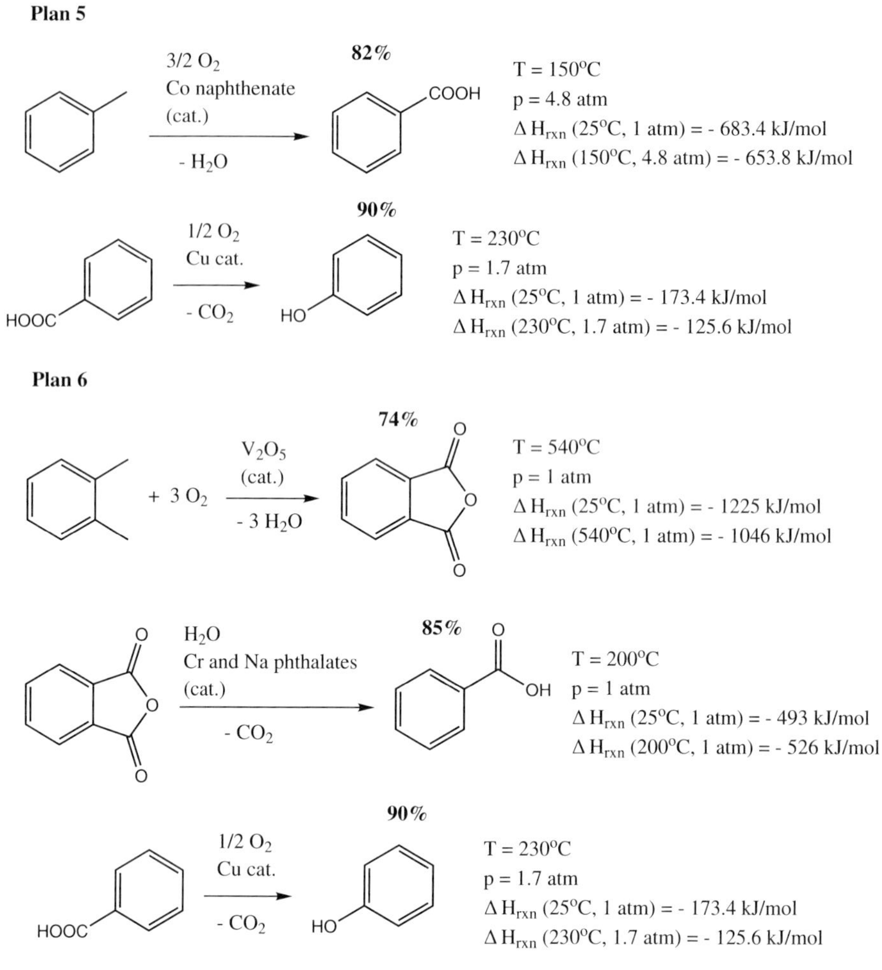

Plan 5

[Toluene] →(3/2 O₂, Co naphthenate (cat.), −H₂O)→ [Benzoic acid, COOH] **82%**
T = 150°C
p = 4.8 atm
ΔH_{rxn} (25°C, 1 atm) = −683.4 kJ/mol
ΔH_{rxn} (150°C, 4.8 atm) = −653.8 kJ/mol

[Benzoic acid, HOOC] →(1/2 O₂, Cu cat., −CO₂)→ [Phenol, HO] **90%**
T = 230°C
p = 1.7 atm
ΔH_{rxn} (25°C, 1 atm) = −173.4 kJ/mol
ΔH_{rxn} (230°C, 1.7 atm) = −125.6 kJ/mol

Plan 6

[o-Xylene] →(+ 3 O₂, V₂O₅ (cat.), −3 H₂O)→ [Phthalic anhydride] **74%**
T = 540°C
p = 1 atm
ΔH_{rxn} (25°C, 1 atm) = −1225 kJ/mol
ΔH_{rxn} (540°C, 1 atm) = −1046 kJ/mol

[Phthalic anhydride] →(H₂O, Cr and Na phthalates (cat.), −CO₂)→ [Benzoic acid, OH] **85%**
T = 200°C
p = 1 atm
ΔH_{rxn} (25°C, 1 atm) = −493 kJ/mol
ΔH_{rxn} (200°C, 1 atm) = −526 kJ/mol

[Benzoic acid, HOOC] →(1/2 O₂, Cu cat., −CO₂)→ [Phenol, HO] **90%**
T = 230°C
p = 1.7 atm
ΔH_{rxn} (25°C, 1 atm) = −173.4 kJ/mol
ΔH_{rxn} (230°C, 1.7 atm) = −125.6 kJ/mol

Scheme 3.12 *Continued*

that plans 4 and 5 have the lowest waste originating from excess reagents (E-excess), whereas the high overall PMI values for plans 6 and 7 arise from significant excess reagent consumption. Plan 3 has the lowest waste contribution arising from reaction by-products, which is consistent with it having the highest atom economy. Based on these material efficiency metrics calculations, plans 4 and 5 are singled out as being most green with respect to minimum total waste generation.

3.4.3
Environmental and Safety–Hazard Impact

BI [10] and SHI [11] may be used to assess and rank the environmental and safety–hazard impacts, respectively, of individual reactions or entire synthesis

Plan 7

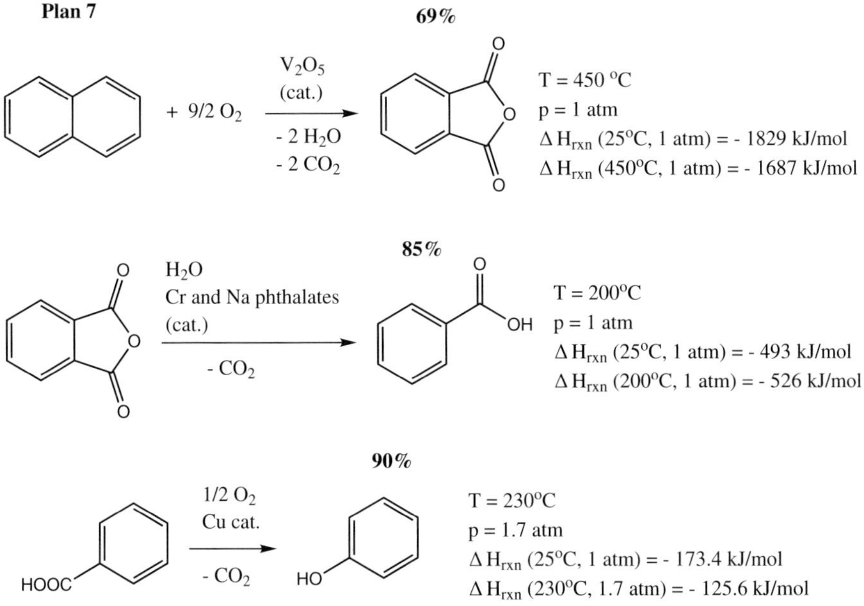

Naphthalene + 9/2 O₂ →[V₂O₅ (cat.), −2 H₂O, −2 CO₂] phthalic anhydride (69%)

T = 450 °C
p = 1 atm
ΔH$_{rxn}$ (25°C, 1 atm) = − 1829 kJ/mol
ΔH$_{rxn}$ (450°C, 1 atm) = − 1687 kJ/mol

phthalic anhydride →[H₂O, Cr and Na phthalates (cat.), −CO₂] benzoic acid (85%)

T = 200°C
p = 1 atm
ΔH$_{rxn}$ (25°C, 1 atm) = − 493 kJ/mol
ΔH$_{rxn}$ (200°C, 1 atm) = − 526 kJ/mol

benzoic acid →[1/2 O₂, Cu cat., −CO₂] phenol (90%)

T = 230°C
p = 1.7 atm
ΔH$_{rxn}$ (25°C, 1 atm) = − 173.4 kJ/mol
ΔH$_{rxn}$ (230°C, 1.7 atm) = − 125.6 kJ/mol

Scheme 3.12 *Continued*

plans. Both metrics are mass weighted quantities with respect to the fractional contribution of each chemical component to the overall mass of input materials used or overall mass of waste produced, depending on whether the analysis is applied to input materials used or waste produced. BI accounts for potential environmental impacts from acidification–basification (AB), ozone depletion (OD), smog formation (SF), global warming (GW), inhalation toxicity (INHT), ingestion toxicity (INGT), bioaccumulation (BA), and abiotic resource depletion (ARD). Similarly, SHI accounts for potential safety–hazard impacts from corrosive properties of gases, liquids, or solids (CG and CL), flammability (F), oxygen balance (OB), hydrogen gas generation (HG), explosive vapor (XV), explosive strength (XS), impact sensitivity (IS), occupational exposure limit (OEL), skin dose (SD), and risk phrases (RP). Values of BI or SHI closer to 1 indicate low environmental or safety–hazard impact and are thus relatively greener than values closer to 0. Associated uncertainties in BI and SHI for a given reaction may be estimated by determining the fraction of input parameters that are missing because they are either unknown or unreliable [13]. The applicability of BI and SHI computations and their rankings is strongly linked to the availability of verified data. According to the framework of this analysis, each input chemical used or waste chemical generated in a chemical reaction requires eight parameters to determine its BI (number of carbon atoms, pK_a of acidic hydrogen atoms, ozone depletion, smog formation, LD50(oral), LC50(inhalation), Henry law constant, and $\log K_{ow}$) and 11 parameters to determine its SHI (number of oxygen atoms

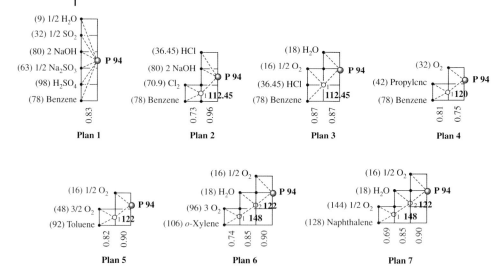

Figure 3.11 Synthesis tree representation of phenol plans shown in Scheme 3.12.

Table 3.5 Summary of global material efficiency metrics for phenol syntheses according to Scheme 3.12.

Plan	Number of steps	RY (%)	AE (%)	E	RME (%)	PMI	Rank
1	1	83	26	6.8	13	7.8	5
2	2	70	35	4.3	19	5.3	4
3	2	75	84	4.2	19	5.2	3
4	2	61	62	1.6	38	2.6	2
5	2	73	60	1.3	43	2.3	1
6	3	56	40	57.6	2	58.6	6
7	3	53	31	58.8	2	59.8	7

Table 3.6 Global E-factor breakdown for phenol syntheses according to Scheme 3.12.

Plan	E_{kernel}	E_{excess}	$E_{rxn\ solvent}$	E_{cat}	E_{workup}	$E_{purification}$	E_{total}
1	3.6	3.1	0	0	0	0	6.8
2	2.6	1.7	0	0	0	0	4.3
3	0.6	3.4	0	0.2	0	0	4.2
4	1.5	0.1	0	0	0	0	1.6
5	1.2	0.1	0	0.01	0	0	1.3
6	3.3	54.0	0	0.4	0	0	57.6
7	4.9	53.5	0	0.4	0	0	58.8

Table 3.7 BI(waste) results for phenol syntheses according to Scheme 3.12.

Plan	BI(waste)	% Uncertainty	Rank	Highest impact contribution
1	0.8761	0	5	INGT for unreacted sodium hydroxide
2	0.8696	0	6	INGT for unreacted sodium hydroxide; INHT for unreacted chlorine gas
3	0.9768	0	3	INGT and INHT for hydrochloric acid catalyst
4	0.7404	0	7	INGT for acetone by-product; GW for cumene, acetone, benzene, and propene
5	0.9724	3	4	INGT for unreacted benzoic acid; GW for unreacted toluene, carbon dioxide by-product, unreacted benzoic acid
6	0.9937	9	1	INGT for vanadium pentoxide catalyst
7	0.9929	8	2	INGT for vanadium pentoxide catalyst; INHT for naphthalene

required to oxidize substance, LD50(dermal), flash point, LC50(inhalation), number of moles of hydrogen gas generated, lower explosion limit, impact sensitivity, Trauzl lead block test, occupational exposure limit, skin dose, and risk R-phrases). A detailed compilation of reliable database sources for all of these parameters has been published [10,11]. In principle, both BI and SHI may be extended to include more potentials such as carcinogenicity, endocrine disruption, and any other biologically relevant environmental impact parameter provided the necessary data exist. However, due to limited availability of such data these were not included in the present analysis. Both BI and SHI have been applied to assess industrial syntheses of aniline [10,11] and β-azidation reactions of α,β-unsaturated ketones and acids [13]. Based on these limited analyses ingestion and inhalation toxicity potentials for liquids and gases, respectively, had the highest contributions to BI and occupational exposure limit and skin dose had the greatest contributions to SHI.

Since environmental harm is mainly caused by waste materials generated in a chemical reaction, BI(waste) is determined. For impacts due to safety and hazard, both SHI(input) and SHI(waste) are important since chemical workers are exposed to the handling of both input and waste materials when carrying out a chemical reaction. Table 3.7 summarizes the BI results for waste materials for the seven phenol syntheses given in Scheme 3.12, and Tables 3.8 and 3.9 summarize the SHI results for waste and input materials. We observe that plans 6 and 7 that are the most material inefficient have the highest BI and SHI scores albeit with the greatest degree of uncertainty, whereas plan 4 that is the most material efficient has the lowest BI and SHI scores. This apparent paradoxical situation arises since both of these indexes are calculated based on mass weighted impact potentials. In plan 4, the proportion of the waste coming from the most offending chemical, acetone, accounts for 38% of the mass of total

Table 3.8 SHI(waste) results for phenol syntheses according to Scheme 3.12.

Plan	SHI(waste)	% Uncertainty	Rank	Highest impact contribution
1	0.9539	2	5	OEL for unreacted benzene and sulfuric acid; CL for unreacted benzene
2	0.9136	4	6	OEL for unreacted benzene and chlorine gas; CL for unreacted benzene; SD for unreacted hydrochloric acid
3	0.9806	1	4	OEL for unreacted benzene; SD for unreacted hydrochloric acid
4	0.8327	2	7	OEL for unreacted benzene; SD for acetone by-product
5	0.9953	4	1	OEL for cobalt naphthenate catalyst
6	0.9941	8	2	OEL for vanadium pentoxide catalyst
7	0.9938	8	3	OEL for vanadium pentoxide catalyst

Table 3.9 SHI(input) results for phenol syntheses according to Scheme 3.12.

Plan	SHI(input)	% Uncertainty	Rank	Highest impact contribution
1	0.8494	2	5	OEL for unreacted benzene and sulfuric acid; CL for unreacted benzene
2	0.8323	4	6	OEL for unreacted benzene and chlorine gas; CL for unreacted benzene; SD for unreacted hydrochloric acid
3	0.8807	1	4	OEL for unreacted benzene; SD for unreacted hydrochloric acid
4	0.6851	2	7	OEL for unreacted benzene; SD for acetone by-product
5	0.9972	4	1	OEL for cobalt naphthenate catalyst
6	0.9944	8	2	OEL for vanadium pentoxide catalyst
7	0.9939	8	3	OEL for vanadium pentoxide catalyst

waste produced. By contrast, in plans 6 and 7 the most offending chemical, vanadium pentoxide catalyst, accounts for only 0.6% of the total waste. This situation was also noted in the analysis of β-azidation reactions of α,β-unsaturated ketones and acids [13]. In that work it was noted that "optimization in the direction of producing benign and safe waste products suggests that not only the waste materials should be composed of inherently lower impact potentials with respect to environment and safety, but also that the mass proportion of the most offending chemical with respect to the overall waste mass profile is kept to a minimum as far as possible." Plan 5 that ranks a close second in material efficiency ranks first in SHI(input) and SHI(waste) scores making it the plan that prevails in two levels of metrics analysis. It is possible to depict the performances of all seven plans visually using radial polygon diagrams that are constructed from the six metrics parameters (RY, AE, RME, BI(waste), SHI(waste), and SHI(input)) that are all quantities ranging between 0 and 1. From these diagrams, the vector magnitude ratio (VMR) for each plan is determined, given by Eq. (3.11), which estimates an unbiased overall score of green performance covering material efficiency, environmental impact, and safety–hazard impact relative to the ideal score of unity.

$$\text{VMR} = \frac{1}{\sqrt{6}}\left[(\text{AE})^2 + (\text{RY})^2 + (\text{RME})^2 + (\text{BI}_w)^2 + (\text{SHI}_w)^2 + (\text{SHI}_{in})^2\right]^{1/2}$$

(3.11)

Figure 3.12 shows the respective radial polygons and corresponding VMR scores for the seven phenol syntheses, as shown in Scheme 3.12. From these diagrams, we observe that plans 5 and 3 rank highest over three levels of metrics analysis with respect to VMR scores consistent with the above discussion.

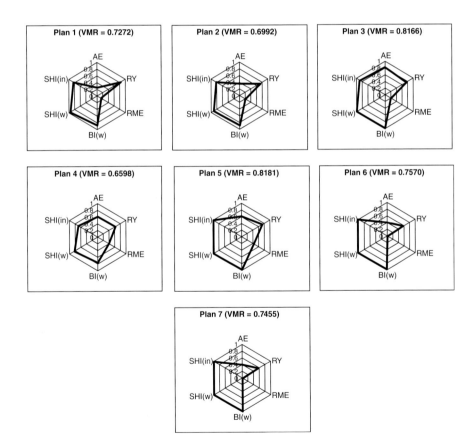

Figure 3.12 Radial polygon diagrams showing VMR scores for the seven phenol plans in Scheme 3.12.

3.4.4
Input Energy

The final layer of metrics analysis that can be applied is the determination of total energy consumption required to carry out all reaction steps in a synthesis plan. Specifically, the total energy required to heat and pressurize all input materials to carry out the entire synthesis plan at the given reaction conditions is determined. In this treatment, the energy consumption to operate all necessary equipment including their intrinsic efficiencies is neglected. Hence, the figures presented should be considered theoretical minimum estimates for input energy consumption. We first set a basis scale of 1 ton for phenol for all plans. Next we determine the corresponding mole scales and masses of all input materials required to produce a final mass of 1 ton of phenol. For each input material in a plan, the enthalpy changes in kJ for heating that substance from 298 K to the reaction temperature and for pressurizing that substance from 1 atm to the reaction pressure are determined on a per mole basis. The sum of these two contributions constitutes the overall enthalpic input energy required for that substance and is given by the parameter q. Processes involving heating have $q > 0$ and those involving cooling have $q < 0$. The q value of a given substance is then multiplied by its corresponding mole scale as prescribed in the synthesis plan in which that substance appears. The procedure is repeated for all input materials in a plan and the overall input energy sum is determined for a 1 ton synthesis of phenol that can then be converted back to units of kJ/mol. As shown in Scheme 3.12 in all chemical reactions T_{rxn} is always larger than 298 K and p_{rxn} is greater than or equal to 1 atm. The enthalpic contribution due to heating a substance depends on the phase transitions it can undergo from standard state conditions (298 K and 1 atm) to reaction conditions. The following cases given by Eqs. (3.12)–(3.20) illustrate the possible calculation scenarios for determining the temperature change contribution to enthalpy. All thermodynamic data were obtained from the DIPPR (Design Institute for Physical Property Data) database [54].

3.4.5
Case I

Heating a liquid at 298 K from 298 K to T_{rxn} (T_{rxn} is above boiling point, T_b) where the liquid undergoes a phase transition from liquid to gas:

$$q = \int_{298}^{T_b} C_{p,liq}(T) dT + \int_{T_b}^{T_{rxn}} C_{p,gas}(T) dT + \Delta H_{vap} \tag{3.12}$$

$$C_{p,liq}(T) = A + BT + CT^2 + DT^3 + ET^4 \tag{3.13}$$

$$C_{p,\text{gas}}(T) = A + B\left[\frac{C/T}{\sinh(C/T)}\right]^2 + D\left[\frac{E/T}{\cosh(E/T)}\right]^2 \qquad (3.14)$$

where ΔH_{vap} is the heat of vaporization, the functions $C_p(T)$ represent the temperature-dependent heat capacity functions at constant pressure for liquids and gases, and the parameters A, B, C, D, and E are constants specific to a given substance.

3.4.6
Case II

Heating a liquid at 298 K from 298 K to T_{rxn} (T_{rxn} is below boiling point, T_b) where the liquid does not undergo a phase transition:

$$q = \int_{298}^{T_{\text{rxn}}} C_{p,\text{liq}}(T)\,dT \qquad (3.15)$$

3.4.7
Case III

Heating a gas at 298 K from 298 K to T_{rxn}:

$$q = \int_{298}^{T_{\text{rxn}}} C_{p,\text{gas}}(T)\,dT \qquad (3.16)$$

3.4.8
Case IV

Heating a solid at 298 K from 298 K to T_{rxn} (T_{rxn} is above both the boiling point, T_b, and the melting point, T_m) where the solid undergoes phase transitions from solid to liquid and from liquid to gas:

$$q = \int_{298}^{T_m} C_{p,\text{sol}}(T)\,dT + \int_{T_m}^{T_b} C_{p,\text{liq}}(T)\,dT + \int_{T_b}^{T_{\text{rxn}}} C_{p,\text{gas}}(T)\,dT + \Delta H_{\text{fus}} + \Delta H_{\text{vap}}$$

$$(3.17)$$

$$C_{p,\text{sol}}(T) = A + BT + CT^2 + DT^3 + ET^4 \qquad (3.18)$$

where ΔH_{fus} is the heat of fusion and $C_{p,\text{sol}}(T)$ is the temperature-dependent heat capacity function at constant pressure for solids.

3.4.9
Case V

Heating a solid at 298 K from 298 K to T_{rxn} (T_{rxn} is above melting point, T_m) where the solid undergoes a phase transition from solid to liquid:

$$q = \int_{298}^{T_m} C_{p,sol}(T)dT + \int_{T_m}^{T_{rxn}} C_{p,liq}(T)dT + \Delta H_{fus} \tag{3.19}$$

3.4.10
Case VI

Heating a solid at 298 K from 298 K to T_{rxn} (T_{rxn} is below melting point, T_m) where the solid does not undergo a phase transition:

$$q = \int_{298}^{T_{rxn}} C_{p,sol}(T)dT \tag{3.20}$$

The enthalpic contribution due to changing pressure was determined by applying the Redlich–Kwong equation of state function to gases [55,56] and the Tait equation to liquids [57,58]. In all cases the magnitude of the temperature change contribution to q far exceeded that of the pressure change contribution. Table 3.10 summarizes the overall energy consumption results for the seven phenol syntheses based on the reaction conditions given in Scheme 3.12. Figure 3.13 shows histograms of total input energy for the production of

Table 3.10 Summary of energy consumption data for phenol syntheses shown in Scheme 3.12.

Plan	Energy consumption to heat input materials (kJ/mol phenol produced)	Energy consumption to pressurize input materials (kJ/mol phenol produced)	Net energy consumption, q (kJ/mol phenol produced)	Rank
1	358	0	358	5
2	202	−22	180	3
3	341	0	341	4
4	122	−35	87	1
5	153	0	153	2
6	16 220	0	16 220	7
7	16 187	0	16 187	6

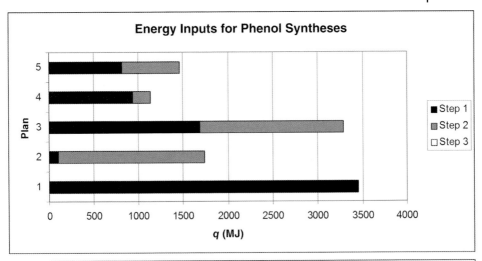

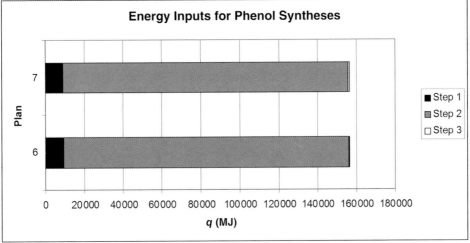

Figure 3.13 Histograms showing total energy input to produce 1 ton of phenol as a function of each reaction step contribution.

1 ton of phenol illustrating the relative contributions from each reaction step in each plan.

Plan 4 requires the least amount of energy input (87 kJ/mol) followed by plan 5 (153 kJ/mol) and plan 2 (180 kJ/mol). By contrast, plans 6 and 7 required the highest energy consumption exceeding 16 000 kJ/mol. From Figure 3.13 we observe that the first step of plan 2 consumes the least energy overall, the second steps of plans 6 and 7 consume the most energy overall, the energy demand of plan 3 is evenly divided between both steps, and the first steps of the remaining plans consume more energy than later steps.

3.4.11
Concluding Remarks and Outlook for Improvements

Taking all three levels of metrics analysis into account, it appears that plan 5, toluene oxidation to benzoic acid followed by thermal decarboxylation, is the overall winning green plan with respect to its VMR score based on material efficiency, BI, and SHI performance, and on its second best place standing for energy demands. Plan 4, the cumene oxidation-Hock rearrangement process, is competitive with respect to energy input and material efficiency and plan 3, benzene chlorination with gaseous hydrochloric acid followed by hydration, is competitive with respect to its VMR score. Both of these plans score high in two out of the three metrics analysis levels.

The present green metrics results are entirely consistent with the fact that over 90% of industrial phenol production is carried out using the cumene oxidation process where industry has focused its attention on optimizing material and energy demands [59,60]. However, this does not exclude the possibility of striving for even more efficient methods to make phenol as there is always room for improvement [61]. Indeed, there is an ongoing interest in developing an efficient and scalable route via direct oxidation of benzene using molecular oxygen [62–64], nitrous oxide [65,66], or hydrogen peroxide [67–72] as oxidant. Such transformations are single-step reactions that begin from a readily available commercial starting material and have high atom economies of 100, 77, and 84%, respectively. All three alternative routes require catalysts such as titanium silicates, zeolites, or metal oxides that may or may not need preparation via energy-intensive calcination processes. The by-products, nitrogen gas or water, are benign which is consistent with an excellent environmental impact profile. A unique advantage of the nitrous oxide reaction is that a high impact waste greenhouse gas from adipic acid production is put to good use as a reagent and is ultimately converted to a gas that already exists in large abundance in the atmosphere. The main drawbacks in using an oxidation approach include reaction control to prevent overoxidation reactions leading to side products such as catechol, benzoquinone, or hydroquinone; and safety with respect to scale-up. Overoxidation can be eliminated by using oxidants in low concentrations; however, this comes at the price of low reaction yields ranging from 3 to 15 to 53% for molecular oxygen, nitrous oxide, and hydrogen peroxide, respectively, on a per pass basis. The corresponding PMI profiles for best performing reactions are 590 [62], 23 [65], and 12 [72]. In terms of determining input energy consumption following the same method as described for previous plans, reaction of benzene with hydrogen peroxide requires 74 kJ, nitrous oxide requires 859 kJ, and molecular oxygen requires 1600 kJ/mol of phenol produced. Clearly, the case of the hydrogen peroxide reaction, typically run at 1 atm and 25–60 °C, has the best energy consumption profile of all methods to make phenol when compared with traditional methods (see Table 3.10).

This case study nicely illustrates the compromise nature of pursuing greener synthesis plans to a desired target molecule, particularly when searching for a plan that best satisfies material, energy, environmental impact, and safety–hazard constraints. In the first edition of this book, we have described in detail quantitative and qualitative guidelines for achieving "good" syntheses with respect to satisfying metrics targets and incorporating synthesis design strategies [73]. The top-ranked plans 3–5 for phenol satisfy the "golden limit" threshold criterion of AE exceeding 62% on a per step basis. Since phenol is a simple molecule structurally, it is made in no more than three steps. From a synthesis design perspective, the transformation from benzene to phenol is a substitution of ArC-H for ArC-OH meaning that many of the elegant strategy options that are implemented in ring construction or multicomponent couplings do not apply in this case. The substitution can be done directly by an oxidative process as discussed above, or indirectly in a stepwise manner using sulfonic acid (plan 1), chlorine (plans 2 and 3), isopropyl (plan 4), or carboxylic acid (plans 5–7) as sacrificial substituents. The direct method of substitution is particularly attractive because of the high atom economies and the fortuitous benign nature of by-products as discussed above.

There has been one report of a biofeedstock route to phenol from glucose via shikimic acid [74] in near-critical water that has been advertised as a nonbenzene route from renewable starting materials (see Scheme 3.13). The primary products arising from dehydration of shikimic acid are *m*- and *p*-hydroxybenzoic acid that subsequently decarboxylate to yield phenol. Although it is appealing from a green chemistry perspective, unfortunately verification of its merits by a detailed green metrics analysis as we have discussed in this chapter is severely hampered since no detailed experimental procedure was given. One key missing piece of information is the scale at which the reaction was conducted, which makes it difficult to gage its material efficiency and energy consumption performance against well-documented proven traditional routes. The only metrics for the shikimic acid to phenol transformation that can be stated are a reaction yield of 53% and an atom economy of 54%.

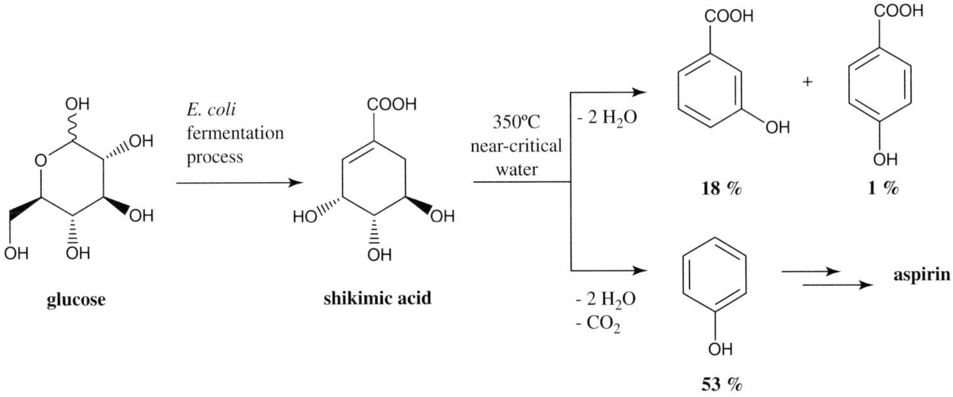

Scheme 3.13 Biotechnological route to phenol from glucose.

Green metrics can offer a tremendous advantage in directing such optimizations in a meaningful way. However, their determination not only must be computationally correct, but also must be carried out in an integrated way so as to make the decision-making process unbiased and fair as far as possible. Also, it is evident that despite best efforts to achieve such a winning green plan it should be understood that its ranking is comparative and not absolute. Its green ranking is only meaningful in the context of other plans according to specific criteria that are clearly defined and parameterized. This means that any claims of greenness must be couched in the language and proof of metrics analysis, otherwise such claims are without merit. Unfortunately, most of the green chemistry literature published to date over the last 20 years does not subscribe to this simple and logical guideline, which fuels suspicion among skeptics of the scientific motivation and aims of the field. Publications in the field are better categorized as advertisements rather than properly documented scientific papers with verifiable experimental procedures. There are two chief issues that challenge further progress in metrics analysis, particularly among novice scientists who genuinely wish to implement these ideas in their own research work. The first is the plethora of disconnected coined names for metrics that unnecessarily complicates the problem and impedes integration of concepts into a unified field. Often different authors use multiple names for the same parameter with a view to carve personal niches in the literature, rather than to actually solve problems. The second is the miscalculation and/or misuse of metrics particularly for determining material efficiencies of synthesis plans composed of sequential reactions that we have already noted in this chapter. A common way to cheat on synthesis plan performance is to begin the metrics analysis from a starting material with an advanced chemical structure thereby artificially truncating the actual number of steps in a plan to a desired target product. Another misuse of metrics is to separately analyze reaction, work-up, and purification stages of a set of transformations beginning from a common reactant and leading to a common target product. And then to select the three best performing stages from separate real examples in order to come up with a hypothetical overall best performing reaction without experimentally verifying if that contrived recipe actually works. Another famous abuse of material efficiency greenness is the announcement of so-called "solvent free" methodologies that eliminate reaction solvents, but use extensive, often undisclosed, volumes of chromatographic solvents in purification procedures. Announcing claims of greenness on the basis of material efficiency performance alone cannot be used to claim global greenness, though it is a key pillar suite of metrics. Rankings among synthesis plans with respect to one set of criteria can change when another set of criteria such as environmental impact, safety, or energy input demands are considered as was observed for the phenol analysis. This inevitably leads to tradeoffs that must be weighed objectively against each other. Finally, upon finding a properly determined and rationalized green plan or process for a chemical it should not be interpreted as the final word on making that target substance as there will always exist future alternate options that await discovery.

References

1. Andraos, J. (2013) Application of green metrics to scalable industrial synthesis plans: approaches to oseltamivir phosphate (Tamiflu®), in *Scalable Green Chemistry: Case Studies from the Pharmaceutical Industry* (ed. S.G. Koenig), CRC Press/Taylor & Francis Group, Boca Raton, FL, pp. 75–104.
2. Andraos, J. (2012) Parameterization and tracking of optimization of synthesis strategy using computer spreadsheet algorithms, in *Handbook of Green Chemistry Volume 7: Green Synthesis* (ed. C.-J. Li), Wiley-VCH Verlag GmbH, Weinheim, pp. 387–414.
3. Andraos, J. (2012) *The Algebra of Organic Synthesis: Green Metrics*, CRC Press/Taylor & Francis Group, Boca Raton, FL.
4. Andraos, J. (2006) On using tree analysis to quantify the material, input energy, and cost throughput efficiencies of simple and complex synthesis plans and networks: towards a blueprint for quantitative total synthesis and green chemistry. *Organic Process Research and Development*, **10**, 212–240.
5. Moores, A. (2009) Atom economy – principles and some examples, in *Handbook of Green Chemistry Volume 1: Homogeneous Catalysis* (ed. R.H. Crabtree), Wiley-VCH Verlag GmbH, Weinheim, pp. 1–15.
6. Andraos., J. and Sayed, M. (2007) On the use of "Green" metrics in the undergraduate organic chemistry lecture and lab to assess the mass efficiency of organic reactions. *Journal of Chemical Education*, **84**, 1004–1010.
7. Earle, M.J., Noe, M., Perosa, A., and Seddon, K.R. (2014) Improved synthesis of tadalafil using dimethyl carbonate and ionic liquids. *RSC Advances*, **4**, 1204–1211.
8. Toniolo, S., Arico, F., and Tundo, P. (2014) A comparative environmental assessment for the synthesis of 1,3-oxazin-2-one by metrics: greenness evaluation and blind spots. *ACS Sustainable Chemistry & Engineering*, **2**, 1056–1062.
9. Andraos, J. (2013) Green chemistry metrics: material efficiency and strategic synthesis design, in *Innovations in Green Chemistry and Green Engineering* (eds P.T. Anastas and J.B. Zimmerman), Springer, New York, pp. 81–113.
10. Andraos, J. (2012) Inclusion of environmental impact parameters in radial pentagon material efficiency metrics analysis: using benign indices as a step towards a complete assessment of "greenness" for chemical reactions and synthesis plans. *Organic Process Research and Development*, **16**, 1482–1506.
11. Andraos, J. (2013) Safety/hazard indices: completion of a unified suite of metrics for the assessment of "greenness" for chemical reactions and synthesis plans. *Organic Process Research and Development*, **17**, 175–192.
12. Mercer, S.M., Andraos, J., and Jessop, P.G. (2012) Choosing the greenest synthesis: a multivariate metric green chemistry exercise. *Journal of Chemical Education*, **89**, 215–220.
13. Andraos, J., Ballerini, E., and Vaccaro, L. (2015) A comparative approach to the most sustainable protocol for the β-azidation of α,β-unsaturated ketones and acids. *Greem Chemistry*, **17**, 913–925.
14. Dicks, A.P. (2012) *Green Organic Chemistry in Lecture and Laboratory*, CRC Press/Taylor & Francis Group, Boca Raton, FL.
15. Dicks, A.P. and Hent, A. (2015) *Green Chemistry Metrics*, Springer, London.
16. Laird, T. (2013) Green chemistry is good process chemistry. *Organic Process Research and Development*, **16**, 1–2.
17. Anastas, P.T. (2015) Green chemistry next: moving from evolutionary to revolutionary. *Aldrichimica Acta*, **48**, 3–4.
18. Constable, D.J.C. (2015) Moving toward a green chemistry and engineering design ethic. *Aldrichimica Acta*, **48**, 7.
19. Warner, J.C. (2015) Where we should focus green chemistry efforts. *Aldrichimica Acta*, **48**, 29.
20. Jimenez-Gonzalez, C., Ponder, C.S., Broxterman, Q.B., and Manley, J.B. (2011) Using the right green yardstick: why process mass intensity is used in the pharmaceutical industry to drive more

21 Dach, R., Song, J.J., Roschangar, F., Samstag, W., and Senanayake, C.H. (2012) The eight criteria defining a good chemical manufacturing process. *Organic Process Research and Development*, **16**, 1697–1706.

22 Andraos, J. (2005) Unification of reaction metrics for green chemistry: applications to reaction analysis. *Organic Process Research and Development*, **9**, 149–163.

23 Roschangar, F., Sheldon, R.A., and Senanayake, C.H. (2015) Overcoming barriers to green chemistry in the pharmaceutical industry – the Green Aspiration Level™ concept. *Greem Chemistry*, **17**, 752–768.

24 Tundo, P. and Andraos, J. (2014) *Green Syntheses*, vol. **1**, CRC Press/Taylor & Francis Group, Boca Raton, FL.

25 Sheldon, R.A. (2007) The E factor: fifteen years on. *Green Chemistry*, **9**, 1273–1283.

26 Andraos, J. and Hent, A. (2015) Useful material efficiency green metrics problem set exercises for lecture and laboratory. *Journal of Chemical Education*, **92**, 1831–1839.

27 Greener, B.S. and Millan, D.S. (2010) Bortezomib (Velcade): a first-in-class proteasome inhibitor, in *Modern Drug Synthesis* (eds J.J. Li and D.S. Johnson), John Wiley & Sons, Inc., New York, pp. 99–110.

28 Millennium Annual Report (2007) http://library.corporate-ir.net/library/80/801/80159/items/297170/MLNM_AR2007.pdf (accessed May 1, 2015).

29 Adams, J., Ma, Y.-T., Stein, R., Baevsky, M., Grenier, L., and Plamondon, L. (1998) Boronic ester and acid compounds. US Patent 5,780,454.

30 Li, Y., Plesescu, M., Sheehan, P., Daniels, J.S., and Prakash, S.R. (2007) Synthesis of four isotopically labeled forms of a proteasome inhibitor, bortezomib. *Journal of Labelled Compounds and Radiopharmaceuticals*, **50**, 402–406.

31 Beenen, M.A., An, C., and Ellman, J. (2008) Asymmetric copper-catalyzed synthesis of r-amino boronate esters from *N-tert*-butanesulfinyl aldimines. *Journal of the American Chemical Society*, **130**, 6910–6911.

32 Ivanov, A.S., Zhalnina, A.A., and Shishkov, S.V. (2009) A convergent approach to synthesis of bortezomib: the use of TBTU suppresses racemization in the fragment condensation. *Tetrahedron*, **65**, 7105–7108.

33 Milo, L.J. Jr., Lai, J.H., Wu, W., Liu, Y., Maw, H., Li, Y., Jin, Z., Shu, Y., Poplawski, S.E., Wu, Y., Sanford, D.G., Sudmeier, J.L., and Bachovchin, W.W. (2011) Chemical and biological evaluation of dipeptidyl boronic acid proteasome inhibitors for use in prodrugs and pro-soft drugs targeting solid tumors. *Journal of Medicinal Chemistry*, **54**, 4365–4377.

34 Ivanov, A.S., Shishkov, S.V., and Zhalnina, A.A. (2012) Synthesis and characterization of organic impurities in bortezomib anhydride produced by a convergent technology. *Scientia Pharmaceutica*, **80**, 67–75.

35 Janca, M. and Dobrovolny, P. (2009) Methods for preparing bortezomib and intermediates used in its manufacture. WO Patent 2009/004350.

36 Palle, R.V., Kadaboina, R., Murki, V., Manda, A., Gunda, N., Pulla, R.R., Hanmanthu, M., Mopidevi, N.N., and Ramdoss, S.K. (2009) Bortezomib and process for producing same. WO Patent 2009/036281.

37 Roemmele, R.C. (2011) Proteasome inhibitors and processes for their preparation, purification and use. WO Patent 2011/087822.

38 Kumar, S., Durvasula, V.V., Rathod, P.D., and Aryan, R.C. (2011) Process for the preparation of bortezomib. WO Patent 2011/098963.

39 Rao, D.R., Kankan, R.N., Pathi, S.L., and Puppala, R. (2014) Process for preparing of bortezamib. WO Patent 2014/041324.

40 Ravi, J.R.R., Kondaveeti, S., Ibhatla, K.S.B.R., Muddasani, P.R., and Nannapaneni, V.C. (2014) Stable and pure polymorphic form of bortezomib. WO Patent 2014/097306.

41 Henschke, J.P., Xie, A., Huang, X.Y., and Chen, Y.F. (2012) Process for preparing and purifying bortezomib. US Patent 2012/0289699.

42 Tao, S. and Li, H. (2013) Method for synthesizing bortezomib. CN Patent 101,812,026.
43 Yang, Y., Lei, Z., Jie, L., Lie, L., Lei, W., Hua, T., Yong, X., and Wei, W. (2012) Method for preparing bortezomib. CN Patent 102,675,415.
44 Li, Z.X. and Zhao, L.X. (2013) Synthetic method of high-purity bortezomib and intermediate thereof. CN Patent 103,012,551.
45 Pickersgill, I., Bishop, J., Koellner, C., Gomez, J.M., Geiser, A., Hett, R., Ammoscato, V., Munk, S., Lo, Y., Chiu, F.T., and Kulkarni, V.R. (2005) Synthesis of boronic ester and acid compounds. WO Patent 2005/097809.
46 Constable, D.J.C., Dunn, P.J., Hayler, J.D., Humphrey, G.R., Leazer, J.L., Jr., Linderman, R.J., Lorenz, K.L., Manley, J., Pearlman, B.A., Wells, A., Zaks, A., and Zhang, T. (2007) Key green chemistry research areas – a perspective from pharmaceutical manufacturers. *Green Chemistry*, **9**, 411–420.
47 Monks, B.M. and Whiting, A. (2013) Direct amide formation avoiding poor atom economy reagents, in *Sustainable Catalysis: Challenges and Practices for the Pharmaceutical and Fine Chemical Industries*, 1st edn (eds P.J. Dunn, K.K. Hii, M.J. Krische, and M.T. Williams), John Wiley & Sons, Inc., Hoboken, NJ.
48 Valeur, E. and Bradley, M. (2009) Amide bond formation: beyond the myth of coupling reagents. *Chemical Society Reviews*, **38**, 606–631.
49 Dev, D., Palakurthy, N.B., Thalluri, K., Chandra, J., and Mandal, B. (2014) Ethyl 2–cyano-2-(2-nitrobenzenesulfonyloxyimino)acetate (o–nosylOXY): a recyclable coupling reagent for racemization-free synthesis of peptide, amide, hydroxamate, and ester. *Journal of Organic Chemistry*, **79**, 5420–5431.
50 Blacker, A.J. and Williams, M.T. (2011) *Pharmaceutical Process Development: Current Chemical and Engineering Challenges*, Royal Society of Chemistry, Cambridge, pp. 251–252.
51 Faith, W.L., Keyes, D.B., and Clark, R.L. (1966) *Industrial Chemicals*, 3rd edn, John Wiley & Sons, Inc., New York.
52 Lowenheim, F.A. and Moran, M.K. (1975) *Faith, Keyes, and Clark's Industrial Chemicals*, 4th edn, John Wiley & Sons, Inc., New York.
53 Andraos, J. (2009) Global green chemistry metrics analysis algorithm and spreadsheets: evaluation of the material efficiency performances of synthesis plans for oseltamivir phosphate (Tamiflu) as a test case. *Organic Process Research and Development*, **13**, 161–185.
54 Design Institute for Physical Property Data (DIPPR) Project 801 (2015) http://www.aiche.org/dippr/projects/801 (accessed February 2015).
55 Redlich, O. and Kwong, J.N.S. (1949) On the thermodynamics of solutions. V. *Chemical Review*, **44**, 233–244.
56 Smith, J.M. and van Hess, H.C. (1987) *Introduction to Chemical Engineering Thermodynamics*, McGraw-Hill Book Co., New York.
57 Danner, R.P. and Daubert, T.E. (1988) *Manual for Predicting Chemical Process Design Data – Data Prediction Manual*, American Institute of Chemical Engineers, New York.
58 Dymond, J.H. and Malhotra, R. (1988) The Tait equation: 100 years on. *International Journal of Thermophysics*, **9**, 941–951.
59 Weber, M., Weber, M., and Kleine-Boymann, M. (2012) *Ullmann's Encyclopedia of Industrial Chemistry*, vol. **26**, Wiley-VCH Verlag GmbH, Weinheim, pp. 503–519.
60 Kirk, R.E., Othmer, D.F., and Grayson, M. (1991) *Encyclopedia of Chemical Technology*, 4th edn, vol. **18**, John Wiley & Sons, Inc., New York, pp. 286–291.
61 Schmidt, R.J. (2005) Industrial catalytic processes – phenol production. *Applied Catalysis A*, **280**, 89–103.
62 Tassinari, R., Bianchi, D., Ungarelli, R., Battistel, E., and D'Aloisio, R. (1998) Process for the synthesis of phenol from benzene. EP Patent 0,894,783.
63 Tsuruya, S. (1999) Method of manufacturing phenol by direct oxidation of benzene. EP Patent 1,024,129.
64 Ricci, M., Bianchi, D., and Bortulo, R. (2009) Towards the direct oxidation of benzene to phenol, in *Sustainable Industrial Processes* (eds F. Cavani, G. Centi, S. Perathoner, and F. Trifiró),

65. Gubelmann, M. and Tirel, P.J. (1989) Process for preparation of phenol. EP Patent 0,341,165 (Rhone-Poulenc).
66. Kharitonov, A.S., Panov, G.I., Ione, K.G., Romannikov, V.N., Sheveleva, G.A., Vostrikova, L.A., and Sobolev, V.I. (1992) Preparation of phenol or phenol derivatives. US Patent 5,110,995.
67. Bianchi, D., Bortolo, R., Tassinari, R., Ricci, M., and Vignola, R. (2000) A novel iron-based catalyst for the biphasic oxidation of benzene to phenol with hydrogen peroxide. *Angewandte Chemie International Edition*, **39**, 4321–4323.
68. Gao, X. and Xu, J. (2006) A new application of clay-supported vanadium oxide catalyst to selective hydroxylation of benzene to phenol. *Applied Clay Science*, **33**, 1–6.
69. Jian, M., Zhu, L., Wang, J., Zhang, J., Li, G., and Hu, C. (2006) Sodium metavanadate catalyzed direct hydroxylation of benzene to phenol with hydrogen peroxide in acetonitrile medium. *Journal of Molecular Catalysis A*, **253**, 1–7.
70. Balducci, L., Bianchi, D., Bortolo, R., D'Aloisio, R., Ricci, M., Tassinari, R., and Ungarelli, R. (2003) Direct oxidation of benzene to phenol with hydrogen peroxide over a modified titanium silicalite. *Angewandte Chemie International Edition*, **42**, 4937–4940.
71. Bianchi, D., Bortolo, R., Buzzoni, R., Cesana, A., Dalloro, L., and D'Aloisio, R. (2004) Integrated process for the preparation of phenol from benzene with recycling of the by-products. US Patent 2004/122264 (Polimeri Europa S.p.A.).
72. Arunabha, D., Sakthivel, S., and Kumar, S.J. (2008) Process for the liquid phase selective hydroxylation of benzene. US Patent 2008/234524 (Council of Scientific and Industrial Research).
73. Andraos, J. (2008) Application of green metrics analysis to chemical reactions and synthesis plans, in *Green Chemistry Metrics: Measuring and Monitoring Sustainable Processes* (eds A. Lapkin and D.C. Constable), Blackwell Scientific, Oxford, pp. 118–119.
74. Gibson, J.M., Thomas, P.S., Thomas, J.D., Barker, J.L., Chandran, S.S., Harrup, M.K., Draths, K.M., and Frost, J.W. (2001) Benzene-free synthesis of phenol. *Angewandte Chemie*, **113**, 1999–2002.

4
Life Cycle Assessment

Concepción Jiménez-González

4.1
Introduction

Life cycle inventory and assessment (LCI/A) is a methodology that allows one to estimate the cumulative environmental impacts associated with manufacturing the chemicals, materials, and equipment used to make a product or deliver a service, thereby providing a comprehensive view of the potential trade-offs in environmental impacts associated with a given activity. The term life cycle refers to the major phases of a process or product, from the extraction of all raw materials to the final manufacture, transportation, use, maintenance, reuse, to its final fate [1,2].

Depending on the objectives of the assessment, the life cycle evaluation of the environmental impacts is likely to have different boundary conditions [3,4]. For instance, it could cover the entire supply chain (cradle-to-grave, or CtG), a single chemical plant (gate-to-gate, or GtG), or downstream production impacts. Life cycle metrics can either be

- direct life cycle inventory (LCI) data, for example, life cycle energy, life cycle mass, life cycle emissions;
- or, they could come from a life cycle impact assessment (LCIA), which measures either individual impacts, such as global warming potential, or aggregates the impacts into a score or index, such as the EcoIndicator 99 method [5].

Life cycle thus provides a framework of more holistic "green metrics" to estimate the environmental footprint of a route, process, or reaction. The use of LCI/A to measure the "greenness" of chemistries has been championed previously elsewhere as a strategic need in the development and use of green metrics [6–11]. Using a life cycle approach to estimate environmental impacts of processes and products provides the practitioner with the following advantages:

- Evaluates the true greenness of a process (e.g., avoids shifting impacts).
- Provides a set of commonly accepted metrics to measure specific impacts.

- Identifies trade-offs between impacts and phases in the life cycle and highlights hidden costs and impacts.
- Identifies "hot spots," or the phases or areas that contribute the most to the "environmental footprint" of an activity to drive opportunities for resource optimization.
- Systematically identifies key environmental impacts at each of the life cycle stages of a product.
- Provides information to decision-makers for strategic planning, priority setting, product or process design or redesign.
- Identifies information gaps.
- Provides scientific data that can be applied to recognized marketing schemes (e.g., eco-labeling, environmental claims, environmental product declarations).

There has been increased interest in LCI/A techniques to evaluate the "greenness" of processes, perhaps primarily driven by the increased global interest in climate change and the drive to estimate the global warming potential impacts of a chemical route. However, it is lamentable that attention has been predominantly directed at global warming potential to the exclusion of other impact categories. Even with this increased uptake, the routine use of LCI/A remains far from being fully embedded as a business process, and continues to undergo an evolution, with different companies being at diverse stages of sophistication in their use of LCI/A as a business decision-making tool [12,13].

4.2
The Evolution of Life Cycle Assessment

Life cycle assessment traces its roots back to the 1960s, where the main interest was in cumulative energy consumption. Later on, global modeling studies published in *The Limits to Growth* [14] and *A Blueprint for Survival* [15] were focused on estimations of population change in the face of limited resources. The predicted depletion of fossil fuels caused more calculations of energy use but the foundations for modern life cycle assessment methodologies began in 1969 with the initiation of an internal study by The Coca-Cola Company who desired to compare the environmental releases associated with two different kinds of beverage containers. Raw materials use was quantified as were the environmental emissions associated with the manufacture of each container. Other companies performed similar inventories in the early 1970s. At that time, most of the data came from publicly available sources and industry-specific data were hard to find. From 1975 to the late 1980s the interest in performing these types of detailed assessments decreased as the oil crises waned and environmental issues focused on toxic chemicals, pollution prevention, and hazardous waste management. When hazardous waste became a global issue, life cycle inventory/ assessment re-emerged as a tool for analyzing systemic and cumulative environmental problems. Still, there was a need for a standardized approach, and there

4.3 LCA Methodology at a Glance

Figure 4.1 Illustrative examples of the historical evolution of life cycle practice.

were many questions about the usefulness and appropriateness of the tool. These shortcomings created a demand for the standardization of existing *ad hoc* approaches. The Society of Environmental Toxicology and Chemistry [16–18] stepped into the void with the publication of standardized approaches and these efforts culminated with the development of international LCA standards through the International Standards Organization (ISO) 14 000 series (14040, 14044). The ISO standards guide LCA practitioners on the elements needed to conduct an assessment, including uncertainty and sensitivity analysis, as well as critical review [19,20]. At this point in time, Product Category Rules (PCR) and Environmental Product Declarations (EPD) are being explored as complements to LCI/A and are being applied more frequently by different industries [21]. Figure 4.1 shows a few illustrative examples of the historical evolution of LCA practice.

As the practice of life cycle inventory/assessment develops and becomes more accepted as a legitimate and helpful scientific discipline, challenges remain in the practicalities of the approach to obtaining the desired information at the appropriate level for the objectives of the assessment, and how to effectively use the results to improve business or development decisions.

4.3
LCA Methodology at a Glance

The LCA process is a systematic approach consisting of four discrete phases: goal definition and scoping, inventory analysis, impact assessment, and interpretation. The next few paragraphs will cover the general principles of the methodology.

4.3.1
Goal and Scope

The goal and scope phase is arguably among the most important phases of a LCI/A. In this stage we define and describe why we are performing a life cycle inventory and/or assessment. This includes defining intended application, the reasons for carrying out the study, the intended audience, and whether or not the results will be used in comparisons to be disclosed to the public. Key aspects to defining goal and scope include the following:

- *Functional unit.* Why and how is this product used? This is particularly important when comparing two or more products or systems
- *System boundaries.* Do we have to cover the entire life cycle? A decision to use a boundary other than a cradle-to-grave boundary needs to be understood and documented. If the boundary is too limited, significant elements may be excluded; if the boundary is too open, resources may be unnecessarily wasted.
- *Allocation.* When the manufacture of any given product produces other valuable secondary outputs, which life cycle environmental impacts belong to which product? The ISO standard dictates that allocation should be avoided whenever possible, and provides guidance on how to address it; for example, by broadening the system boundaries (i.e., system expansion method). For more information on this topic, see Ref. [22].

4.3.2
Inventory Analysis

This is the backbone of a life cycle assessment, as the inventory identifies and quantifies the resources used (e.g., energy, water, and materials) and the environmental emissions associated with the product or service (e.g., air emissions, solid waste disposal, waste water discharges, etc.). Most of the time spent in completing an LCI/A is typically spent in the inventory phase, given the needs for data collection, validation, documentation, and potentially refining system boundaries and assumptions.

There are several documented methodologies that can be used to calculate life cycle inventory information, either in full, on a gate-to-gate basis, or for specific systems such as energy production, waste treatment, and others [23–28]. There are also commercially available software packages and databases such as Eco-Invent [29], SimaPro [30], DEAM [31], UMBERTO [32], GABI [33], and others that contain life cycle inventory information for generic processes, energy production, waste treatment, and some other subsystems. The literature also contains life cycle information that can be used in LCIs, and there are several good review articles of the methods used in different applications [34–36].

Transparency in the development of life cycle inventory data are critically important since it is very unlikely that accurate information may be obtained from a single source, and calculated data will be derived from multiple inputs.

Comparative studies of life cycle inventory data obtained for the same product or service collated from different LCI databases have found a lack of transparency, double-counting of primary sources, and significant discrepancies [37].

4.3.3 Impact Assessment

At this stage, the life cycle inventory results are used to assess the potential human and ecological effects of the process or activity under study. The standard ISO 14044 has defined guidelines for each of the elements that comprise the impact assessment.

- *Category Selection and Definition.* This determines which impacts would be evaluated (e.g., cumulative energy demand, acidification potential, etc.), the models to be used (e.g., CML, EDIP, etc.), the chosen framework (e.g., Nordic guidelines, SETAC, corporate guidelines, etc.), the cause and effect chains, and the methods to measure them. Selecting and defining these categories will depend on the goal and scope.
- *Classification.* In this step, the inventory results (e.g., total kilogram of carbon dioxide emissions, etc.) are matched with specific impact categories (e.g., global warming potential, etc.).
- *Characterization.* This step estimates the environmental impact for each category (e.g., global warming potential, etc.) using science-based conversion factors.
- *Normalization.* Allows to express impact results in a way that meaningful comparisons may be drawn (e.g., normalizing by regional impact, benchmarking by process, etc.)
- *Grouping.* For this element we think about how to sort the results; as, for example, into local, regional, and global impacts. This may be done either for presenting the results or as part of completing the assessment.
- *Weighting.* One can add some relative judgment to the impacts. For instance, we may subjectively give an impact category greater importance because of its potential to cause a particularly bad outcome; for example, as in the case where we may weight a human cancer outcome more highly than crop damage.
- *Quality Assessment.* The intent here is to obtain a better understanding of the reliability of the results by applying appropriate statistical and other analytical methods (e.g., sensitivity analysis, data quality indicators [38], uncertainty analysis, etc.).

4.3.4 Interpretation

During this part of the life cycle assessment, the results of the inventory analysis and impact assessment are evaluated for the products, processes, or services under study. In principle, this may sound simple, and in theory we would have a system where clear conclusions can be drawn from both the inventory and the impact assessment. However, it is relatively common that with a long list of impacts to evaluate, one can find that process A might have a better profile than process B in some impacts, but worse in the rest of the impacts. One must, therefore, come to terms with these trade-offs between environmental impacts during the decision-making process. Whether the trade-offs are important or not generally depends on the specific priorities of the decision makers, how large of a difference there is among the trade-offs, and how much uncertainty there is in the data quality or the overall assessment. Some considerations to keep in mind during the interpretation phase are as follows:

- The goal and scope of the LCA: One cannot easily extrapolate from one system to another unless the data and science support it.
- Boundaries and boundary conditions are extremely important. In comparative LCI/As these are fundamental, but it is even more important when talking about stand-alone LCI/As, as the differences in boundary conditions can result in different outcomes.
- The degree of uncertainty in the data and the associated results. This is especially important when comparing outputs – how significant are the differences?
- Sensitivity of the results to variations in specific parameters. If other databases or assessment methods were used, would the results remain valid?
- Trade-offs: Are all the important life cycle environmental impacts considered or do hidden trade-offs exist in the assessment results?
- Applicability of the data: In some instances, spatial and temporal differences in the data might cause important differences.
- Organizational and societal values: Which impacts are valued above others and do these align with the inherent priorities of the impact assessment methods employed?
- Micro- and macro-economic factors: Would the total cost of the activity, from a holistic viewpoint, change the results? Are there market forces affecting the assessment?
- Societal values: Are there cultural impacts that might not be immediately apparent in the metrics used for the assessment?

4.3.5
LCI/A Limitations

Although life cycle inventory and assessment is a very powerful tool, as is the case with any methodology, there are limitations in its application. First of all, performing a full LCI/A can be resource and time intensive. This is particularly true if there is not enough data readily available, or if one's knowledge of the system under study is limited. In many instances available, data are limited and for those data that are readily at hand, there is likely to be greater uncertainty associated with these data than what one might be willing to accept. In almost all LCI/A studies, one will inevitably be faced with having to make some assumptions about the source data or will be forced to utilize averaged data (i.e., background data).

These inherent limitations in LCI/A studies necessitate that one maintain transparency about the assumptions and the data quality for any data used in the study. Transparency ensures that there is an explicit demarcation of the sources, assumptions, the information utilized, and the data gaps within a life cycle inventory or assessment. LCI/A is an iterative methodology and very seldom is it the case that an LCI/A result is fixed; thus, data transparency will greatly help to bolster the credibility and acceptance of the study while making it easier to integrate more accurate information as new data become available.

In addition to problems or issues with data limitations, when seen within the broader green metrics context, LCI/A is a methodology that needs to continue to evolve with the state of the art and science. While LCI/A is arguably the most holistic sustainability methodology at the moment, social or economic aspects have historically been out of the scope of traditional LCI/As. However, it should be noted that other tools such as Social LCA and Total Cost Accounting (Life Cycle Costing) tend to cover these aspects.

4.3.6
Critical Review

As we saw in the previous section, LCI/A methodology has several limitations, many of them linked to the transparency of the underlying data, the assumptions, and the methodology used to undertake the assessment. In an attempt to ensure the credibility of LCI/A methodology, and to answer concerns about objectivity, there has been a movement toward increased standardization of LCI/A methodologies and critical peer reviews. Besides enhancing credibility, critical peer review helps to maintain data confidentiality and integrity by allowing a reviewer to validate the study as they review the entire assessment. This allows the final results to be communicated with confidence without having to show the confidential information. Finally, when the results of the LCI/A may possibly impact policy or market decisions, inviting interested parties and stakeholders as part of the critical review is an advisable option.

ISO standards contemplate three types of reviews: review by internal experts, review by external experts, and review by interested parties. Reviews performed by interested parties are normally done as a panel of experts, whereas internal and external expert reviews tend to be done by a single reviewer. Not all LCI/As are subject to critical review, but it is definitely advisable to consider it in the planning. In addition, not all critical reviews are the same. The ISO standards recognize two comprehensive and formal critical reviews:

- The ISO standard full review, which is mandatory in cases where comparative assessments are done for a public audience in mind. The primary focus of this review is on the methodology, its consistency with the ISO standard, and its scientific validity.

- The review of EPD, where the primary focus is on the underpinning data. In these cases, customers are expected to draw conclusions and comparison based on potentially nontransparent (commercially sensitive or proprietary), aggregated data.

There are very specific guidelines for these two types of mandatory critical reviews. In contrast to these reviews, the focus of the internal expert or external expert reviews will need to be decided as part of the goal and scope of the LCI/A in question.

4.3.7
Streamlined Life Cycle Assessment

Performing a full LCI/A can be very time consuming, and such a study normally requires an extensive collection of data and a large database. When the results of an LCI/A are intended for process or product development decision-making, producing a full LCI/A is not always feasible due to limited data availability during process development. Waiting to obtain a complete and well-validated data set that could stand up to the full rigor of an LC/IA would very likely mean that assessment results would be produced too late to be practical. At the same time, the most efficient and cost effective time to introduce sustainability considerations into a process, product, or activity is precisely during development.

One way to address this dilemma is to use a simplified or streamlined life cycle assessment method. According to SETAC, a streamlined LCA applies LCA methodology and covers the same aspects, but at a higher level. Thus, instead of using site-, time-, mode- and technology-specific data throughout the assessment, a streamlined LCA would tend to use generic background data to conduct a simplified assessment that focuses on the most important phases or impacts. One would then assess the reliability of the results with uncertainty and sensitivity analysis.

Most simplified LCAs consist of three phases: an initial screening to determine the main points of focus; a simplification phase that uses the results of the screening to determine which areas can be simplified and which need more in-depth assessments; and finally a quality analysis phase that assesses the reliability of the results. As is common with a full LCA, a streamlined LCA can be used iteratively to identify those parts of the system that require further assessment, or it can be used to determine indicators, as a prior step to performing a full LCI/A, or as a complementary study to a full LCA, or as a basis to evaluate similar systems in a faster manner. As might be expected, the quality analysis phase is a very important part of any streamlined LCA because as one streamlines the study the uncertainty increases, and there is an increased chance of obtaining significantly different results than would be obtained with a full LCI/A. However, there are many occasions where the risk associated with a streamlined LCA is acceptable, as in the case of process and product development, or when there is insufficient data to perform a full LCI/A, or when the timelines required to complete a full LCI/A would hinder the integration of any results into the improvement assessment, or when the effort is better spent in the most important subsection of the LCI/A, and so on. Given the increased importance of streamlined LCAs, some general guidance has been proposed by Hunt *et al.* for how to conduct streamlined LCAs and for how to select a streamlining in such a way as to reduce the potential for error [39].

4.4
Measuring Greenness with LCI/A – Applications

The application of LCI/A metrics is still not a widespread practice. However, the use is definitely more widespread now than a decade ago, and its use has been highlighted many times as a key element of an effective green chemistry program [40–43]. As LCI/A practice has evolved, the emphasis of the areas of interest has also shifted from the initial exploration of the concept with a few case studies, assessing either full production routes or manufacturing subsystems, moving to LCI/As of products, EPD, and environmental footprints of entire companies. Part of this evolution has been the balance between a full set of impacts and the use of single indicators and streamlined tools for general screening [44,45].

Nowadays, practitioners could use LCA in an array of different applications: through case studies to better understand the wider environmental implications of processes, to compare different chemical routes, or to compare the use of different unit operations, to name a few. Table 4.1 contains a few illustrative examples where LCI/As have been used as a measure of greenness.

In simple terms, one can think about two typical applications of LCI/A. First, life cycle inventory assessment may be seen as a means to compare the environmental footprint of different alternatives to make development or improvement decisions. For instance, when comparing reaction A with reaction B, in which either:

- A is better than B (A > B)
- B is better than A (B > A), or
- A and B are essentially similar (A ≈ B).

Second, life cycle inventory assessment is also used as a comprehensive methodology for understanding the major contributors to the environmental footprint of a product, therefore, allowing for environmental improvements across larger systems, such as in the case of supply chain optimization. This section will cover several examples of the application of LCI/A.

4.4.1
Probing case studies

It is not surprising that the initial applications of LCI/A would start with some probing case studies, partly to increase the understanding of the process or product, and partly to understand what type of questions an LCI/A could answer.

Table 4.1 Selected illustrative examples of application of LCI/A.

Application	Examples	References
Probing case studies through LCA of specific molecules	• Pfizer – evaluation of chemical routes for Sertraline • GSK – cradle-to-gate LCIA in the synthesis of an Active Pharmaceutical Ingredient (API) • Pfizer – evaluation of the route for Pregablin • Cradle-to-gate LCI of vancomycin hydrochloride • LCIA of Biorefineries	[46] [47] [48] [49] [50,51]
Route or process comparison	• Chemical versus an enzymatic route for an intermediate • LCIA comparing Hoffmann La-Roche and GSK APIs • Adipic acid route comparisons • Caprolactam traditional versus catalytic routes • Bioplastics versus fossil plastics • Basic chemicals – bio based versus fossil based • Continuous versus batch processing	[52] [53,54] [55,56] [57] [58–63] [64–67] [68–72]
Materials assessment	• Solvent selection – GSK, AstraZeneca, others • GSK Reagent Selection guides, including LCA • Enzymes LCA assessment • Biofuels • Polysaccharides • Ionic liquids • Nanomaterials • Packaging	[73–76] [77] [78,79] [80] [81] [82–84] [85–88] [89–92]
Subsystem assessment	• Waste treatment: Pfizer, Novartis, BMS, GSK • Disposable versus reusable processing equipment • Exergetic LCA for technology options • LCA for biocatalytic processes assessment	[93,94] [95–99] [100,101] [102]
Product LCA	• GSKs cradle-to-gate LCA for a device product • BASF Eco-efficiency analysis of B12 and indigo dye • Novartis – limited LCA of two products • Apple products (iPhones, iPads, others)	[11] [103,104] [105] [106]
Foot-printing and single indicators	• GSK enterprise and product carbon and water footprint • UK's National Health System Green House Gas Accounting for Pharmaceuticals and Medical Devices • ABPI/Carbon Trust tool blister pack carbon footprint • Greenhouse gas emissions for USA Maize production • Land-use in biofuels and crops • Water footprinting	[107–109] [110] [111] [112] [113–117] [118,119]

Source: Reproduced with permission of Ref. [120]. Copyright 2014, The Royal Society of Chemistry (RSC).

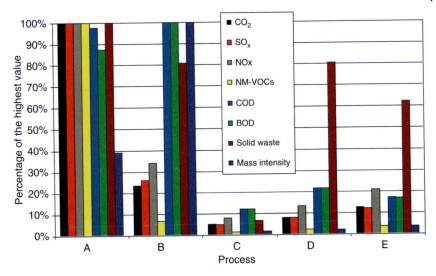

Figure 4.2 Comparison of selected life cycle inventory results for several processes (A, B, C, D, and E) for the production of an API. NM-VOCs = non-methane volatile organic compounds, BOD = biological oxygen demand, COD = chemical oxygen demand. (Reproduced with permission from Ref. [46].)

One example is the LCI/A that was sponsored by Pfizer to evaluate several processes at different stages of development for the production of sertraline, the active pharmaceutical ingredient (API) for the antidepressant Zoloft®, and its synthetic precursor, tetralone [46]. Figure 4.2 shows a few illustrative life cycle inventory results for several processes from the production of sertraline at the time of the assessment. This comparison demonstrates how LCI results may be used to contrast environmental profiles. This study provided insights that may be applied to synthetic chemistry route and process technology selection. About a decade later, a life cycle inventory/assessment evaluation was performed for three different manufacturing routes to Pfizer's API, pregablin [48]. The study found that a majority of the life cycle environmental impacts were attributable to raw material manufacture, ranging from about 48 to 61% for the three processes evaluated. Life cycle environmental impacts associated with waste disposal and recovery fared second, with a tighter range from 31 to 34%. Life cycle environmental impacts associated with energy use ranged from 7.6 to about 20% of the total.

GlaxoSmithKline (GSK) performed a cradle-to-gate LCI/A to evaluate environmental impacts and identify hotspots in the synthesis of another API [47]. The results of the LCI/A revealed that solvent usage contributes significantly to the environmental footprint of the synthetic chemical processes used to manufacture APIs, as shown in Figure 4.3. This assessment was also the basis for the development of a streamlined life cycle assessment tool known as FLASC® that simplified

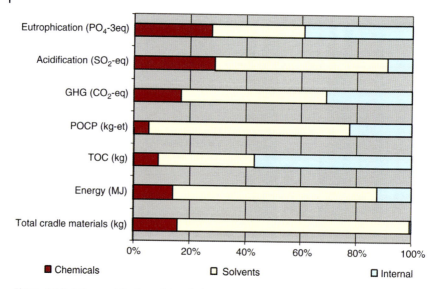

Figure 4.3 Relative contribution of chemicals, solvents, and internal processes on the environmental life cycle impacts of an API. TOC = total organic carbon, POCP = photochemical ozone creation potential, GHG = greenhouse gases (global warming potential). (Reproduced with permission from Ref. [47]. Copyright 2004, Springer.)

and accelerated the use of life cycle assessment within GSK [121]. The GSK methodology was also the basis for a streamlined tool for eco-footprinting that has now been applied across the pharmaceutical industry through the American Chemical Society's Green Chemistry Institute Pharmaceutical Roundtable [122].

Another example of a cradle-to-gate LCI/A is the study of the production of vancomycin hydrochloride [49]. This LCI/A was performed as part of larger study of the environmental impacts associated with the introduction of biocide-coated medical textiles for the prevention of nosocomial infections. The study found that the low-yield fermentation process accounts for 47% of the total life cycle environmental impacts associated with energy use and accounts for a majority of the impacts associated with raw materials production and use in the process. Over 75% of the total cradle-to-gate energy consumption is associated with steam used for assuring initial sterility of the reactor and its contents.

4.4.2
Chemical Route Comparison

This is a subsection of the A versus B application of LCI/A. Practitioners, scientists, and engineers are interested in using LCI/A comparisons for decision-making purposes, although sometimes it is not straightforward, given the uncertainties, time, and resources involved. Unfortunately, very few assessments in the literature report the uncertainty and sensitivity ranges, but one can cite examples of application for initial comparisons. There are some instances of the

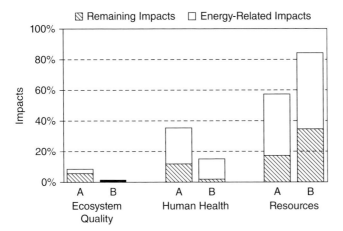

Figure 4.4 Energy-related and remaining impact composition of the EI99 scores relative to the total scores for A and B. (Reproduced with permission from Ref. [53]. Copyright 2010, Springer.)

use of streamlined tools to expedite the comparisons. One example is the assessment comparing the environmental profile of two APIs, one from Hoffmann-La Roche and one from GSK [53,54]. The assessment found that the environmental profiles of the two APIs were generally comparable, given the differences in their molecular complexity and associated routes of manufacture, and that optimizing material and energy use were the most effective routes of reducing the environmental footprint of these systems, as shown in Figure 4.4.

LCI/A has also been used to compare unit operations and processes operating in either batch or continuous flow conditions. Microreactors have also been assessed from a life cycle perspective, either comparing the technology to other alternatives where microreactors showed less energy-related impacts [68], or to use the comparison to provide direction on process development for several different process scenarios and chemical syntheses [69–72]. Figure 4.5 shows a comparison of the screening results for a few batch and continuous processes, and Figure 4.6 shows an illustration of more detailed global warming potential results for the comparison of batch and continuous processes under different conditions.

There have now been quite a few LCI/A comparisons between chemical and biological processes for the production of chemicals and products. While no sweeping conclusions can be made, there are some indications that biological processes may exhibit a smaller footprint, assuming that the processes are optimized, and that materials with large life cycle environmental footprints are not used in excess. For instance, a limited life cycle assessment was performed to compare biological and petrochemical processes for adipic acid production. The study found that the cumulative energy demand and global warming potential associated with the manufacture of adipic acid may be reduced by moving from the petrochemical process to a combined biological and chemical process. The highest calculated reduction potential cumulative energy demand was estimated between 29 and 57%, depending on the reagents used and the concentrations in the fermentation

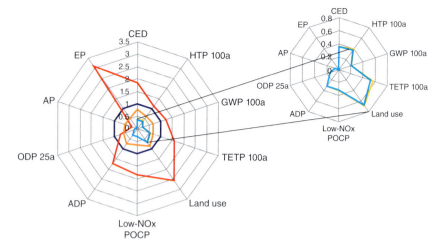

Figure 4.5 Environmental screening of selected results to obtain a more holistic view on the overall environmental impacts of the processes. CED, cumulative energy demand; HTP, human toxicity potential; GWP, global warming potential; TETP, terrestrial ecotoxicity potential; Low NOx; POCP, photo oxidant creation potential caused by NOx emissions at ground level; ADP, abiotic resource depletion potential; ODP, ozone depletion potential; AP, acidification potential; E, eutrophication potential. (Reproduced with permission from Ref. [71]. Copyright 2013, American Chemical Society.)

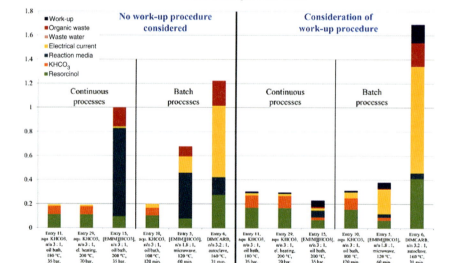

Figure 4.6 GWP of selected results of the Kolbe–Schmitt synthesis in batch and continuous processes, with and without consideration of workup. (Reproduced with permission from Ref. [71]. Copyright 2013, American Chemical Society.)

broth [55,56]. In another example of bulk or commodity chemical production that compared fermentation to traditional petrochemical routes, an assessment was made of caprolactam routes and compared differences in the global warming potential. This study found that the biocatalyzed route had a smaller GWP [57].

A cradle-to-gate life cycle comparison of a chemical and a biocatalytic process to produce 7-aminocephalosporic acid (7-ACA) was performed [52]. Impact estimations were performed using the GSK Fast Life Cycle Assessment of Synthetic Chemistry (FLASC^{TM}) tool [119] and employed a modular gate-to-gate methodology [23]. The comparison revealed that the chemical process has larger overall environmental impacts, with the chemical process having about 60% more life cycle impacts associated with energy use, about 16% more mass (excluding water) utilization, has doubled the greenhouse gas (GHG) impact and about 30% higher photochemical ozone creation potential (POCP) and acidification potentials. There have also been quite a few studies performed comparing bio-based chemicals and fossil-based plastics.

4.4.3
Material Assessment

Some of the first LCA assessments were completed for packaging materials, starting with the plastics eco-profiles first published in 1993 [90] and which are now accessible on the internet (Figure 4.7). These provide LCI information and

Figure 4.7 Screenshot of the Eco-Profiles Web site, which provides LCI and LCA information on plastics.

EPD for plastics [123]. The LCI/A of packaging materials has evolved significantly and there are now well-established Product Category Rules, material information, a set of position papers, and guidance developed for different types of packaging.

LCI/A has also been applied in material assessment and selection for chemical processes, exemplified by both GSK and AstraZeneca, who have incorporated life cycle considerations into their solvent assessment and selection guides [73,74,124]. Figure 4.8 shows examples of GSK's solvent selection guide, including (a) the summary guide for medicinal chemistry and (b) a comparison of ethers, where one can see the relative LCI/A score when making decisions. Capello *et al.* [76] have developed a comprehensive framework for the environmental assessment of solvents, which included a very thorough life cycle assessment evaluation that not only provides insights for material selection, but also provides data to select the best treatment or recovery options accounting for life cycle impacts, as shown in Figure 4.9. Additional applications in academia include process assessments performed by Slater and Savelski [75], which are based on the amounts and characteristics of the solvents used, including health, safety, and environmental life cycle parameters obtained from commercial LCA software.

Similarly to solvent selection, LCA has also been incorporated in guides that aid in the selection of more desirable reagents for a given transformation, as in

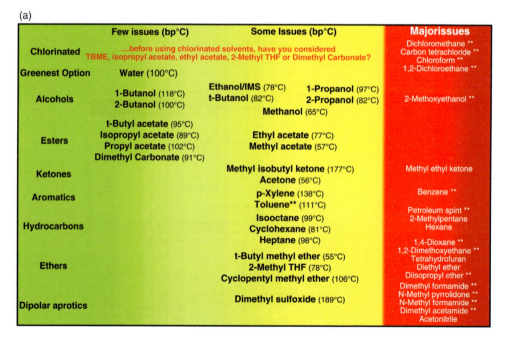

Figure 4.8 (a) A summary guide for medicinal chemistry and (b) a comparison of ethers, where one can see the relative LCA score when making decisions taken from GSK's Solvent Selection Guide. (Reproduced with permission from The Royal Society of Chemistry. http://pubs.rsc.org/en/Content/ArticleLanding/2011/GC/c0gc00918k#!divAbstract.)

4.4 Measuring Greenness with LCI/A – Applications

(b)

Classification	Solvent	Cas number	Melting point °C	Boiling point °C	Waste	Environ-mental Impact	Health	Flamm-ability & Explosio	Reactivity/ Stability	Life Cycle Score
Ether	t-Amyl methyl ether	994-05-8	-80	86	5	5	5	5	9	8
	t-Butylmethyl ether	1634-04-4	-109	55	4	5	5	3	9	8
	Cyclopentyl methyl ether	5614-37-9	-140	106	6	4	4	5	8	4
	t-Butyl ethyl ether	637-92-3	-74	70	5	5	4	4	9	8
	2-Methyltetrahydrofuran	96-47-9	-137	78	4	5	4	3	6	4
	Diethyl ether	60-29-7	-116	35	4	4	5	2	4	6
	Bis(2-methoxyethyl) ether	111-96-6	-68	162	4	5	2	8	4	6
	Dimethyl ether	115-10-6	-141	-25	3	5	7	1	4	7
	1,4- Dioxane	123-91-1	12	102	3	4	4	4	5	6
	Tetrahydrofuran	109-99-9	-108	65	3	5	6	3	4	4
	1,2-Dimethoxyethane	110-71-4	-58	85	4	5	2	4	4	7
	Diisopropyl ether	108-20-3	-86	68	4	3	8	1	1	9

Figure 4.8 (Continued)

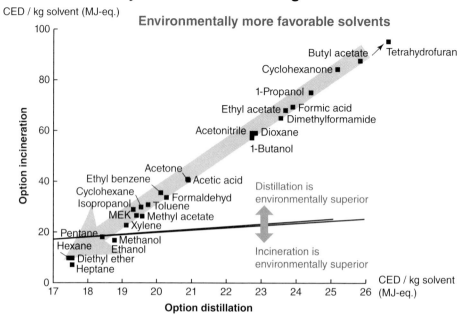

Figure 4.9 Life cycle assessment of the treatment options, incineration, and distillation developed by Capello et al. Reproduced with permission from Ref. [76]. Copyright 2007, Royal Society of Chemistry.)

GSK's reagent selection guides, which include LCA as one aspect that needs to be assessed to ensure the reagent with fewer impacts is considered and utilized in synthetic routes [77]. Another example is found in the life cycle assessments that Novozymes routinely undertakes for their enzymes [78], the results of which have been corroborated by independent studies evaluating the LCA of enzyme use in chemical synthesis processes [79]. Recently, there has been growing interest in evaluating the environmental life cycle impact of chemicals and materials, particularly as they are considered as potential replacements for existing commonly used chemicals and materials. Examples of this include LCAs of biofuels, ionic liquids, or nanomaterials.

4.4.4
Product LCAs

As part of the growing standardization of LCA methodologies, the ISO has set standards for providing the public with environmental information about products, including LCA, through an EPD. ISO 14025 defines an EPD as containing quantified environmental data in preset categories based on the ISO 14040 series of standards, but the EPD may include additional environmental information [21]. PCR are documents that define the rules and requirements for EPDs within a certain product category.

Of course, some LCAs of products have been performed to better understand the life cycle environmental impacts and advance research outside of the framework of EPD. For example, Novartis has performed limited LCAs for a few products [106] such as the cradle-to-grave LCA of Exelon, a drug for Alzheimer's disease. In this case, the LCA was performed using primary energy requirements as a baseline to better understand the relative environmental impacts each life cycle stage contributed to the overall life cycle impacts. The study found that the main contributor to life cycle environmental impacts associated with energy was distribution (38%), followed by packaging (34%), API production (12%), formulation (5%), and end-of-life impacts (1%). It is interesting to note that the large impact associated with distribution is a consequence of this drug being manufactured in Switzerland and distributed by plane, mainly to the US market. A similar cradle-to-grave LCA was performed for tegretol, a medicine to treat epilepsy. In this case, it was found that a majority of the impacts came from the API production (62%), followed by distribution (20%), packaging (14%), and formulation (4%).

GSK has also performed several LCA assessments of formulated products. One published example is the LCA for a formulated product for treating asthma that is administered through a device [11]. In this assessment, raw materials used to make the device were shown to have the largest contribution to the overall life cycle impacts. Figure 4.10 shows the relative contributions of material, production, and transportation.

Another example of using life cycle assessment in the evaluation of products is found in the eco-efficiency and SEEBalance® assessments developed by

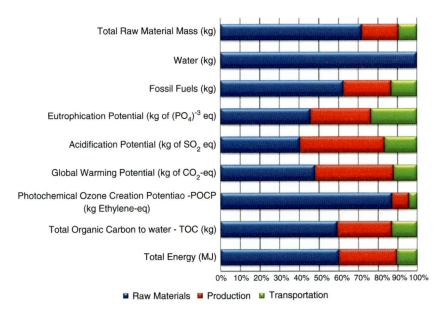

Figure 4.10 Cradle-to-gate LCA of a pharmaceutical product in a device showing relative contributions of raw materials, production, and transportation for several impacts. (Reproduced with permission from The Royal Society of Chemistry.)

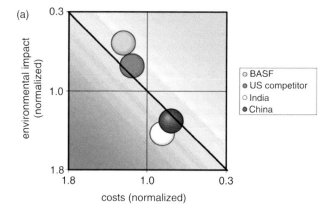

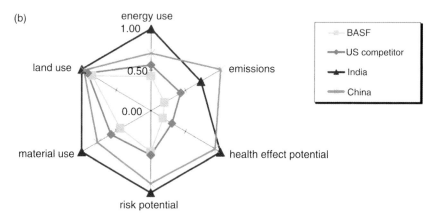

Figure 4.11 Example of an assessment using BASF's Eco-efficiency analysis.

BASF [104]. In the BASF methodology, the Eco-efficiency assessment attempts to measure both environmental and economic factors within a life cycle assessment framework. BASF has been using these assessments since 1996 on numerous products and processes as a means to achieve constant improvement in environmental and economic performance.

Some examples where the tool has been applied include: the selection of the best synthetic route to indigo based on an assessment of four different technologies for Indigo production; the assessment of two methods for cereal crop preservation to serve as a basis for discussing chemical additives in animal feeds; the analysis of two furniture board processes, which aided in making technology investment decisions; and others.[1] Figure 4.11 shows an example of an eco-efficiency analysis performed comparing the life cycle environmental impacts and costs for BASF and several competitors to produce ibuprofen. According to

---

1) http://www.basf.com/group/sustainability_en/index, (accessed November 5, 2008).

the BASF Web site, results of the eco-efficiency analysis are intended to inform business decisions about the following:

- Whether or not further improvements are possible or if the assessment can be used as a selling point when the analysis confirms a high degree of eco-efficiency.
- Whether and to what extent the economic or ecological footprint can be improved when the analysis shows that a product or process has inadequate eco-efficiency but it is possible to improve it.
- Whether to abandon or substitute a product or process, when the analysis shows that it has poor eco-efficiency and cannot be improved with a reasonable investment.
- Whether to direct further research and development to the most eco-efficient alternative when a product or process is in the early stages of research and there are several credible paths for further development to follow.

Another example of a company using life cycle thinking to evaluate products is the environmental reports available for Apple® products [107]. These reports may be downloaded from the company Web site, and include information on climate change, energy efficiency, material efficiency, restricted substances, and recycling. Table 4.2 presents the global warming potential data provided for the iPhone 6 Plus.

4.4.5
Footprinting

In recent years, there has been a trend toward companies performing limited assessments that evaluate only a single life cycle environmental impact, for example, global warming potential. This has been undertaken by individual companies, but has also gained traction with governments and NGOs, as in the case of the United Kingdom's National Health System's "Green House Gas Accounting for Pharmaceuticals and Medical Devices" or the ABPI/Carbon Trust's tool "blister pack carbon footprint."

This trend is mainly being driven in part by the complexity, resources, and time involved in performing a full LCA following ISO standards, partly by the drive toward using life cycle thinking in business decision-making by businesses that may not have the appropriate in-house LCA capability, and partly to ease communication with the layperson about the trade-offs or to reduce the number

Table 4.2 Some of the LCA data provided in the environmental report of the iPhone 6 Plus.

Global warming potential	Life cycle phase contributions
110 kg CO_2-eq	Production – 81%
	Use – 14%
	Transport – 4%
	Recycling – 1%

Table 4.3 Carbon and water footprint data as disclosed in GSK's 2014 responsibility report.

Life cycle stage	Global warming potential [million ton CO_2-eq/yr] (% of total)	Water use [million m^3] (% of total)
Raw materials	5.6 (40%)	1200 (Nd)
Operations	1.6 (11%)	15 (Nd)
Logistics	0.2 (2%)	Nd
Product use	6.4 (46%)	Nd (13%)
Disposal	0.2 (2%)	Nd
Total	15 (100%)	Nd

Nd = not disclosed.

of metrics that need to be explained. With the increased focus on climate change impacts, it is not surprising that most of the initial single-metric work has been on carbon footprinting. As societal focus shifts to concerns about other large global environmental impacts, one can see examples of water footprint calculators (water is the new carbon) and research on land use changes, particularly as focused on biomaterials increases.

One example of a company using a single metric is GSK, which has used carbon footprint analysis across its entire global supply chain since 2010 to determine the main areas of concern directly under the control of the company [107]. Most recently, the company has undertaken a water footprint exercise to better understand water usage, although a full picture of the water footprint has not been disclosed. Table 4.3 shows the latest carbon and water footprint data as disclosed in GSK's 2014 responsibility report [109]. GSK has also performed over 40 product carbon footprints to obtain insights into product and process enhancements. For instance, the carbon footprint of Ventolin showed that GSK needed to focus on reducing the release of the propellant HF134a into the atmosphere as it contributed to more than 95% of the product's carbon footprint (28 kg CO_2-equivalent per device). On the other hand, the assessment of the Advair Diskus dry powder inhaler showed the need to focus on the plastics used to make the device, and on the energy used in GSK's operations.

Given the uptake of single-metric studies, advocates of a holistic sustainability point of view favor a multiobjective assessment that provides a better and more systematic understanding of environmental impacts and helps one to identify significant trade-offs that may help individuals, organizations, and society avoid unintended consequences if and when changes are made to a system.

Limited assessment approaches relying on single-metrics trade resolution and a holistic point of view for the ability to incorporate LCA insights into industrial decision-making, which is a calculated and what some argue, a practical trade-off. Although the debate will surely continue, it is not an either or proposition. The key to continued use of these limited approaches is the recognition that working at a lower resolution and with higher uncertainty may not be applicable to all scenarios, and it is necessary to maintain an appropriate level of transparency regarding assumptions and limitations of a single-metric approach. Current advances in more sophisticated data analytics and visualization approaches may help bridge this balance between transparency and simplicity.

4.5
Final Remarks

Industry use of life cycle assessment has continued to increase in the last few years as an evolutionary process to evaluate "greenness" of processes, perhaps primarily driven by the increased global interest in climate change. The demand for high-quality life cycle information is increasing as governments, industrial groups, consumer organizations, and NGO groups ask carbon footprint questions, insist on PCR, and EPD, and who generally want LCI/A information on commercial products.

Even with this increased demand and uptake, the routine use of life cycle assessment is not fully embedded in most companies as a formal business process, and its use continues to evolve, with different companies being at diverse stages of implementation and maturity.

As demand grows, so does the need for quality LCI/A data and standardized techniques. At this point, it is fair to say that the demand is exceeding supply on both fronts, and that the quantity and quality of data will need to be continuously enhanced to meet that demand.

This chapter presented examples of applying LCA to various fields and depths. Although the usefulness of LCA metrics can be seen clearly, one may also quickly see the diversity of approaches and applications. Part of this diversity is explained by the fact that LCI/A approaches and assessment methods inevitably vary with the type of question the LCA is trying to answer, with the goal and scope definition being perhaps the most important step of an LCA that leads to the greatest diversity. However, this diversity also highlights the importance of how one presents and uses LCI and LCIA outputs to provide useful insights and drive the desired behavior.

Jimenez-Gonzalez and Overcash recently published important observations on the practice of LCI/A over the last two decades, which are summarized as follows [120]:

- Corporations with advanced life cycle capabilities have a competitive advantage when they integrate broad environmental benefits and consequences of products into decision-making for R&D and for process improvement.
- The conversion of LCI data to LCI/A data increases the variability thus making comparisons of alternatives more challenging.
- LCI/A data are not available for many chemicals in LCI databases (~1200 chemicals), with missing data rarely identified in the LCI/A results. Efforts such as the Environmental Genome Initiative (http://environmentalgenome.org/) aim to fill a substantial part of that gap.
- Lack of transparency reduces the credibility of life cycle studies (e.g., publication of only relative percent, LCI data not available).
- Uncertainty and sensitivity analysis are still not routinely used in LCA, thus potentially masking or overestimating the importance of certain impacts.

The observations highlight a field in evolution, with many applications and also room to continue evolving and improving. Ongoing developments in data analytics and visualization provide an opportunity to address LCIA's limitations and drive its simplification and adoption. In spite of its current limitations and needed improvements, LCA is certainly a powerful tool for assessment of sustainability and greenness that can be used to aid the decision-making toward greener products and processes.

References

1 United States Environmental Protection Agency (2008) Life Cycle Assessment: Principles and Practice. EPA/600/R-06/060, National Risk Management Research Laboratory, Cincinnati, OH, USA (2006). Available at http://www.epa.gov/ORD/NRMRL/lcaccess/pdfs/600r06060.pdf (accessed November 1, 2008).

2 Wenzel, H., Hauschild, M., and Alting, L. (1997) *Environmental Assessment of Products. Volume 1: Methodology, Tools and Case Studies in Product Development*, Chapman and Hall, London.

3 International Organization of Standardization (2006) ISO 14040– Environmental Management – Life Cycle Assessment – Principles and framework, ISO, Geneva, Switzerland.

4 International Organization of Standardization (2006) ISO 14044– Environmental Management–Life Cycle Assessment–Requirements and Guidelines, ISO, Geneva, Switzerland.

5 Pré Consultants (2000) The Eco-indicator 99 – a damage oriented method for life cycle impact assessment – Methodology Report and Manual for Designers Pré Consultants.

6 Grossmann, E.I. (2004) Challenges in the new millennium: product discovery and design, enterprise and supply chain optimization, global life cycle assessment. *Computers and Chemical Engineering*, **29**, 29–39.

7 Manley, J.B., Anastas, P.T., and Cue, B.W. (2008) Frontiers in green chemistry: meeting the grand challenges for sustainability in R&D and manufacturing. *Journal of Cleaner Production*, **16**, 743–750.

8 Lapkin, A. and Constable, D.J.C. (2008) *Green Chemistry Metrics: Measuring and Monitoring Sustainable Processes*, Wiley-Blackwell, New Jersey, 344 p.

9 Henderson, R.K., Constable, D.J.C., and Jiménez-González, C. (2010) Green Chemistry Metrics, in *Green Chemistry in the Pharmaceutical Industry* (eds P. Dunn, A. Wells, and M.T. Williams), Wiley-VCH Verlag GmbH, Weinheim.

10 Jiménez-González, C., Ponder, C.S., Broxterman, Q.B., and Manley, J.B. (2011) Using the right green yardstick: why process mass intensity is used in the pharmaceutical industry to drive more sustainable processes. *Organic Process Research and Development*, **15** (4), 912–917.

11 Jiménez-González, C., Constable, D.J.C., and Ponder, C. (2012) Evaluating the "Greenness" of chemical processes and products in the pharmaceutical industry – a green metrics primer. *Chemical Society Reviews*, **41**, 1485–1498.

12 Nygren, J. and Antikainen, R. (2010) Use of life cycle assessment (LCA) in global companies. Reports of the Finnish Environment Institute 16. Helsinki.

13 Watson, W.J.W. (2012) How do the fine chemical, pharmaceutical, and related industries approach green chemistry and sustainability? *Green Chemistry*, **14**, 251–259.

14 Meadows, D.H. et al. (1972) *The Limits to Growth: A Report for the Club of Rome's Project on the Predicament of Mankind*, Universe Books, New York, p. 205.

15 Goldsmith, E. and Allen, R. (1972) A blueprint for survival. The Economist, **2** (1)

16 SETAC (1993) *Guidelines for Life Cycle Assessment: A 'Code of Practice* (eds F. Consoli, D. Allen, I. Boustead, J. Fava, W. Franklin, A.A. Jensen, N. Oude, R. Parrish, R. Perriman, D. Postlethwaite, B. Quay, J. Seguin, and B. Vigon), Society of Environmental Toxicology and Chemistry.

17 SETAC (1997) *Life Cycle Impact Assessment: The State-of-the-Art* (eds L. Barnthouse, J. Fava, K. Humphreys, R. Hunt, L. Laibson, S. Noessen, J.W. Owens, J.A. Todd, B. Vigon, K. Wietz, and J. Young), Society of Environmental Toxicology and Chemistry.

18 SETAC (2004) *Life-Cycle Management* (eds D. Hunkeler, G. Rebitzer, M. Finkbeiner, W.-P. Schmidt, A.A. Jensen, H. Stranddorf, and K. Christiansen), Society of Environmental Toxicology and Chemistry.

19 International Organization of Standardization (1997) ISO 14040 – Environmental Management – Life Cycle Assessment – Principles and Framework, Geneva, Switzerland.

20 International Organization of Standardization (2006) ISO 14044 – Environmental Management – Life Cycle Assessment – Requirements and Guidelines, Geneva, Switzerland.

21 International Organization of Standardization (2006) ISO 14025– Environmental Labels and Declarations – Type III Environmental Declarations – Principles and Procedures, ISO, Geneva, Switzerland.

22 Baumann, H. and Tillman, A.M. (2004) *The Hitch Hiker's Guide to LCA*, Professional Publishing House.

23 Jiménez-González, C., Kim, S., and Overcash, M. (2000) Methodology of developing gate-to-gate life cycle analysis information. *International Journal of Life Cycle Assessment*, **5** (3), 153–159.

24 Jiménez-González, C. and Overcash, M. (2000) Energy sub-modules applied in life cycle inventory of processes. *Journal of Clean Products and Processes*, **2**, 57–66.

25 Jiménez-González, C., Overcash, M., and Curzons, A. (2001) Treatment modules – a partial life cycle inventory. *Journal of Chemical Technology and Biotechnology*, **76**, 707–716.

26 Capello, C., Hellweg, S., Badertscher, B., Betschart, H., and Hungerbühler, K. (2007) Environmental assessment of waste-solvent treatment options Part 1: The ecosolvent tool. *Journal of Industrial Ecology*, **11**, S. 26–38.

27 Doka, G. (2003) Life Cycle Inventories of Waste Treatment Services, Ecoinvent report No.13. Swiss Centre for Life Cycle Inventories, December.

28 Kim, S. and Dale, B. (2005) Life cycle inventory information of the united states electricity system. *The International Journal of Life Cycle Assessment*, **10** (4), 294–304.

29 Ecoinvent: Swiss Centre for Life Cycle Inventories (2008) www.ecoinvent.org (accessed May 1, 2017).

30 SimaPro, PRé Consultants (2017) www.pre-sustainability.com/simapro (accessed May 1, 2017).

31 DEAM™ Database, EcoBilan (2017) http://www.ghgprotocol.org/Third-Party-Databases/DEAM (accessed May 1, 2017).

32 Institute for Environmental Informatics (2017) UMBERTO software ifu, Hamburg GmbH. Available at https://www.ifu.com/en/umberto/ (accessed May 1, 2017).

33 GaBi Software PE International. www.gabi-software.com/ (accessed May 1, 2017).

34 Kralisch, D., Ott, D., and Gericke, D. (2015) Rules and benefits of life cycle assessment in green chemical process and synthesis design: a tutorial review. *Green Chemistry*, **17**, 123.

35 Tufvesson, L.M., Tufvesson, P., Woodley, J.M., and Börjesson, P. (2013) Life cycle assessment in green chemistry: overview of key parameters and methodological concerns. *International Journal of Life Cycle Assessment*, **18**, 431–444.

36 Suh, S. and Huppes, G. (2005) Methods for life cycle inventory of a product. *Journal of Cleaner Production*, **13**, 687–697.

37 Jiménez-González, C. and Overcash, M. (2000) Life cycle inventory of refinery products: review and comparison of commercially available databases. *Environmental Science and Technology*, **34** (22), 4789–4796.

38 Weidema, B.P. and Wesnæs, M.S. (1996) Data quality management for life cycle inventories – an example of using data

39 Hunt, R.G., Boguski, T.K., and Weitz, K., and Sharma, A. (1998) Examining LCA streamling techniques. *The International Journal of Life Cycle Assessment*, **3** (1), 36–42.

40 Leahy, D.K., Tucker, J.L., Mergelsberg, I., Dunn, P.J., and Kopach, M.E., and Purohit, V.C. (2013) Seven important elements for an effective green chemistry program: an IQ consortium perspective. *Organic Process Research and Development*, **17**, 1099–1109.

41 Lima-Ramos, J., and Tufvesson, P., and Woodley, J.M. (2014) Application of environmental and economic metrics to guide the development of biocatalytic processes. *Green Processing and Synthesis*, **3**, 195–213.

42 Sneddon, H. (2014) Embedding sustainable practices into pharmaceutical R&D: what are the challenges? *Future Medicinal Chemistry*, **6** (12), 1373–1376.

43 Andraos, J. and Dicks, A.P. (2012) Green chemistry teaching in higher education: a review of effective practices. *Chemistry Education Research and Practice*, **13**, 69–79.

44 Finnveden, G., Hauschild, M.Z., Ekvall, T., Guinée, J., Heijungs, R., Hellweg, S., Koehler, A., and Pennington, D., and Suh, S. (2009) Recent developments in life cycle assessment. *Journal of Environmental Management*, **91**, 1–21.

45 Ridoutt, B. *et al.* (2015) Making sense of the minefield of footprint indicators. *Environmental Science and Technology*, **49**, 2601–2603.

46 Jiménez-González, C. (2000) Life cycle assessment in pharmaceutical applications, Ph.D. thesis, North Carolina State University, Raleigh, NC.

47 Jiménez-González, C., Curzons, A.D., and Constable, D.J.C., and Cunningham, V.L. (2004) Cradle-to-gate life cycle inventory and assessment of pharmaceutical compounds. *The International Journal of Life Cycle Assessment*, **9** (2), 114–121.

48 Savelski, M.J., Slater, C.S., Dunn, P.J., Knoechel, D.J., and Visnic, C.M. (2011) Use of life cycle assessment to evaluate the sustainable manufacture of the active pharmaceutical ingredient pregabalin. Presentation at the AIChE 2011 Annual Meeting, October 17, 2011, Minneapolis.

49 Ponder, C. and Overcash, M. (2010) Cradle-to-gate life cycle inventory of vancomycin hydrochloride. *Science of the Total Environment*, **408**, 1331–1337.

50 Cherubini, F. and Jungmeier, G. (2010) LCA of a biorefinery concept producing bioethanol, and chemicals from switchgrass. *The International Journal of Life Cycle Assessment*, **15**, 53–66.

51 Cherubini, F. and Ulgiati, S. (2010) Crop residues as raw materials for biorefinery systems – a LCA case study. *Applied Energy*, **87**, 47–57.

52 Henderson, R.K., Jiménez-González, C., Preston, C., Constable, D.J.C., and Woodley, J. (2008) EHS & LCA assessment for 7-ACA synthesis: a case study for comparing biocatalytic & chemical synthesis. *Industrial Biotechnology*, **4** (2), 180–192.

53 Wernet, G., Conradt, S., Isenring, H.P., Jiménez-González, C., and Hungerbühler, K. (2010) Life cycle assessment of fine chemical production: a case study of pharmaceutical synthesis. *The International Journal of Life Cycle Assessment*, **15** (3), 294–303.

54 Conradt, S. (2008) Life cycle assessment of a pharmaceutical compound, M.Sc. thesis, ETH-Zürich, Zürich.

55 Wang, Q., Vural-Gürsel, I., Shang, M., and Hessel, V. (2013) Life cycle assessment for the direct synthesis of adipic acid in microreactors and benchmarking to the commercial process. *Chemical Engineering Journal*, **234**, 300–311.

56 van Duuren, J.B.J.H., Brehmer, B., Mars, A.E., Eggink, G., dos Santos, V.A.P.M., and Sanders, J.P.M. (2011) A limited LCA of bio-adipic acid: manufacturing the nylon-6,6 precursor adipic acid using the benzoic acid degradation pathway from different feedstocks. *Biotechnology and Bioengineering*, **108**, 1298–1306.

57 Roes, A.L. and Patel, M.K. (2011) Ex-ante environmental assessments of novel technologies – improved caprolactam catalysis and hydrogen storage. *The*

Journal of Cleaner Production, **19**, 1659–1667.
58 Akiyama, M., Tsuge, T., and Doi, Y. (2003) Environmental life cycle comparison of polyhydroxyalkanoates produced from renewable carbon resources by bacterial fermentation. *Polymer Degradation and Stability*, **80**, 183–194.
59 Harding, K.G., Dennis, J.S., von Blottnitz, H., and Harrison, S.T.L. (2007) Environmental analysis of plastic production processes: comparing petroleum-based propylene and polyethylene with biologically-based poly-β-hydroxybuturic acid using life cycle analysis. *Journal of Biotechnology*, **130**, 57–66.
60 Kim, S. and Dale, B.E. (2005) Life cycle assessment study of biopolymers (polyhydroxyalkanoates) derived from no-tilled corn. *The International Journal of Life Cycle Assessment*, **10** (3), 200–210.
61 Madival, S., Auras, R., Singh, S.P., and Narayan, R. (2009) Assessment of environmental profile of PLA, PET and PS clamshell containers using LCA methodology. *The Journal of Cleaner Production*, **17**, 1183–1194.
62 Weiss, M., Patel, M., Heilmeier, H., and Bringezu, S. (2007) Applying distance to-target weighing methodology to evaluate the environmental performance of bio-based energy, fuels, and materials. *Resources, Conservation and Recycling*, **50**, 260–281.
63 Vink, E.T.H., Ràbaga, K.R., Glassner, D.A., and Gruber, P.R. (2003) Applications of life cycle assessment to NatureWorks™ polylactide (PLA) production. *Polymer Degradation and Stability*, **80**, 403–419.
64 BREW (2006) Medium and long-term opportunities and risks of the biotechnological production of bulk chemicals from renewable resources – the potential of white biotechnology. The BREW Project. Utrecht, The Netherlands.
65 Urban, R.A. and Baksi, B.R. (2009) 1,3-Propanediol from fossil versus biomass: a life cycle evaluation of emissions and ecological resources. *Industrial and Engineering Chemistry Research*, **48**, 8068–8082.
66 Adlercreutz, D., Tufvesson, P., Karlsson, A., and Hatti-Kaul, R. (2010) Alkanolamide biosurfactants: techno-economic evaluation of biocatalytic versus chemical production. *Industrial Biotechnology*, **6** (4), 204–211.
67 Ekman, A. and Börjesson, P. (2010) Life cycle assessment of mineral oilbased and vegetable oil-based hydraulic fluids including comparison of biocatalytic and conventional methods. *The International Journal of Life Cycle Assessment*, **16**, 297–305.
68 Jiménez-González, C., Curzons, A., Constable, D.J.C., Overcash, M., and Cunningham, V. (2001) How do you select the 'greenest' technology? Development of Guidance for the Pharmaceutical Industry. *Clean Products and Processes*, **3**, 35–41.
69 Huebschmann, S., Kralisch, D., Hessel, V., Krtschil, U., and Kompter, C. (2009) Environmentally benign microreaction process design by accompanying (simplified) life cycle assessment. *Chemical Engineering and Technology*, **32**, 1757–1765.
70 Huebschmann, S., Kralisch, D., Loewe, H., Breuch, D., Petersen, J.H., Dietrich, T., and Scholz, R. (2011) Decision support towards agile eco-design of micro-reaction processes by accompanying (simplified) life cycle assessment. *Greem Chemistry*, **13**, 1694–1707.
71 Kressirer, S., Kralisch, D., Stark, A., Krtschil, U., and Hessel, V. (2013) Agile green process design for the intensified Kolbe–Schmitt synthesis by accompanying (simplified) life cycle assessment. *Environmental Science and Technology*, **47**, 5362–5371.
72 Kralisch, D., Streckmann, I., Ott, D., Krtschil, U., Santacesaria, E., Di Serio, M., Russo, V., De Carlo, L., Linhart, W., Christian, E., Cortese, B., de Croon, M.H.J.M., and Hessel, V. (2012) Transfer of the epoxidation of soybean oil from batch to flow chemistry guided by cost and environmental issues. *ChemSusChem*, **5**, 300–311.

73 Jiménez-González, C., Curzons, A.D., Constable, D.J.C., and Cunningham, V.L. (2005) Expanding GSK's solvent selection guide–application of life cycle assessment to enhance solvent selections. *Clean Technologies and Environmental Policy*, **7**, 42–50.

74 AstraZeneca (2014) http://www.astrazeneca.com/Responsibility/The-environment/Product-environmental-improvement (accessed April 21, 2014).

75 Slater, C.S. and Savelski, M.J. (2009) Towards a Greener Manufacturing Environment. *Innovations in Pharmaceutical Technology*, **29**, 78–83.

76 Capello, C., Fischer, U., and Hungerbuhler, K. (2007) What is a green solvent? A comprehensive framework for the environmental assessment of solvents. *Greem Chemistry*, **9**, 927–934.

77 Adams, J.P., Alder, C.M., Andrews, I., Bullion, A.M., Campbell-Crawford, M., Darcy, M.G., Hayler, J.D., Henderson, R.K., Oare, C.A., Pendrak, I., Redman, A.M., Shuster, L.E., Sneddon, H.F., and Walkera, M.D. (2013) Development of GSK's reagent guides – embedding sustainability into reagent selection. *Greem Chemistry*, **15**, 1542.

78 Nielsen, P.H., Oxenbøll, K.M., and Wenzel, H. (2007) Cradle-to-gate environmental assessment of enzyme products produced industrially in Denmark by novozymes A/S. *The International Journal of Life Cycle Assessment*, **12** (6), 432–438.

79 Kim, S., Jiménez-González, C., and Dale, B.E. (2009) Enzymes for pharmaceutical applications – a cradle-to-gate life cycle assessment. *The International Journal of Life Cycle Assessment*, **14** (5), 392–400.

80 Concawe, EUCAR, EC Joint Research Centre (2007) Well-to-wheels analysis of future automotive fuels and powertrains in the European context. Available at http://iet.jrc.ec.europa.eu/about-jec/sites/iet.jrc.ec.europa.eu.about-jec/files/documents/wtw3_wtw_report_eurformat.pdf (accessed May 9, 2015).

81 Shen, L. and Patel, M.K. (2008) Life cycle assessment of polysaccharide materials: a review. *Journal of Polymers and the Environment*, **16**, 154–167.

82 Kralisch, D., Reinhardt, D., and Kreisel, G. (2007) Implementing objectives of sustainability into ionic liquids research and development. *Greem Chemistry*, **9**, 1308–1318.

83 Anastas, P.T. (2010) *Handbook of Green Chemistry - Green Solvents* (eds P. Wasserscheid and A. Stark), Wiley-VCH Verlag GmbH, Weinheim.

84 Reinhardt, D., Ilgen, F., Kralisch, D., Konig, B., and Kreisel, G. (2008) Evaluating the greenness of alternative reaction media. *Greem Chemistry*, **10**, 1170–1181.

85 Griffiths, O.G., O'Byrne, J.P., Torrente-Murciano, L., Jones, M.D., Mattia, D., and McManus, M.C. (2013) Identifying the largest environmental life cycle impacts during carbon nanotube synthesis via chemical vapour deposition. *The Journal of Cleaner Production*, **42**, 180–189.

86 Anctil, A., Babbitt, C.W., Raffaelle, R.P., and Landi, B.J. (2011) Material and energy intensity of fullerene production. *Environmental Science and Technology*, **45**, 2353–2359.

87 Walser, T., Demou, E., Lang, D.J., and Hellweg, S. (2011) Prospective environmental life cycle assessment of nanosilver T-shirts. *Environmental Science and Technology*, **45**, 4570–4578.

88 Miseljic, M. and Olsen, S. (2014) Life-cycle assessment of engineered nanomaterials: a literature review of assessment status. *Journal of Nanoparticle Research*, **16**, 1–33.

89 World Packaging Organization (2015) Position Paper LCA – Life Cycle Assessment. Available at www.worldpackaging.org (accessed May 10, 2015).

90 PlasticsEurope (2011) Eco-profiles and Environmental Declarations of the European Plastics Manufacturers. Version 2.0. Available at http://www.plasticseurope.org/documents/document/20110421141821-plasticseurope_eco-profile_methodology_version2-0_2011-04.pdf (accessed May 23, 2015).

91 Corrugated Packaging Alliance (2010) Corrugated Packaging Life-cycle assessment Summary Report. Prepared

by Pre Americas and Five Winds International. Available at http://www.corrugated.org/upload/LCA%20Summary%20Report%20FINAL%203-24-10.pdf (accessed May 10, 2015).

92 United Nations Environmental Program (2015) An Analysis of life cycle assessment in packaging for food and beverage applications. Available at http://www.lifecycleinitiative.org/wp-content/uploads/2013/11/food_packaging_11.11.13_web.pdf. (accessed May 10, 2015).

93 Jiménez-González, C., Overcash, M., and Curzons, A. (2001) Waste treatment modules – a partial life cycle inventory. *Journal of Chemical Technology and Biotechnology*, **76**, 707–716.

94 Raymond, M.J., Slater, C.S., and Savelski, M.J. (2010) LCA approach to the analysis of solvent waste issues in the pharmaceutical industry. *Greem Chemistry*, **12**, 1826–1834.

95 Sinclair, A., Leveen, L., Monge, M., Lim, J., and Cox, S. (2008) The environmental impact of disposable technologies. *BioPharm International*, **21** (11), S4–S15.

96 Mauter, M. (2009) Environmental life cycle assessment of disposable bioreactors. *BioProcess International*, **7** (4), S18–S29.

97 Whitford, W.G. (2010) Single-use systems as principal components in bioproduction. *BioProcess International*, December 2010, 34–44.

98 Thomas, P. (2011) Green gets granular: single-use vs. traditional biopharm process trains. Pharmaceutical Manufacturing. Available at http://www.pharmamanufacturing.com/articles/2011/060/.

99 Scott, C. (2011) Sustainability in bioprocessing. *BioProcess International*, **9** (10), 25–36.

100 Van der Vorst, G., Van Langenhove, H., De Paep, F., Aelterman, W., Dingenenb, J., and Dewulf, J. (2009) Exergetic life cycle analysis for the selection of chromatographic separation processes in the pharmaceutical industry: preparative HPLC versus preparative SFC. *Greem Chemistry*, **11**, 1007–1012.

101 De Soete, W., Dewulf, J., Cappuyns, P., Van der Vorst, G., Heirman, B., Aelterman, W., Schoeters, K., and Van Langenhove, H. (2013) Exergetic sustainability assessment of batch versus continuous wet granulation based pharmaceutical tablet manufacturing: a cohesive analysis at three different levels. *Greem Chemistry*, **15**, 3039.

102 Lima-Ramos, J., Tufvesson, P., and Woodley, J.M. (2014) Application of environmental and economic metrics to guide the development of biocatalytic processes. *Green Processing and Synthesis*, **3**, 195–213.

103 Saling, P., Kicherer, A., Dittrich-Krämer, B., Wittlinger, R., Zombik, W., Schmidt, I., Schrott, W., and Schmidt, S. (2002) Eco-efficiency analysis by BASF: the method. *IJLCA*, **7** (4), 203–218.

104 Saling, P. and Kicherer, A. (2006) Assessment of biotechnology-based chemicals, in *Renewables-Based Technology – Sustainability Assessment*, John Wiley & Sons, West Sussex, pp 299–314.

105 De Jonge, A.M. (2003) Limited LCAs of pharmaceutical products: merits and limitations of an environmental management tool. Social responsible. *Environmental Management*, **10**, 78–90.

106 Apple (2014) Apple environmental responsibility report. Available at https://www.apple.com/environment/reports/last (accessed May 10, 2015).

107 GlaxoSmithKline Corporate responsibility report 2010. Available at www.gsk.com/responsibility (accessed May 10, 2015).

108 GlaxoSmithKline Corporate responsibility report 2013. Available at www.gsk.com/responsibility (accessed May 10, 2015).

109 GlaxoSmithKline Corporate responsibility report 2014. Available at www.gsk.com/responsibility (accessed May 10, 2015).

110 Environmental Resources Management (ERM) and UK National Health System (NHS) Sustainable Development Unit (2012) Greenhouse Gas Accounting Sector Guidance for Pharmaceutical Products and Medical Devices. Summary

Document, November 2012, pp 26. Available at http://www.ghgprotocol.org/files/ghgp/Summary-Document_Pharmaceutical-Productand-Medical-Device-GHG-Accounting_November-2012.pdf (accessed April 29, 2014).

111 The Association of the British Pharmaceutical Industry (ABPI) and Carbon Trust (2013) Blister pack carbon footprint tool. Available at http://www.abpi.org.uk/our-work/mandi/Pages/sustainability.aspx (accessed April 21, 2014).

112 Kim, S., Dale, B., and Keck, P. (2014) Energy requirements and greenhouse gas emissions of maize production in the USA. *BioEnergy Research*, **7** (2), 753–764.

113 Kim, S., Dale, B.E., Heijungs, R., Azapagic, A., Darlington, T., and Kahlbaum, D. (2014) Indirect land use change and biofuels: mathematical analysis reveals a fundamental flaw in the regulatory approach. *Biomass and Bioenergy*, **71**, 408–412.

114 Börjesson, P. and Tufvesson, L.M. (2010) Agricultural crop-based biofuels – resource efficiency and environmental performance including directland use changes. *The Journal of Cleaner Production*, **19**, 108–120.

115 Fargione, J., Hill, J., Tilman, D., Polasky, S., and Hawthorne, P. (2008) Land clearing and the biofuel carbon debt. *Science*, **319**, 1235–1238.

116 Searchinger, T., Heimlich, R., Houghton, R.A., Dong, F., Elobeid, A., Fabriosa, J., Tokgoz, S., Hayes, D., and Yu, T.H. (2008) Use of U.S. croplands for biofuels increases greenhouse gases through emissions from land use change. *Science*, **319**, 1238–1240.

117 Müller-Wenk, R. and Brandâo, M. (2010) Climatic impact of land use in LCA– carbon transfers between vegetation/soil and air. *The International Journal of Life Cycle Assessment*, **15**, 172–182.

118 Water footprint (2015) Water footprint network. Available at http://waterfootprint.org/en/last (accessed May 23, 2015).

119 National Geographic (2015) Water footprint calculator. Available at http://environment.nationalgeographic.com/environment/freshwater/change-the-course/water-footprint-calculator/ (accessed May 23, 2015).

120 Jiménez-González, C. and Overcash, M.R. (2014) The evolution of life cycle assessment in pharmaceutical and chemical applications – a perspective. *Green Chemistry*, **16**, 3392–3400.

121 Curzons, A., Jiménez-González, C., Duncan, A., Constable, D.J.C., and Cunningham, V. (2007) Fast life-cycle assessment of synthetic chemistry tool, FLASC™ tool. *The International Journal LCA*, **12** (4), 272–280.

122 Jiménez-González, C., Ollech, C., Pyrz, W., Hughes, D., Broxterman, Q.B., and Bhathela, N. (2013) Expanding the boundaries: developing a streamlined tool for eco-footprinting of pharmaceuticals. *OPRD*, **17**, 239–246.

123 Plastics Europe (2015) Eco-profiles. Available at http://www.plasticseurope.org/plasticssustainability/eco-profiles.aspx (accessed May 23, 2015).

124 Shuster, L. (2015) GSK's solvent sustainability guide: the next generation. Presentation at the 19th Annual Green Chemistry and Engineering Conference. Bethesda, MD.

5
Sustainable Design of Batch Processes
Tânia Pinto-Varela and Ana Isabel Carvalho

5.1
Introduction

The new global markets' environment is characterized by the production of high-value and low-volume products, subject to highly uncertain demands. Consequently, agility in reacting to variability in customer orders, demand forecasts, material availability, and production throughput is a prerequisite in almost all production facilities. This fact has triggered an increased interest in batch processing because of the ability to rapidly adjust production volumes. In general, batch processes involve multipurpose facilities where a variety of products are produced by sharing all the available resources. Different sequences of production may lead to the same final product and the same operation may be performed in different unit operations at different timelines [1]. Such plants can be easily reconfigured or adapted to allow modifications in production and/or to cover a wide range of operating conditions within the same plant configuration.

The design of multipurpose batch processes involves determination of the number, type, and capacities of the resources required in the process, as well as definition of its operability (periodic or nonperiodic), so as to produce a specific set of products. This task should be done while guaranteeing a set of predefined conditions and optimizing a given objective. Due to the inherent agility of the multipurpose resources utilization, where the same resource can be used to perform different tasks, operational scheduling considerations need to be taken into account at the design stage. The processes with those characteristics enable flexibility without any significant equipment changes, which is essential in the current global markets. Summarizing, batch processes are characterized by the following:

- Each task is operated independently.
- Batch tasks characterize the manufacturing operations.
- The resources are shared (man-power, utilities, etc.).
- Manufacturing process runs in short time bursts.
- The quantity or scale of manufacture does not justify continuous operation.
- Flexibility is maximized through different ways of connecting equipment.

Handbook of Green Chemistry Volume 11: Green Metrics, First Edition. Edited by David J. Constable and Concepción Jiménez-González.
© 2018 Wiley-VCH Verlag GmbH & Co. KGaA. Published 2018 by Wiley-VCH Verlag GmbH & Co. KGaA.

- Multipurpose equipment that can be used to process different tasks at different time.

However, these batch processing characteristics are no longer as suitable as they once were. Process sustainability has become increasingly important, mainly triggered by the European legislation, which imposes reductions in the use of energy and water. There are also increasing issues associated with nonrenewable resources depletion. Consumers are also increasing their purchases of products that are considered to be more environmentally preferable, forcing companies to include this topic in their primary agenda. To succeed in this highly competitive culture, producers must keep pace with constantly shifting needs and expectations from their customers for innovative products and shorter lead times. These issues have prompted the batch processing industry to develop an innovation culture that leads to the manufacture of a broad range of high-value products. The following additional challenges are faced by the batch process industry:

- Complex manufacturing processes involving mixing/splitting and blending operations.
- Customer pressure for higher service and quality levels at lower cost.
- Conflicting production strategies: make-to-stock versus make-to-order.
- Pressure to reduce costs and enhance operational agility.
- Conflicting goals: economic, environmental, and social.
- Pressure to use processes that have fewer environmental impacts associated with them.
- Escalating, highly variable resource costs.
- Intense global competition.

These characteristics of batch processes have paved the way to new research studies toward an integrated approach/framework to sustainable process design and industrial ecology. Therefore, this chapter aims to present a generic framework for sustainable batch process design, where sustainability metrics and methods, to deal with each step of batch design, are proposed. This framework is a pioneer in the assessment of sustainability in batch processes, which turns this framework into a useful guideline for researchers and practitioners, toward the development of sustainable batch designs.

This chapter is organized as follows. In Section 5.2 an overview of the state of the art on batch process methodologies and on sustainability metrics is presented. Then, in Section 5.3 a detailed description of Sustainable Batch Design Framework (SBD-FRAME) is presented. Two case studies are presented in Section 5.4 to highlight the application of the framework. Finally, conclusions are drawn in Section 5.5.

5.2
State of the Art

This section underpins the importance of batch process design (Section 5.2.1) reviewing the methodologies used in batch design and is followed by a literature review on the assessment areas required for a more sustainable design (Section 5.2.2).

5.2.1
Design and Retrofit of Batch Processes

Batch processes are key enablers in imparting flexibility and variability to the production of end products, and in helping to make rapid adjustments to shifting market demands, but bring more complexity to integrated processing systems. The increase in marketing products globally, coupled with increasing market competitiveness, forced the batch chemical processing industry to include sustainability as a key strategic decision in batch process design. Despite this urgent requirement, there are still few publications that describe or document full sustainability integration in batch process design. However, several authors have begun addressing batch process design, employing different methodologies and improving some sustainability pillars, especially the economic pillar.

The first peer-reviewed publication to summarize batch process design research was published in 1990 by Reklaitis [2] where process integration and design under deterministic and uncertainty aspects were addressed. Beyond that, Reklaitis presented a conceptual decomposition of batch processing design and identified several areas for further research. The decomposition process was defined by four decision levels. The first level characterizes the processing network, which involves the definition of the product recipe and its tasks. The aim of the second level is to select the best operating strategy where the decision is about the operating mode to be used for the production of different products. The third level is associated with the allocation of process and storage equipment to tasks, while the fourth, and last level, is characterized by the design of all processing, storage, and connectivity equipment in the plant. Following considerable research, the areas identified for further research were task network determination, selection of operating strategy, short campaign operation, mixed operating structures, retrofit design, heat integration, and enabling technology. Later Barbosa-Povoa [1] published a critical review for grass root and retrofit of batch design for multiproduct and multipurpose plants. This work focused on the integration of basic design and scheduling as well as on the detailed aspects at the design level such as plant topology, layout, operational characteristics, and uncertainty.

In this chapter, the focus will be on works published after 2007 describing methodologies that may be applied to the sustainable design of batch plants. For previous work, please refer to the review of Reklaitis [2] and Barbosa-Povoa [1]. After 2007, several methods have been applied to batch process design, namely, mathematical programming, metaheuristics, heuristics, path flow decomposition, and multicriteria analysis, among others.

The optimization of batch processes, through mathematical programming, enables one to achieve better levels of process efficiency, when considering coupled design and scheduling. Several authors have used mathematical programming for the detailed design, retrofit, and scheduling of batch plants, considering demand uncertainty through mixed-integer linear programming (MILP) and a two-stage stochastic approach (Pinto-Varela *et al.* [3]) and (Fumero *et al.* [4]). Pinto-Varela *et al.* [3] considered the design and retrofit of a multipurpose cyclic batch process through a generic and holistic representation, maximal State-Task Network (mSTN), while Fumero *et al.* [4] explored the batch sequence in a campaign and its assignment to units when parallel units are used. The design aspects were explored simultaneously with environmental goals by Al-Mutairi and El-Halwagi [5]. An optimization formulation has been developed for the case of project schedule while allowing design retrofitting changes that include new environmental units, modification of design, and operating conditions in the process (without new process units). Seid and Majozi [6] presented a robust scheduling formulation for the synthesis and design of multipurpose batch plants. The formulation performance was compared with some benchmark examples and was found to have better computational performance. The integration of variations in prices, product demand limits, costs, and raw materials availability at the design level was developed by Fumero *et al.* [7]. The formulation selects the optimal topology (design and number) but also the detailed production planning in each time period.

Despite the diversity and complexity associated with batch processes, the use of utilities and resources are common aspects found in the majority of batch processes. Therefore, it has become crucial for the sustainability of the planet to include an integrated system's approach that includes process integration and an efficient utilization of resources, for instance, through the integration of renewable energies. A consideration of these issues triggered the authors Tokos *et al.* [8], Halim and Srinivasan [9], Moreno-Benito *et al.* [10], and Chaturvedia *et al.* [11] to develop their research in process integration design, with the aim of water and energy minimization in batch processes. Tokos *et al.* [8,12] explored process integration through MILP and mixed-integer nonlinear programming (MINLP) formulation, respectively. In the former work, the authors explored the combination of electricity, heating, and cooling production for the selection of an optimal polygeneration system. The proposed heat integration process and the selected cogeneration system improved a company's economic performance and reduced its environmental impact. The latter formulation explored multilevel strategies for retrofitting a large-scale water system, which integrates water-using operations and wastewater treatment units in different production sections within the same network. Halim and Srinivasan [9], to overcome a literature gap, proposed a scheduling mathematic formulation (MILP) incorporating simultaneous energy and water minimization. The formulation is decomposed into three parts – scheduling, heat integration, and water reuse optimization – and solved sequentially. Moreno-Benito *et al.* [10] developed an optimization-based approach for integrating process synthesis and plant allocation decisions, such as feed-forward control trajectories of processing conditions or the selection of operating modes and equipment items. A modeling strategy based on Mixed-Logic Dynamic Optimization and two-stage stochastic formulation is used to maximize the expected profit taking into account several uncertain scenarios. In 2016, Chaturvedia et al. [11] presented an algebraic/graphical method for heat integration, exploring the products storage, allowing the direct heat integration delay, and providing an opportunity for energy conservation while avoiding the use of an intermediate fluid.

The Path Flow decomposition approach was followed by Carvalho et al. [13], Bumann et al. [14], Banimostafa et al. [15], and Casola et al. [16] in their research. Path flow decomposition is a technique, based on the graph theory. The flowsheet is described with nodes (operation units) and with arcs (flows between units), and then a decomposition of all possible paths for the compounds entering and leaving the system is performed through different established algorithms. This approach screens the batch design alternatives using environmental, economic, and safety indicators, in order to identify the bottlenecks of the process and propose new design alternatives. The screen and identification of sustainable design alternatives considering operational, environmental, economic, and safety issues in the process was first presented by Carvalho et al. [13]. Bumann et al. [14] extended the concept using a new path flow indicator category that focuses on unit occupancy time and the multiobjective process assessment in order to reveal sustainable retrofitting actions. Banimostafa et al. [15] and Casola et al. [16] explored the same approach systematically, screening design alternatives generation in the pharmaceutical industry. In the same year, Banimostafa et al. [17] enriched the path flow decomposition, with hazard assessment and life cycle analysis path flow indicators, proposing a classification for coupling these new path flow indicators with heuristics for process alternative generation. The method highlights solvent recovery or substitution as important retrofitting actions, generates diverse process layout structures to achieve this task, and evaluates them from a cost, hazard, and life cycle assessment point of view.

Traditionally, the design of batch process has been undertaken based on individual criterion and applying only economic objectives, such as cost minimization or profit maximization. However, the increasing market competition, the customers' change expectations, the value of goods and services, combined with advances in technology and fast access to information demanded an integrated approach toward more agile production. Simultaneously, society has been developing an increasing level of awareness for environmental sustainability, and companies have been realizing that economic objectives ought no longer to be their sole concern. Companies realized that environmental impacts resulting not only from their structures but also from their operation need to be minimized. This fact leads to new challenges, which drive the application of methodologies able to tackle several objectives. While several multiobjective methodologies have been applied to fulfill this need, the nature, dimension, and complexity of these problems usually lead to large mixed-integer linear program formulations that come associated with a high computational burden.

In order to overcome this difficulty, several authors – Dietz *et al.* [18], Aguilar-Lasserre *et al.* [19], Mokeddem and Khellaf [20], and Chibeles-Martins *et al.* [21] – applied meta-heuristic and heuristics, capturing the advantage of those methods to define the Pareto optimal solutions in a single simulation run. In addition, fuzzy logic associated with multiobjective integer linear programming approach has been applied by Pinto-Varela *et al.* [22]. In 2008, Dietz *et al.* [18] used a multiobjective genetic algorithm (MOGA) to explore the design and retrofit of a batch plant, considering the investment cost and environmental impact minimization from the operating conditions. The work developed by Aguilar-Lasserre *et al.* [19] explores the batch design with imprecision demands using fuzzy logic concepts in the MOGA. The net present value (NPV), the production delay/advance with respect to a fixed date, and a flexibility criterion that measures the capacity of the plant to manufacture an additional production were the criteria used as objective functions in Pareto nondominated solutions characterization. A multicriteria methodology, analytic hierarchy process (AHP), was used as a decision-support tool for the final batch plant design selection. A similar approach was followed by Mokeddem and Khellaf [20], where the nondominating sorting genetic algorithm (NSGA-II) followed by a multicriteria tool, the PROMETHEE II, was used to help the decision-maker select the optimal compromise solution. In the following year, several research projects based on simulated annealing, fuzzy, and mixed integer nonlinear and linear programming were explored. Pinto-Varela *et al.* [22] includes the design and scheduling of a multipurpose nonperiodic process, considering the maximization of the total revenue as well as the minimization of the total cost. A symmetric fuzzy linear programming (SFLP) approach was applied and compared with a mixed-integer linear programming Pareto nondominate solution through an e-constraint approach. Chibeles-Martins *et al.* [21] explored the design and scheduling of a multipurpose batch process. To overcome the increased time it would take due to the problem complexity, they used a meta-heuristic approach, based on simulated annealing (SA) for the design and scheduling of batch plants. The Pareto nondominated solution was characterized using revenue maximization and cost (capital investment and operational) minimization and compared with the exact Pareto nondominated solution defined by the e-constraint. Tokos *et al.* [23] used a mixed-integer nonlinear programming approach to characterize a water system retrofit by estimating both the economic and environmental impacts of the water network design. The environmental impact was evaluated via benchmarking. The economic evaluation included the water network total cost, the freshwater cost, annual investment costs of the storage, piping and local treatment unit installation, and wastewater treatment cost. The Pareto front was designed using the normal-boundary intersection (NBI) method. This approach can be used for the separate integration of production sections, but also for joint integration of the sections via temporal decomposition. Yue and You [24] addressed multiproduct and multipurpose scheduling problems having economic and environmental concerns. The economic objective was the profit rate with respect to the production time horizon, while the environmental objective is evaluated by its environmental impact per functional unit based on a life cycle assessment. The profit maximization and environmental minimization are contradictory objectives, requiring the use of ϵ-constraint method as a multiobjective approach, allowing the characterization of Pareto nondominant solution. Pinto-Varela [25] explores the multiobjective approach, based on a MILP and ϵ-constraint method to define the Pareto nondominant solution for the optimal design and retrofit of multipurpose batch facilities operating in a nonperiodic and periodic mode. The decision-maker has a range of plant topologies, facilities designs, and storage policies associated with a scheduling operating mode that minimizes the total cost of the system while maximizing the production, subject to total product demands and operational restrictions.

With modern technological advances, it is now possible to solve problems and models that were impossible to solve some years ago; however, the development of efficient computational tools is still required, leading to an open area of research. Borisenko et al. [26] recently proposed a parallel algorithm for a real-world application to identify optimal designs for multiproduct batch plants. Two parallelization strategies with shared-memory and distributed-memory were explored. Carvalho et al. [27] presented a sustainable process design software tool, SustainPro, to help select design alternatives that are evaluated using environmental impact assessment tools and safety indices. Pinto-Varela et al. [28] presented a continuous improvement tool, BatchRetroLC, that include a series of retrofit activities that have to be conducted during the life cycle of a typical industrial batch process. BatchRetroLC aims to guide decision-makers in continuous improvement actions, leading to a sequence of iterative improvement steps. BatchRetroLC integrates an indicator-based methodology, SustainPro, with optimization models developed for the batch retrofit problem, OptimRetHeat.

5.2.2
Sustainability Assessment

The sustainable development concept has become widespread after the definition presented by WCED [29] in the Brundtland Report, where sustainable development was defined as "the development that meets the needs of the present without compromising the ability of future generations to meet their own needs." After two decades, Elkington [30] translated this idea into three main aspects: economic, environmental, and social. The economic pillar "concerns the organisation's impacts on the economic conditions of its stakeholders and on the economic systems at local, national, and global levels" [31]. This pillar has been the key pillar of company's pursuing so that profitability and competitiveness in the global markets can be achieved [32,33]. Economic performance is the highest priority and has therefore been the most developed pillar for assessing sustainability in batch process design. Several economic

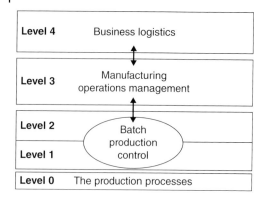

Figure 5.1 ISA 95 framework.

indicators, covering different aspects of processes, have been proposed [34]. Economic indicators should be the primary focus in batch process design in order to ensure the reliability of the proposed processes. ISA 95 is a standard that aims to integrate business logistics systems into manufacturing operations systems [35]. This standard has been applied to batch processes, giving insights into key business drivers derived from external relationships with different stakeholders and internal company activities. The business drivers are the most critical areas for performance assessment. Figure 5.1 presents a diagram, summarizing the ISA 95 areas.

Based on an extensive literature review of economic indicators used for sustainability assessments of batch processes, Table 5.1 contains a summary of indicators encompassed by the ISA 95 standards.

The environmental pillar "concerns an organization's impacts on living and non-living natural systems, including ecosystems, land, air, and water" [31]. The external drivers imposed by governments and international market competition are forcing companies to reduce process effluents and emissions and therefore their environmental impacts [44]. Several methods have been presented in the literature to assess the environmental impact of processes [45]. The International Organization for Standardization (ISO) has proposed a standard for the assessment of the environmental impacts of a product, service, or process over its whole life cycle, which is called the Life Cycle Assessment (LCA) [46]. LCA attempts to quantify all inputs and outputs needed for all entities of the life cycle [47], identifying and quantifying the environment impacts. Carvalho et al. [45] presented a literature review on environmental impact methods (25 methods), which can be employed to assess environmental impacts in chemical/biochemical processes. CML [48], Impact 2002+ [49], WAR [50], ReCiPe [51], and TRACI [52] are among the most applied LCA methods, which can be used to assess environmental sustainability in chemical processes. The existing LCA methods present several impact categories, which were grouped by Carvalho et al. [45] into categorical groups (groups obtained by aggregation of all impact categories that are allied, that have similar meaning, or that complement each

Table 5.1 Economic assessment areas to evaluate sustainability in ISA 95 level.

	ISA 95 level	Economic indicators	Definition	References
External drivers	Business planning and logistics – level 4	Financial support from the government	Support on incentives, taxes reductions, among other aspects that governments can provide	GRI [31]; Tanzil [36]
		Future liabilities	Potential obligation that may be incurred by the company	WBCSD [37]; Powell [38]; Azapagic [39]
		Market presence	Market share	WBCSD [37]; Al-Sharrah [40]; GRI [31]
		Indirect economic impacts	Other economic impacts that are not related to the core business of the company	Powell [38]; GRI [31]; Tanzil [36]
		Procurement practices	Costs associated with the procurement activities	Al-Sharrah [40]; GRI [31]
		Product value	Value of the product in the market	Powell [38]
Internal drivers	Business planning and logistics – level 4	NPV	Net present value, which is the actualized profit after an investment	Ruiz-Mercado et al. [41]; Shokrian [42], Shadiya [43]
		Payback	Time for recovering the investment	Ruiz-Mercado et al. [41]
		ROI	Return of investment	Ruiz-Mercado et al. [41]
		Turnover ratio	Financial ratio between annual income statement divided by the average balance of an asset per year	Ruiz-Mercado et al. [41]
	Manufacturing operations management – level 3	Final product revenue	Revenues derived from the sale of final products	WBCSD [37]; Al-Sharrah [40]; GRI [31]; Tanzil [36]; Azapagic [39]
		Eco-products revenue	Revenues derived from eco-products	Ruiz-Mercado et al. [41]; Tanzil [36]
		Production cost	Operational costs related to production	Ruiz-Mercado et al. [41]; WBCSD [37]; Azapagic [39]; Shokrian [42]

(*continued*)

Table 5.1 (Continued)

ISA 95 level	Economic indicators	Definition	References
	Raw materials cost	Raw material expenses (acquisition, treatment, etc.)	Ruiz-Mercado et al. [41]; WBCSD [37]; Powell [38]; Shokrian [42]; Shadiya [43]
	Waste cost	Waste treatment/disposal cost	WBCSD [37]; Shokrian [42]; Powell [38]; Shadiya [43]
	Energy cost	Cost with energy utilities	Ruiz-Mercado et al., [41]; WBCSD [37]; Shadiya [43]
	Water cost	Cost with water utilities	Ruiz-Mercado et al., [41]; WBCSD [37]
	Profit	Difference between revenues and costs, including taxes	WBCSD [37]; Shokrian [42]; Shadiya [43]; Azapagic [39]

other in assessing a certain effect). The categorical groups were further aggregated into areas of improvement, called classes – ecological, resources, and human health classes. The ecological and the human health classes evaluate the impact of a process' emissions on ecosystems and human heath, while the resource class evaluates the extraction of compounds from Nature. Table 5.2 presents a summary of the environmental categorical groups most often applied to life cycle environmental assessments.

Finally, the social pillar represents the "Quality of societies. It signifies the nature–society relationships, mediated by work, as well as relationships within the society" [53]. Social indicators are difficult to propose, since they are built on human intuition, being influenced by the developer's experience and know-how. This subjective task usually leads to qualitative and semiqualitative indicators [54]. Chemical processes involve some major social aspects that are critical to evaluate, namely, materials and process safety [54]. Meckenstock et al. [55] identified that health and safety issues are the most relevant social aspects to be considered at the process level and therefore batch processes are not an exception. Safety indicators have been extensively presented in the literature for evaluating processes' performance. Khan and Abbasi [56] presented the Hazard Identification and Ranking System (HIRA), Heikkila [57] proposed the Inherent Safety Index (ISI), Gupta and Edwards [58] introduced

Table 5.2 Environmental assessment areas to assess environmental impacts on processes.

	Impact categorical group		Impact categorical group		Impact categorical group
Resources	Abiotic resources	Ecological	Acidification	Human health	Carcinogenic
	Biotic resources		Global warming		Causalities
	Element reserves		Air pollution		Human health
	Energy		Bioaccumulation		Human toxicity
	Land use		Damage to flora		Ionizing radiation
	Natural resources		Ecotoxicity		Respiratory effects
	Nonrenewable energy		Eutrophication		Life expectancy
	Nonrenewable, metals		Extinction of species		Morbidity
	Nonrenewable, nuclear		Hazard substances		Noncarcinogenic
	Nonrenewable, primary forest		Volatile organic compounds		Nuisance
	Resources consumption		Malodorous air		
	Water		Oxygen consumption		
	Renewable energy		Ozone depletion		
	Renewable, biomass		Particulate matter		
	Renewable, geothermal		Photochemical oxidation		
	Renewable, solar		Heavy metals		
	Renewable, wind		Waste heat		
	Recycling effect				

Source: Adapted from Ref. [45].

the Inherently Safer Design (ISD) Index, and Khan and Amyotte [59] brought into discussion the Integrated Inherent Safety Index (IISI). Despite the relevance of safety and health issues, some social indicators are important to consider. Labor practices and decent work (acceptable working conditions) are a cornerstone for process design assessment [31,60]; however, process design might influence external stakeholders through the whole life cycle of the product and therefore aspects covering process impact in the community, corruption, stakeholders' participation, human rights, and customer satisfaction are other areas that should be considered [61]. Table 5.3 summarizes the social areas classified according to the ISA 95 standards that may be used to assess batch processes.

Table 5.3 Economic assessment areas to evaluate sustainability in ISA 95 level [57,59,61].

	ISA 95 level	Social indicators
External drivers	Business planning and logistics – level 4	Business impact in the community
		Corruption in business
		Stakeholders' participation
		Human rights
		Consumer's H&S
		Product management and costumer's satisfaction
Internal drivers	Business Planning & Logistics- Level 4	Number of employees hired
		Benefits/promotions
		Rates of injuries
		Training hours
	Manufacturing operations management – level 3	Flammability index
		Heat of reaction
		Toxicity
		Equipment safety
		Layout
		Temperature and pressure
		Explosivity
		Corrosivity

5.3
Framework for Design and Retrofitting in Batch Processes

This section characterizes the proposed framework called Sustainable Batch Design Framework (SBD-FRAME), shown in Figure 5.2. This framework intends to guide practitioners in sustainable batch process design, retrofit, and the evaluation of sustainability across different design steps. The framework presents the most suitable areas for assessment and identifies the most appropriate methods to apply in order to obtain more sustainable batch processes.

State-of-the-art batch process design has been decomposed into four steps by Reklaitis [2]. However, this methodology was not intended to achieve sustainable batch process design and therefore improved methodology is required. As presented, ISA 95 is an industry-accepted framework that practitioners view as a good procedure for businesses to follow in order to increase their speed and responsiveness and thereby remain competitive in today's demanding markets [62]. In order to achieve a methodology for sustainable batch process design, design systems need to be fully integrated utilizing a common workflow model and operational data. In response to this requirement, SBD-FRAME presents a new conceptual sustainable batch design that incorporates all required

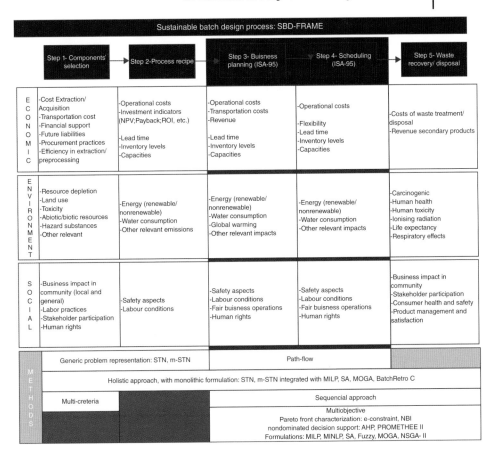

Figure 5.2 Flow diagram of SBD-FRAME.

steps in a holistic life cycle analysis. SBD-FRAME has been decomposed into five steps, which can be defined as follows:

1) *Compound specification*: Here the compounds used to produce different products are determined, and their selection should follow green chemistry principles [63]. The upstream supply chain decisions are a cornerstone in the selection of different compounds. The impacts of natural resource depletion, land use, external stakeholders (e.g., suppliers), social conditions of the employees extracting and processing compounds, among other factors, should be considered in this step so that a sustainable decision is taken. The selection of compounds is part of early-stage design and therefore should already take into account the impacts of this decision on the production process. Principles of green engineering should be considered in a preliminary way, so that selection of compounds is already contemplating future engineering decisions. This integrated decision will provide a holistic analysis,

considering supply chain upstream sustainability impacts and future process sustainability implications.

2) *Process recipe*: In this design step, the products recipes are defined and therefore the tasks are allocated to each product. This step is related to the process definition for each final product. When considering one isolated product, this is quite a simple task; however, when considering multiproducts, batch product definition becomes a complex task. At this stage of batch design, downstream implications (e.g., process synthesis and final product safety) are more important in making sustainable design decisions, since compounds have already been selected.

3) *Business planning*: At this stage, all strategic and tactical level decisions are going to be considered. Here the required equipment, sizing of the equipment, human resources allocation, and storage materials, among other aspects, are determined. In this step, it is necessary to define the enterprise resource planning (ERP), which is the planning bridging the external stakeholders, through their logistics activities, and the internal activities at the operational level. These are medium- to long-term decisions and therefore sustainability assessment should consider this aspect.

4) *Scheduling*: This step accounts for the operational decisions on a daily basis such as the sequence of daily processing activities. This operational planning will allocate required resources to the activities and will clearly define the daily stock requirements and the amount to be kept in stock.

5) *Waste recovery/disposal*: Previous models describing batch process design do not incorporate a critical phase; that is, waste recovery and disposal any batch process design has to consider, if it aims to be more sustainable. This step should consider how product and by-product recovery can take place, since this might be a problem in batch process design. Recycling in these types of processes might involve maintaining stock and therefore this may not be as straightforward a process as it is in continuous processing. Recycling requires detailed planning, resources, and coordination. Moreover, if recovery is not a possibility, disposal should be assessed ensuring a sustainable opportunity for the reduction and/or appropriate treatment of effluents and emissions.

SBD-FRAME proposes for each design step a set of economic, environmental, and social aspects that should be considered in order to evaluate sustainable design alternatives.

5.3.1
Economic Assessment

In the early-stage design (step 1) where compounds are to be selected, economic aspects should evaluate the relationship of the company with the suppliers, assessing the cost of raw materials and logistic costs (transportation, taxes, etc.). It is also important at this stage to evaluate if there is any available financial support for specific production paths involving some compounds. This financial support may be critical to achieving a profitable batch process. Future liability is another issue that should be considered when comparing different compounds. Different compounds mean different reactions, which lead to different processes and subproducts. Some liabilities might arise with one compound versus another and therefore it needs to be taken into consideration in the early design stage. Procurement practices are critical in early-stage design; a companies' process can be sustainable only as far as their suppliers ensure sustainability practices. Supplier's certification in terms of quality, environmental, and social protection should be contemplated in this analysis.

In the second step of sustainable batch design, when the final product composition is determined along with the batch process design, operational costs and investment costs should be considered in the economic assessment for different recipe options. In order to be profitable, batch processes need to be efficient and flexible and technoeconomic aspects are important considerations. Lead times, inventory levels, and capacities should be considered when assessing economic performance of step 2. These aspects are most relevant in batch processing and they are what differentiates batch from continuous in terms of sustainability performance.

In the business planning (step 3) and at the scheduling (step 4), operational costs are essential to be measured. Coordination between external business entity activities and internal activities should be ensured at the minimum operational cost as possible. At this point, revenues are also required to be assessed. The product will enter the market and demand should be fulfilled, so that market share can be kept. The inherent flexibility of batch processes should be in view during this planning step and therefore lead times, inventory levels, capacities, and resource allocation should be clearly evaluated.

Finally, when assessing final disposal or recovery (step 5), investment and operational costs of waste treatment plants should be evaluated. Moreover, cost of waste disposal/treatment should be considered. Revenues from valuation of secondary products or subproducts should be investigated, creating a circular economy, where the waste of one process will be the feedstock for another process. Industrial symbiosis should be considered for its potential to create greater economic value to the main company and thereby generate more sustainable processes.

5.3.2
Environmental Assessment

In the early stage of batch design (step 1) resource depletion, land use, biotic, and abiotic resources are critical to analyze. This design phase should be mainly concerned with the environmental impacts of critical upstream activities, that is, the extraction and preproduction processes. Compound toxicity is a key concern when selecting compounds. This aspect will be reflected in the health and safety issues associated with the process and will be seen as part of product responsibility for internal and external stakeholders. Other relevant aspects are, for instance, ozone depletion and global warming. These and other aspects, as appropriate, should also be analyzed depending on the extraction and preprocessing processes required for the different compounds. This means that depending on the process emissions, different environmental impact categories can be added to account for the emissions of that specific process (see Table 5.2 for more details on the impact categories).

In step 2, the definition of the recipe, there are two relevant issues that should be considered for every batch synthesis process in order to minimize their environmental impacts: renewable/nonrenewable energy and water consumption.

In step 3, business planning, it is critical to analyze not only the batch process environmental aspects but also environmental impacts associated with product distribution. Global warming is the most critical impact category in this step, since transportation is a very high contributor to global greenhouse gases emissions. Finally, for step 4, planning the operational schedule, environmental impacts can be minimized through process integration and intensification. Therefore, environmental impacts associated with energy and water play a critical role in the scheduling definition.

Finally, the environmental assessment of waste treatment/disposal (step 5) should mainly consider its impact on human health (e.g., human toxicity, human health, life expectancy, respiratory effects, among others), because in this step the by-products and final products are going to be released to the surrounding community. For waste recovery, the impact categories to be assessed are dependent on the type of waste treatment that is being considered. Based on that, other relevant environmental aspects should be explored in order to account for the environmental impacts associated with the treatment process.

5.3.3
Social Assessment

With compound selection (step 1), it is critical to evaluate the impact of the process on society. Raw material extraction activities are sometimes done in ways that adversely impact social conditions and therefore several aspects should be monitored. Labor practices, stakeholders' participation, local communities, and human rights are social issues to include in any early-stage sustainability assessment.

When defining the chemicals that go into the process (step 2), safety aspects become the cornerstone of the sustainability assessment. In the business planning and logistics level (step 3), labor conditions and human rights are the most important aspects to consider. Several abuses have been reported regarding safety and human rights not being respected (e.g., excess of hours to a single driver). When moving to a detailed schedule at a more operational level (step 4), it is essential that safety aspects and labor conditions become the top priority in terms of social responsibility. Then, when disposing and selling the by-products and final product (step 5), the business impact on the community should once again be assessed, bridging the operations with the final customers. Consumer health and safety is another aspect that should be assessed in social sustainability, as well as product management and satisfaction.

5.3.4 Methodologies

For many years, different levels associated with the batch design process were explored following a sequential approach [2], mainly due to computational difficulties in solving complex problems. Despite computational improvements, it was still not possible to define the process global optimum, thereby triggering the development of an integrated holistic approach. The integrated approach requires a very high computational capability, not only to explore the more detailed and realistic problem but also to deliver an optimum solution in real time. To overcome the difficulty, metaheuristics were explored.

Nowadays, it is more important than ever to use an integrated holistic approach in the design and scheduling of batch processes. Accordingly, with the SBD-FRAME, the compound selection has impact not only in all the subsequent aspects of the batch design process but also in the environmental impacts that result from it.

As presented in the SBD-FRAME, Figure 5.2, the integration of all framework steps (1 through 5) can be achieved for a grassroots design through mathematical approaches like MILP, MINLP, or by using a more relaxed approach through metaheuristics approaches like SA, MOGA, and NSGA II. The latter approaches have the advantage of being able to develop the results very quickly; however, on the other hand, they are unable to reach the optimum value or give any guidance about the solution quality. The BatchRetroLC, a sequential approach integrating mathematical formulation with path-flow decomposition, takes into account all of the levels of the SD-FRAME for the retrofit of batch plants. The use of a generic representation for the detailed problem characterization through STN or m-STN, in steps 1 and 2 of SBD-FRAME, integrated with the resolution approach previously mentioned, triggers a faster and more precise solution. The multiobjective approach is mainly focused between steps 3 and 5. There are economic and environmental goals that most of the time are contradictory. The algorithm used can be mathematic or metaheuristic. However, the inherent characteristics of metaheuristics (Fuzzy, SA, MOGA, NSGA-II), which can be followed by multicriteria approach (AHP, PROMETHEE II), overcome the computational burden associated with the mathematic formulations. The aim of path-flow decomposition is to develop a screening analysis, which is most relevant at steps 3 and 4 of the SBD-FRAME. The selection of compounds (step 1) can be seen using a holistic approach or as an independent step, through the use of a multicriteria approach. Regardless of which approach is used, each level of the SBD-Frame can use each of the monolithic approaches independently, if necessary.

5.4
Case Studies

Two case studies are presented in order to illustrate how the SBD-FRAME framework guides users in their design of more sustainable batch processes. First, an insulin production case study is presented where a path-flow methodology is employed (Section 5.4.1). Second, a detailed design of a multipurpose batch process plant is described where heat integration and economic savings in utilities are considered (Section 5.4.2).

5.4.1
Retrofit Sustainable Batch Design

The insulin process described by Petrides *et al.* [64] will be analyzed using the SBD-FRAME framework. Figure 5.3 presents the flowsheet for the insulin production process.

This process may be divided into four sections:

1) *Fermentation: Escherichia coli* cells produce the Trp-LE'-MET-proinsulin, the precursor of insulin, in the cellular biomass.
2) *Primary recovery:* In a high-pressure homogenizer, the cells are broken to release the inclusion bodies. The mixture is centrifuged, washed with solvents, and the inclusion bodies are recovered at a high purity.
3) *Series of reaction:* A sequence of chemical reactions are performed to convert the raw materials into insulin.
4) *Purification:* A set of operations is performed to purify the insulin. Multimodal chromatography operations exploit differences in size, molecular charge, and hydrophobicity to isolate biosynthetic insulin. The crystallization of insulin is the last operation of the process.

The objective of this case study was to improve the original process at the planning level. For that purpose, a method based on path-flow decomposition and heuristics, through indicators analysis, has been employed. Following SBD-FRAME to improve the business and logistics (step 3) and the scheduling step (step 4), path-flow decomposition methodologies can be employed to identify bottlenecks in the system and propose new design alternatives. Here, the methodology of *SustainPro*, presented by Carvalho *et al.* [13,27], will be applied. *SustainPro* assists users in screening and identifying new sustainable design alternatives for batch process design and follows a seven-step procedure: *Step 1:* Collect the steady-state data. *Step 2:* Transform equipment flowsheets in an operational flow diagram. *Step 3:* Flowsheet decomposition. *Step 4:* Calculate the sustainability and safety metrics. *Step 5:* Indicator Sensitivity Analysis (ISA) algorithm. *Step 6:* Process sensitivity analysis. *Step 7:* Generation of new design alternatives.

5.4 Case Studies | 143

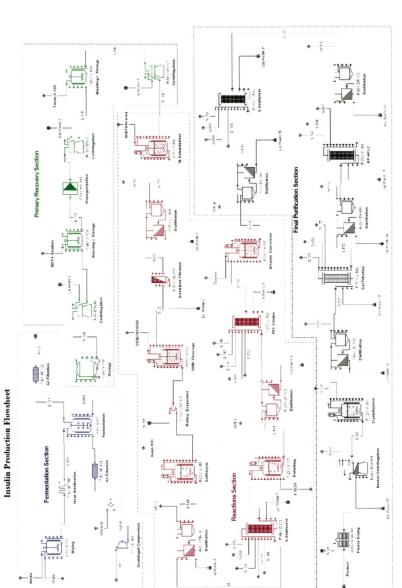

Figure 5.3 Flowsheet of insulin process simulated in SuperPro Design.

Each step is described in more detail as follows:

Step 1: Collect the steady-state data

In this step, mass and energy balance data, time, recipes, equipment, capacities, prices and other relevant data are collected. For the insulin process, the SuperPro Designer [65] software package was used to obtain the data and the prices and costs were taken from Petrides *et al.* [64].

Step 2: Transform equipment flowsheet in an operational flow diagram

In this step, the flowsheet is decomposed into detailed scheduled operations, which is called the operational flow diagram. The equipment flowsheet presented in Figure 5.3 was decomposed in the sequence of operations, achieving an operational flow diagram with 92 operations, 169 streams, and 38 compounds.

Step 3: Flowsheet decomposition

In this step, a flowsheet decomposition, identifying all open- and closed-paths is performed. The closed-paths (CP) are the process recycles; this means the flowpaths start and end in the same unit of the process. An open-path (OP) consists of an entrance and an exit of a compound in the process. The operations flow diagram has been decomposed and 418 closed-paths, 1022 open-paths, and 3344 accumulation paths were achieved.

Step 4: Calculate the indicators

A set of mass and energy indicators that trace the paths of the components is calculated in order to identify batch process bottlenecks in terms of energy, waste, and value. More information about the indicators can be found in Carvalho *et al.* [13, 66]. The most relevant mass indicators were selected and are listed for each section in Table 5.4.

The material value-added (MVA) indicator gives the value generated between the start and the end points of the path. The most sensitive indicator for the insulin production process is the MVA, for the open-paths shown in Table 5.4. They have highly negative values, which means that value is being lost through the process, showing the presence of waste. The most relevant batch indicators were selected and they are presented in Table 5.5.

The operation time factor (OTF) indicator points to the fraction of time that a given operation spends in relation to the total time taken by the whole process. Operations V-102R, V-103(P8)R, V-105R, and V-111R present high OTF indicator values when compared with the other operations, exposing the criticality of those operations in terms of lead time. The most relevant indicators are analyzed for the different section; for more detailed information, see Ref. [13].

Fermentation: The fermentation operation (V-102R) has a high OTF value that translates to a lead time bottleneck. To improve the fermentation process, it will be necessary to take into consideration the parameters that influence the reaction rate.

Table 5.4 Most relevant mass indicators for insulin production.

Section	OP	Path	Component	Flowrate (kg/h)	MVA (10³$/yr)	EWC (10³$/yr)	TVA (10³$/yr)
Fermentation	OP 37	S4-S26	Water	25 881,8	−22 560	68,90	−22 629
Primary recovery	OP 552	S28-S34	TRIS	251,1	−11 934	1,14	−11 935
Reactions	OP 620	S79-S80	Urea	10 399,9	−205 917	0,00	−205 917
	OP 591	S54-S60	Formic acid	10 837,6	−137 334	0,34	−137 334
	OP 613	S77-S80	Urea	4564,4	−90 375	0,00	−90 375
	OP 657	S62-S69	HCL	2987,9	−74 542	0,01	−74 542
Final purification	OP 1016	R V108-S138	Insulin	1,6	−639 591	0,00	−639 591
	OP 1009	R V108-S149	Insulin	1,4	−564 122	0,00	−564 122
	OP 1011	R V108-S159	Insulin	1,3	−507 708	0,00	−507 708
	OP 1005	R V108-S121	Insulin	0,9	−343 498	0,00	−343 498
	OP 822	S137-S138	Acetonitrile	2905,0	−80 527	0,00	−80 527

MVA: Material value vdded; EWC: energy waste cost; TVA: total value added.

Primary recovery: The most critical factor in this part of the process is related to the Tris base waste, given its very negative MVA value. This solvent is used as a buffer to facilitate the separation of the cell body debris from the inclusion bodies. Tris base is not recycled and therefore goes to waste.

Series of reactions: The solvents (urea, WFI, formic acid, HCL, NaCl) involved in this section present very negative values of MVA, because they are not recovered and recycled within the process. In order to improve the batch process toward a more sustainable solution, recovery of the solvents and recycling should be considered. V-103 R and V-105 R present high values of OTF, showing a lead time bottleneck of the system.

Table 5.5 Most relevant batch indicators for insulin production.

Section	Operation	OTF
Fermentation	V-102R	0065
Primary recovery	DS-101(P9)	0022
Reaction	V-103(P8) R	0029
	V-105 R	0043
Final purification	V-111 R	0043

Purification: The insulin paths presented in Table 5.4 have very low flowrates, although their MVA values are very negative. The insulin price is very high and any loss of insulin leads to high value losses. To recover more insulin, two options could be considered: (i) improve the existing separation process; (ii) add a new separation operation.

Step 5: Indicator Sensitivity Analysis algorithm

A sensitivity analysis is performed on the most sensitive indicators to determine those indicators that will yield the most sustainable improvements in the process. These indicators should be the top priority for improvement. The sensitivity analysis was performed using the ISA algorithm, which has been applied to most relevant indicators (some listed in Table 5.4). The ISA algorithm established the MVA indicator of OP591 for formic acid as the target indicator. For batch indicators (see Table 5.5), the most sensitive indicator is the OTF of ammonia in the fermentation operation (V-102R).

Step 6: Process sensitivity analysis

The target indicators operational parameters are screened using a secondary sensitivity analysis in order to determine the parameters that most influence the target indicators. For the MVA- OP591, it was found that the most significant operational parameter is the flowrate of OP591. The operational parameter that has the most influence on the batch target indicator (OTF) is the concentration of ammonia (NH_3) in the reaction.

Step 7: Generation of new design alternatives

To improve MVA, recycling of the formic acid was considered to reduce OP591 flowrate. A separation operation was therefore added to the process in order to recover this compound. Applying the process separation algorithm of Jaksland et al. [67], the proposed separation process was pervaporation. Nakatani et al. [68] found that membranes such as asymmetric aromatic imide polymers are available to purify/recover formic acid from water and thereby making this separation technically viable.

To reduce the OTF, the concentration of ammonia needs to be increased. By implementing a 2% increase in the ammonia concentration, a 0.2% fermentation time reduction was achieved. To further improve the lead time, a new enzyme should be considered.

This new batch design, which includes formic acid recycling and a reduction in the processing time, was assessed using SBD-FRAME (steps 3–5). The results of this analysis showed that the profit increased by 1.98% and the water and energy metrics per value added improved by 2%. In addition, the material metrics improved by 2 and 4%, respectively, per kilogram of final product and per value added. Finally, the environmental impact was reduced by 31.7%. The rest of the performance criteria parameters have remained constant (safety indices included). The sustainability assessment values are presented in Table 5.6.

Table 5.6 Sustainability assessment following SBD-FRAME.

Metrics	Initial	Final	Improvement
Total net primary energy usage rate (GJ/y)	26 727	26 727	0%
% Total net primary energy sourced from renewables	0.72	0.72	0%
Total net primary energy usage per kg product (kJ/kg)	292 397	292 397	0%
Total net primary energy usage per unit value added (kJ/$)	4.55×10^{-4}	4.46×10^{-4}	1.94%
Total raw materials used per kg product (kg/kg)	43 029	42 083	2.20%
Total raw materials used per unit value added (kg/$)	6.70×10^{-5}	6.42×10^{-5}	4.10%
Hazardous raw material per kg product (kg/kg)	4932.35	3986.22	19.18%
Net water consumed per unit mass of product (kg/kg)	7162.22	7162.22	0%
Net water consumed per unit value added (kg/$)	1.11×10^{-5}	1.09×10^{-5}	1,94%
Safety index [57]	20	20	0%
Life cycle analysis (WAR algorithm) [50]	23 709	16 188	31.7%
Profit ($/yr)	7.42×10^{9}	7.56×10^{-9}	1.98%

5.4.2
Design of Batch Process

The optimal plant design with heat integration and utilities rationalization is defined following the SBD-Frame framework through an integrated and holistic approach. A generic and detailed representation methodology, the maximal State-Task Network (m-STN) is applied and for a simpler recipe representation, a State-Task Network (STN) is applied [69]. Through this representation, all levels of the SBD-Frame framework are considered holistically.

In this case study, the holistic approach considers the product recipes through m-STNs representation, the production of all products over a single campaign structure, the plant flowsheet including all possible unit operations with the available heat transfer equipment structures, and the ability to account for heat integration. This holistic approach encompasses component selection, the recipe definition, business planning (steps 2–4), and scheduling, with the aim to increase the sustainability of batch plant design.

The m-STN representation for the design of batch facilities with heat integration is generated automatically from process and plant data. The planning horizon (H) is divided into a number of elementary time steps of fixed duration (δ) where all events are allowed to occur at the interval boundaries and not between them. A MILP is used, where the binary variables are introduced to characterize operational and topological selections, and the continuous ones define the

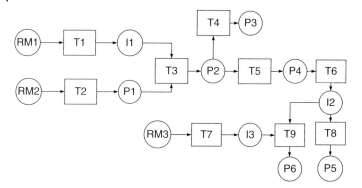

Figure 5.4 Product recipe through State-Task Network representation.

equipment capacities, as well as the amounts of material within the overall process.

In this case study, the design and scheduling of a biochemical multipurpose batch plant with direct integration, operating cyclically over a 40 h period, is addressed. The design and scheduling must consider the three different raw materials (RM1 through RM3) to produce six final products (P1 through P6) using alternative production paths. The production is characterized by nine task (T1 through T9): four are endothermic (T1, T5, T6, and T9) and four are exothermic (T2, T4, T7, and T8). The processing tasks can operate either consuming utilities outside the system or using an integrated operation (defined as H1_Ti tasks $i = 1, \ldots, 8$), which allows heat exchange. Task 3 follows a zero-wait storage policy, consumes an unstable intermediate material (I1) and P1, which is also final product. The recipes to produce all products and task's characterization are shown in Figure 5.4 and Table 5.7, respectively. Three heat transfer units are used for heat integration, the reactors have multipurpose characteristics, but the storage vessels are dedicated.

Table 5.7 Integration task characterization.

Nonintegrated operation		Integrated operation		
Tasks	Utility	Tasks	Exo/endothermic	Utility
T1	Steam	H1_T1	Endothermic	Self-sufficient
T2	Water	H1_T2	Exothermic	Water
T4	Water	H1_T4	Exothermic	Self-sufficient
T5	Steam	H1_T5	Endothermic	Steam
T6	Steam	H1_T6	Endothermic	Self-sufficient
T7	Water	H1_T7	Exothermic	Water
T8	Water	H1_T8	Exothermic	Self-sufficient
T9	Steam	H1_T9	Endothermic	Steam

To increase flexibility in the design in response to demand and integration uncertainties, profit maximization is explored by considering final product production within specified ranges of material units (m.u.): [100,400] for P1, [100,150] for P2, [150,400] for P3, [100,350] for P4, [100,400] for P5, and [100,400] for P6.

Profit is used as the indicator of optimal plant design, including heat integration and utilities rationalization. The aim of this problem is to optimize the economic performance of the plant, measured in terms of capital expenditure and operating costs and revenues through a consideration of the optimal plant configuration (i.e., the number and types of equipment and their connectivity, including the auxiliary equipment and associated circuits); the schedule, making use of selected resources to achieve the desired production (i.e., timing of all tasks, storage policies, batch sizes, amounts transferred, and allocation of tasks to equipment); the heat transfer requirements (i.e., use of external utilities and/or direct plant integration characterized by different equipment structures); the optimal direct plant integration (i.e., use of processing fluids and/or external fluids as heat transfer medium); and the utilities requirements to satisfy all demand. The environmental indicator used is the reduction in energy consumption that comes through heat integration. The detailed formulation, equipment, and operational characteristics are available in Ref. [70].

The optimal plant topology obtained through mixed-integer linear programming is shown Figure 5.5. A serpentine was chosen as auxiliary equipment to

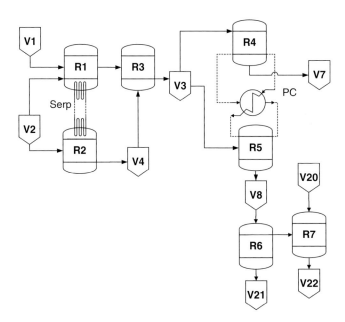

Figure 5.5 Optimal batch plant topology.

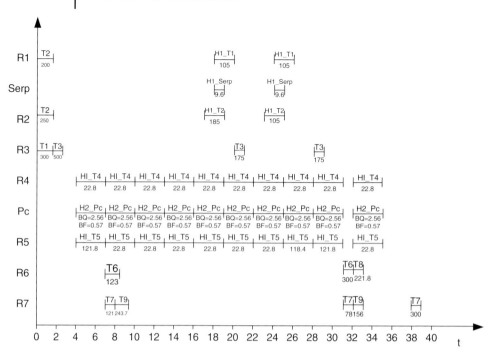

Figure 5.6 Optimal plant scheduling.

perform the heat integration exchange between tasks H1_T1/H1_T2, with an optimal transfer area of 35.43 m², while for tasks H1_T4/H1_T5 a heat-exchanger was used, with an optimal area of 3.12 m².

The reactors selected were R1, R2, R3, R4, R5, R6, and R7, with capacity of 200, 250, 500, 80, 122, 300, and 300 m.u., respectively. The reactors R3, R1, R6, and R7 present a multipurpose operation, while the remaining units are allocated to a single task, shown in Figure 5.6. The utilities savings through heat integration is shown by the sets H1_T1/H1_T2 and H1-T4/H1_T5. Nevertheless, T1 and T2 also operate in a nonintegrated form, like tasks T6, T7, T8, and T9 presented in Figure 5.6. The utility consumption profiles are shown in Figures 5.7 and 5.8, steam and water, respectively.

5.5
Conclusions

A framework for the design and retrofit of more sustainable batch plants, which is called Sustainable Batch Design – Frame (SBD-FRAME) has been proposed (Figure 5.9). The aim of the framework is to guide researchers and practitioners

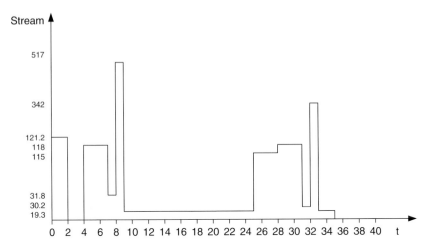

Figure 5.7 Steam consumption profile (m.u./h).

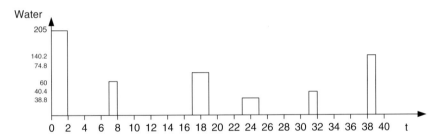

Figure 5.8 Water consumption profile (m.u./h).

through methods selection and other important aspects to consider in the definition of a sustainable batch process. The framework is intended to provide a holistic approach to integrating different process systems engineering methods into sustainable batch design. SBD-FRAME presents a new integrated view of batch design by including economic, environmental, and social aspects. Use of the framework was illustrated through two case studies where different methods are applied (path flow decomposition and mathematical models) and where holistic approaches covering different steps of batch process design and sustainability metrics are combined.

SBD-FRAME is a powerful framework to promote a systematic and holistic approach toward making batch processes more sustainable and it is a pioneer in the definition of a sustainability assessment guideline through batch design and retrofit.

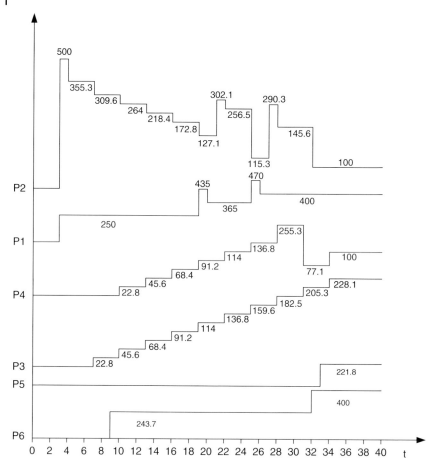

Figure 5.9 Final products profiles (m.u).

References

1 Barbosa-Povoa, A.P. (2007) A critical review on the design and retrofit of batch plants. *Computers & Chemical Engineering*, **31** (7), 833–855.
2 Reklaitis, G.V. (1990) Progress and issues in computer-aided batch process design, in *Foundations of Computer-Aided Process Design* (eds J.J. Siirola, I.E. Grossmann, and G. Stephanopoulos), CACHE-Elsevier, New York, pp. 241–275.
3 Pinto-Varela, T., Barbosa-Povoa, A., and Novais, A.Q. (2009) Design and scheduling of periodic multipurpose batch plants under uncertainty. *Industrial & Engineering Chemistry Research*, **48** (21), 9655–9670.
4 Fumero, Y., Corsano, G., and Montagna, J.M. (2011) Detailed design of multiproduct batch plants considering production scheduling. *Industrial & Engineering Chemistry Research*, **50** (10), 6146–6160.
5 Al-Mutairi, E.M. and El-Halwagi, M.M. (2010) Environmental-impact reduction through simultaneous design, scheduling, and operation. *Clean*

Technologies and Environmental Policy, **12** (5), 537–545.

6. Seid, E.R. and Majozi, T. (2013) Design and synthesis of multipurpose batch plants using a robust scheduling platform. *Industrial and Engineering Chemistry Research*, **52** (46), 16301–16313.

7. Fumero, Y., Corsano, G., and Montagna, J.M. (2013) A mixed integer linear programming model for simultaneous design and scheduling of flowshop plants. *Applied Mathematical Modelling*, **37** (4), 1652–1664.

8. Tokos, H., Pintaric, Z.N., and Glavic, P. (2010) Energy saving opportunities in heat integrated beverage plant retrofit. *Applied Thermal Engineering*, **30** (1), 36–44.

9. Halim, I. and Srinivasan, R. (2011) Sequential methodology for integrated optimization of energy and water use during batch process scheduling. *Computers & Chemical Engineering*, **35** (8), 1575–1597.

10. Moreno-Benito, M., Espuna, A., and Puigjaner, L. (2014) Flexible batch process and plant design using mixed-logic dynamic optimization: single-product plants. *Industrial & Engineering Chemistry Research*, **53** (44), 17182–17199.

11. Chaturvedia, N.D., Manana, Z.A., Alwia, S.R.W., and Bandyopadhyay, S. (2016) Maximising heat recovery in batch processes via product streams storage and shifting. *Journal of Cleaner Production*, **112** (4), 2802–2812.

12. Tokos, H., Pintaric, Z.N., Yang, Y., and Kravanja, Z. (2011) Multilevel strategies for the retrofit of a large industrial water system. *AIChE Journal*, **29**, 1165–1169.

13. Carvalho, A., Matos, H.A., and Gani, R. (2009) Design of batch operations: systematic methodology for generation and analysis of sustainable alternatives. *Computers & Chemical Engineering*, **33** (12), 2075–2090.

14. Bumann, A.A., Papadokonstantakis, S., Fischer, U., and Hungerbuehler, K. (2011) Investigating the use of path flow indicators as optimization drivers in batch process retrofitting. *Computers & Chemical Engineering*, **35** (12), 2767–2785.

15. Banimostafa, A., Papadokonstantakis, S., and Hungerbuehler, K. (2012) Retrofit design of a pharmaceutical batch process improving green process chemistry & engineering principles. 11th International Symposium on Process Systems Engineering, Parts A and B (eds I.A. Karimi and R. Srinivasan), vol. 31, pp. 1120–1124.

16. Casola, G., Yoshikawa, S., Nakanishi, H., Hirao, M., and Sugiyama, H. (2015) Systematic retrofitting methodology for pharmaceutical drug purification processes. *Computers & Chemical Engineering*, **80**, 177–188.

17. Banimostafa, A., Papadokonstantakis, S., and Hungerbuehler, K. (2015) Retrofit design of a pharmaceutical batch process considering "green chemistry and engineering principles". *AIChE Journal*, **61** (10), 3423–3440.

18. Dietz, A., Azzaro-Pantel, C., Pibouleau, L., and Domenech, S. (2008) Strategies for multiobjective genetic algorithm development: application to optimal batch plant design in process systems engineering. *Computers & Industrial Engineering*, **54** (3), 539–569.

19. Aguilar-Lasserre, A.A., BautistaBautista, M.A., Ponsich, A., and Gonzalez Huerta, M.A. (2009) An AHP-based decision-making tool for the solution of multiproduct batch plant design problem under imprecise demand. *Computers & Operations Research*, **36** (3), 711–736.

20. Mokeddem, D. and Khellaf, A. (2009) Optimal solutions of multiproduct batch chemical process using multiobjective genetic algorithm with expert decision system. *Journal of Automated Methods & Management in Chemistry*. doi: http://dx.doi.org/10.1155/2009/927426.

21. Chibeles-Martins, N., Pinto-Varela, T., Barbosa-Povoa, A.P., and Novais, A.Q. (2011) A simulated annealing approach for the bi-objective design and scheduling of multipurpose batch plants. *Computer-Aided Chemical Engineering*, **29**, 865–869.

22. Pinto-Varela, T., Barbosa-Povoa, A.P.F.D., and Novais, A.Q. (2010) Fuzzy-like optimization approach for design and scheduling of multipurpose non-periodic facilities. *Computer-Aided Chemical Engineering*, **28**, 937–942.

23 Tokos, H., Pintaric, Z.N., and Yang, Y. (2012) Bi-objective MINLP optimization of an industrial water network via benchmarking. 11th International Symposium on Process Systems Engineering, Parts A and B. (I.A. Karimi and R. Srinivasan), 31, 475–479.

24 Yue, D. and You, F. (2013) Planning and scheduling of flexible process networks under uncertainty with stochastic inventory: MINLP models and algorithm. *AIChE Journal*, **59** (5), 1511–1532.

25 Pinto-Varela, T. (2015) Scheduling and design of multipurpose batch facilities periodic versus nonperiodic operation mode through a multi-objective approach, in *Synthesis, Design and Resource Optimization in Batch Chemical Plants* (eds T. Majozi, E.R. Seid, and J.-Y. Lee), CRC Press Taylor & Francis.

26 Borisenko, A., Kegel, P., and Gorlatch, S. (2011) Optimal design of multi-product batch plants using a parallel branch-and-bound method. *Parallel Computing Technologies*, **6873**, 417–430.

27 Carvalho, A., Matos, H.A., and Gani, R. (2013) SustainPro-A tool for systematic process analysis, generation and evaluation of sustainable design alternatives. *Computers & Chemical Engineering*, **50**, 8–27.

28 Pinto-Varela, T., Carvalho, A., and Barbosa-Povoa, A. (2014) Framework to batch process retrofit – a continuous improvement approach. 24th European Symposium on Computer Aided Process Engineering, Parts A and B. (J.J. Klemes, P.S. Varbanov, and P.Y. Liew), 33, 1357–1362.

29 WCED (1987) Report of the World Commission on Environment and Development: Our Common Future, Oslo.

30 Elkington, J. (1998) *Cannibals with Forks: The Triple Bottom Line of 21st Century Business*, New Society Publishers, p. 407.

31 Global Report Initiative (2014) G4 Sustainability Reporting Guidelines – Reporting Principles and Standard Disclosures, pp. 25, 52.

32 Doane, D. and MacGillivray, A. (2001) Economic sustainability: the business of staying in business. The Sigma Project.

33 Stamford, L. and Azapagic, A. (2011) Sustainability indicators for the assessment of nuclear power. *Energy*, **36** (10), 6037–6057.

34 Ruiz-Mercado, G.J., Smith, R.L., and Gonzalez, M.A. (2011) Sustainability indicators for chemical processes: I. Taxonomy. *Industrial & Engineering Chemistry Research*, **51**, 2309–2328.

35 ISA-Org (2015) https://www.isa.org/isa95/ (accessed September 2015).

36 Tanzil, D. and Beloff, B.R. (2006) Assessing impacts: overview on sustainability indicators and metrics. *Environmental Quality Management*, **15**, 41–56.

37 Verfaillie, H.A. and Bidwell, R. (2000) Measuring eco-efficiency: a guide to reporting company performance. World Business Council for Sustainable Development.

38 Powell, J.B. (2010) Sustainability metrics, indicators and indices for the process industries, in *Sustainable Development in the Process Industries: Cases and Impact* (eds J. Harmsen and J.B. Powell), John Wiley & Sons, Inc., New York.

39 Azapagic, A., Millington, A., and Collett, A. (2006) A methodology for integrating sustainability considerations into process design. *Chemical Engineering Research and Design*, **84** (A6), 439–452.

40 Al-Sharrah, G., Elkamel, A., and Almanssoor, A. (2010) Sustainability indicators for decision-making and optimisation in the process industry: the case of the petrochemical industry. *Chemical Engineering Science*, **65**, 1452–1461.

41 Ruiz-Mercado, G.J., Gonzalez, M.A., and Smith, R.L. (2013) Sustainability indicators for chemical processes: III. Biodiesel case study. *Industrial & Engineering Chemistry Research*, **52**, 6747–6760.

42 Shokrian, M., High, K.H., and Sheffert, Z. (2014) Screening of process alternatives based on sustainability metrics: comparison of two decision-making approaches. *International Journal of Sustainable Engineering*. doi: 10.1080/19397038.2014.958601.

43 Shadiya, O.O. and High, K.A. (2013) Designing processes of chemical

products for sustainability: incorporating optimization and the sustainability evaluator. *Environmental Progress & Sustainable Energy*, **32** (32), 762–776.

44 Seuring, S. (2012) A review of modeling approaches for sustainable supply chain management. *Decision Support Systems*, **54** (4), 1513–1520.

45 Carvalho, A., Mendes, A.N., Mimoso, A.F., and Matos, H.A. (2014) From a literature review to a framework for environmental impact assessment index. *Journal of Cleaner Production*, **64**, 36–62.

46 ISO14040: (2006) https://www.iso.org/standard/37456.html (accessed April 22, 2017).

47 Ashby, A., Leat, M., and Hudson-Smith, M. (2012) Making connections: a review of supply chain management and sustainability literature. *Supply Chain Management: An International Journal*, **17** (5), 497–516.

48 Guinée, J. (2002) Handbook on life cycle assessment: operational guide to the ISO standards. *The International Journal of Life Cycle Assessment*, **6** (5), 255.

49 Humbert, S., Margni, M., and Jolliet, O. (2005) IMPACT 2002⊠: User Guide. Draft for version 2.1. EPFL e École Polytechnique Fédérale de Lausanne.

50 Young, D.M. and Cabezas, H. (1999) Designing sustainable processes with simulation: the Waste Reduction (WAR) algorithm. *Computers & Chemical Engineering*, **23** (10), 1477–1491.

51 Goedkoop, M., Heijungs, R., Huijbregts, M., Schryver, A.D., Struijs, J., and Van Zelm, R. (2009) ReCiPe 2008, a life cycle impact assessment method which comprises harmonised category indicators at the midpoint and the endpoint level. Report I: Characterisation.

52 Bare, J. (2011) TRACI 2.0: the tool for the reduction and assessment of chemical and other environmental impacts 2.0. *Clean Technologies and Environmental Policy*, **13** (5), 687–696.

53 Littig, B. and Griessler, E. (2005) Social sustainability: a catchword between political pragmatism and social theory. *International Journal of Sustainable Development*, **8**, 65–79.

54 Jayswal, A., Li, X., Zanwar, A., Lou, H.H., and Huang, Y. (2011) A sustainability root cause analysis methodology and its application. *Computers and Chemical Engineering*, **35**, 2786–2798.

55 Meckenstock, J., Barbosa-Póvoa, A.P., and Carvalho, A. (2015) The wicked character of sustainable supply chain management: evidence from sustainability reports. *Business Strategy and the Environment*. doi: 10.1002/bse.1872.

56 Khan, F.I. and Abbasi, S.A. (1998) Multivariate hazard identification and ranking system. *Process Safety Progress*, **17** (3), 157–170.

57 Heikkila, A.M. (1999) Inherent Safety in Process Plant Design: An Index-Based Approach. VTT Publications 384, Technical Research Centre of Finland, Espoo, Finland.

58 Gupta, J.P. and Edwards, D.W. (2003) A simple graphical method for measuring inherent safety. *Journal of Hazard Mater*, **104**, 15–23.

59 Khan, F.I. and Amyotte, P.R. (2004) Integrated inherent safety index (I2SI): a tool for inherent safety evaluation. *Process Safety Progress*, **23** (2), 136–148.

60 IChemE (2002) *Sustainable Development Progress Metrics*, IChemE Sustainable Development Working Group, IChemE Rugby, UK.

61 Simões, M. (2014) Social key performance indicators – assessment in supply chains. Master thesis, Instituto Superior Técnico, Lisboa, Portugal.

62 AspenTech (2008) Performance management for batch processes. www.aspentech.com (accessed July 2015).

63 Jiménez-González, C. and Contable, D.J.C. (2011) *Green Chemistry and Engineering: A Practical Design Approach*, John Wiley & Sons, Inc., Hoboken, NJ. ISBN: 978-0-470-17087-8.

64 Petrides, D., Sapidou, E., and Calandranis, J. (1995) Computer-aided process analysis and economic evaluation for biosynthetic human insulin production: a case study. *Biotechnology and Bioengineering*, **48** (5), 529–541.

65 SuperPro Designer (2008) Example files: Biosynthetic Human Insulin Production.
66 Carvalho, A., Gani, R., and Matos, H. (2008) Design of sustainable chemical processes: Systematic retrofit analysis generation and evaluation of alternatives. *Process Safety and Environmental Protection*, **86** (5), 328–346.
67 Jaksland, C., Gani, R., and Lien, K. (1996) An integrated approach to process/product design and synthesis based on properties–process relationship. *Computers and Chemical Engineering*, **20**, 151–156.
68 Nakatani, M., Sumiyama, Y., and Kusuki, Y. (1994) Pervaporation method of selectively separating water from an organic material aqueous solution through aromatic imide polymer asymmetric membrane. Patent EP0391699.
69 Pinto, T., Barbosa-Povoa, A., and Novais, A.Q. (2008) Design of multipurpose batch plants: a comparative analysis between the STN, m-STN, and RTN representations and formulations. *Industrial & Engineering Chemistry Research*, **47** (16), 6025–6044.
70 Pinto, T., Novais, A.Q., and Barbosa-Povoa, A.P.F.D. (2003) Optimal design of heat-integrated multipurpose batch facilities with economic savings in utilities: a mixed integer mathematical formulation. *Annals of Operations Research*, **120** (1–4), 201–230.

6
Green Chemistry Metrics and Life Cycle Assessment for Microflow Continuous Processing

Lihua Zhang, Qi Wang, and Volker Hessel

6.1 Introduction

6.1.1 Green Chemistry and Green Engineering in the Pharmaceutical Industry

Green chemistry concerns the design of chemical products and processes that reduce or eliminate the use and generation of hazardous substances, increase mass and energy efficiency, and make use of sustainable resources [1]. It primarily means new chemical synthesis strategies, chemistry based on renewable resources, or waste reduction by improved atom economy as well as yield and selectivity improvements. As a counterpart, green engineering is the development and commercialization of industrial processes that are economically feasible and reduce the risk to human health and the environment [2]. It deals with the design of novel equipment and new production methods, using such as multifunctional and hybrid reactors, new operating modes, or microflow processing for process intensification [3]. Thus, the development aims of green chemistry and engineering are to reduce and prevent wastes, environmentally harmful emissions and hazards in both product and process design, and finally to achieve integrated economic, ecological, and social sustainability. Principles of green chemistry and engineering are widely used as general guidelines to assess the greenness and sustainability of a process or products [4,5]. The combination of chemistry and engineering principles could become a key function for greener and more sustainable chemical development.

In 2005, the American Chemical Society (ACS), Green Chemistry Institute (GCI), and several global pharmaceutical corporations established the ACS GCI Pharmaceutical Roundtable (the Roundtable) to encourage the integration of green chemistry and engineering into the pharmaceutical industry, and to catalyze the implementation of both into the business of drug discovery, development, and production [6,7]. Pharmaceutical green chemists and engineers must strive for higher efficiency and enhanced chemical process economics with

Handbook of Green Chemistry Volume 11: Green Metrics, First Edition. Edited by David J. Constable and Concepción Jiménez-González.
© 2018 Wiley-VCH Verlag GmbH & Co. KGaA. Published 2018 by Wiley-VCH Verlag GmbH & Co. KGaA.

reduced environmental burden during chemical synthesis. Therefore, greener, safer, and more sustainable chemistries and processes should be emphasized and implemented throughout every level of the pharmaceutical industry, for example see Ref. [8]:

- Designing efficient processes that minimize the resources (mass and energy) needed to produce the desired product.
- Considering the environmental and health and safety profile of the materials (toxicity, degradability) used in the process.
- Considering the environmental life cycle impacts of the process.
- Considering the economic viability of the process.
- Considering the waste generated in the process.

This involves almost all aspects of chemical syntheses and chemical processes, for example, from raw materials, solvents, and reagents to products, as well as considerations on economy, energy consumption, safety and environmental impact, and so on. Determining whether the selected chemical parameters, processes, or reactions are authentically green and sustainable requires appropriate assessment methods.

According to the principles of green chemistry and green engineering, green metrics should be based on and incorporate the following [9]:

- resource efficiency – mass, energy, waste, atom economy
- the environmental, health, and safety profile of the materials used and the process
- overall life cycle assessment considerations

6.1.2
Green Metrics and Life Cycle Assessment

Against this background, some detailed green metrics enable first qualitative assessments and are often used as screening tools. This embraces, for example, the E-factor [10,11], atom economy (AE) [12], mass-related concepts, for example, mass intensity/process mass intensity/(MI/PMI) [7], mass productivity (MP), mass intensity (MI), carbon efficiency (CE), and reaction mass efficiency (RME) [13,14], as well as solvent rate [15].

These metrics have the advantage that all required data are easily available and can be calculated without special expertise, which is needed for the more sophisticated evaluation methods. However, exactly due to this simplicity, these metrics only consider the amount of starting materials, solvents, reactants, or waste. However, details about the impact on the environment during their supply or disposal are not included. As a consequence, they do not allow conclusions about a potential environmental profile or any comprehensive comparison between different environmental effects.

On the other hand, life cycle assessment (LCA) [16–18] can be used as a holistic evaluation tool to analyze the environmental impacts caused throughout the entire life cycle of a product, process, or activity, from cradle-to-grave [19], or in a shortened version, from cradle-to-(factory)-gate [20]. The LCA methodology is standardized in ISO 14040 [21] and 14044 [22] and is defined as the "compilation and evaluation of the inputs, outputs and potential environmental impacts of a product system throughout its life cycle." The holistic approach includes the whole life cycle of a product or process starting from extraction of resources, the production of all materials, energies used in the system, the usage of products, repair and maintenance, and disposal or recycling or reuse. Along with LCA analysis, life cycle costing analysis (LCC) is also emerging as a preferred decision-making tool in the process industry. This is to be used from the beginning of research and development (R&D) activities until pilot plant processing. LCC is a tool to determine the most cost-effective option among different competing alternatives to purchase, own, operate, maintain and, finally, dispose of an object or process, when each is equally appropriate to be implemented on technical grounds[1]. In this way, the combination of LCA and LCC can give an accurate and holistic overall picture for assessing the environmental and economic performance of chemical processes and products.

In the following section, green metrics, LCC, and LCA methodologies will be reviewed and their applicability for assessing greenness and sustainability in the pharmaceutical and fine chemicals industries will be described.

6.1.3
Continuous Processing at Small Scale

Production in the fine chemicals and pharmaceutical industry is dominated by batch processes, because of the flexibility multipurpose batch chemical manufacturing facilities offer in getting new products to market as fast as possible. However, the purity of products, formation of byproducts, difficulties with scaling up, and possible contamination are areas of ongoing improvement but in some cases the limits of the technology have been approached. On top of that, traditional batch production is facing challenges due to increased product diversity, customer demands, uncertain markets, shortened product lifetime, and fast technology developments [23]. Therefore, radical process improvements are needed that will lead to even more productive and cost-efficient routes, while at the same time maintaining or even improving product quality [24]. As demonstrated over many years in the petro- and bulk chemical manufacturing industry and also in other major manufacturing industries, such as electronic, automotive, and steel manufacturing, continuous processing is an efficient means for making chemical processes more sustainable, ecoefficient, and economical. With the availability of modern process intensification equipment, it made sense to consider continuous manufacturing as a major opportunity for the future green chemistry or green engineering in the fine chemicals and pharmaceutical industry [25,26].

---

1) https://en.wikipedia.org/wiki/Life-cycle_cost_analysis (accessed April 25, 2017).

	BATCH PROCESSING	CONTINUOUS PROCESSING
PROS	Easy verification of product quality; Regulations; Suited to the shifts change; Easier in case of low volume production; Operations like crystallization, drying, precipitation are easily done in batch; Flexible and versatile – important for small volume productions.	Reduced plant size; High processing efficiency; Reduced manufacturing costs; Easier scale up; Better process control – temperature, mass transfer, quality; Unusual operating conditions (Novel Process Windows) – reactions with highly concentrated or explosive mixtures made possible; Reduced inventory, waste and energy requirements;
CONS	Difficult scaling up; Difficult to achieve homogeneous process conditions – changeable temperature, velocity and, concentration profiles inside the equipment; Poor mixing; Safety, health, and environmental issues.	Complex to change processing mode once license has been obtained; Difficult processing of solids – clogging, difficult cleaning; Time and money intensive adaptation to strict batch-favored regulations; Not cost-effective in case of small scale applications when there already is in-ground capital for batch.

Figure 6.1 Batch versus continuous operation. (Reproduced with permission from Ref. [24]. Copyright 2014, American Chemical Society).

In 2007, the ACS-GCI Roundtable developed a list of 10 key green chemistry research areas. Through voting among industrialists, continuous processing has been ranked as top-1 priority within the prime green engineering measures [6]. Continuous processing is well known as a means of improving the safety profile for a given manufacturing process while at the same time increasing materials efficiency[2]. As a result, there has been increasing interest and recognition that it represents a promising alternative to conventional batch manufacturing processes in the fine chemical and pharmaceutical industry [27–29]. Several advantages of continuous processing include the following [6]:

- Economics: lower cost of production
- Quality: improved product quality and consistency
- Safety: enhanced process safety
- Environmental: significant reduction in adverse impacts

On the other hand, despite such high efficiency, large-scale continuous processing is inflexible because of its process specialization compared to modular batch processing facilities that provide greater manufacturing flexibility. However, in recent years there has been increasing development of modular, small-scale continuous equipment that offers the benefits of modularity and flow. The main advantages and drawbacks related to batch and continuous operation modes were summarized by Denčić et al. [24] and are shown in Figure 6.1. Naturally, as pointed out by Kralisch et al. [30], merely transferring batch manufacturing protocols to continuous manufacturing does not necessarily lead to more ecological sound processes.

2) http://www.gcande.org/going-green-with-continuous-chemistry/ (accessed June 3, 2017).

To pave the way toward improved continuous processing that meets the needs of fine chemicals and pharmaceutical industry, recent efforts have focused on the best ways to implement small-scale continuous technology [31–33]. Micro-/millistructured reaction technologies have emerged as serious alternatives to conventional macroscopic methods over the past several years [34]. The typical diameter of the channels in these devices ranges from about 1 mm (milli) down to a few 100 µm (micrometer). The internal volumes of such small-scale reactors, mixers, heat exchangers, and so on can thus range from the milliliter to the microliter scale. While microflow reactors follow the same engineering rules as their large-scale counterparts, their high performance is, first of all, a simple-scale effect. However, due to several orders of magnitude miniaturization, this effect is still considered by many chemical engineers as being somewhat "out of the box" and, therefore, offers many opportunities for greater use and implementation. The heat- and mass transfer in such small-scale processes is greatly enhanced because of the small path lengths and high surface-to-volume ratios. Given that reaction efficiencies for many reactions are highly dependent on adequate heat and mass transfer (e.g., for fast mixing-masked or highly exothermic reactions), the global outcome can be higher yield/selectivity and increased safety. In the bigger picture, reduced manual handling, flexibility in production capacity, easy reproducibility, improved product quality, and shorter development times are inherent benefits of continuous production [35]. Moreover, small-scale processing is a way to circumvent many of the scale-up challenges commonly encountered with batch processing. In the simplest picture, the processing is simply "copied" many times through numbering-up. Realistically speaking, however, before one considers that, a smart scale-out toward miniflow technology needs to be evaluated, as numbering-up, while being a simple approach to scale-up, may need complex flow controls for distribution and monitoring of the whole assembly [36].

The small size of the processing volume also opens the door to new processing windows, which were either formerly not accessible or forbidden due to safety restrictions. This has been coined by Hessel as chemical intensification [37,38] as one approach within the overarching concept of Novel Process Windows [37–43]. The engineering counterpart is process-design intensification that is concerned with process integration and process simplification (see Refs [37,38,42]). While these are common principles for macroscale reactors, the small size and high performance of the reactors gives a new level of compactness and integration and enables a far-reaching process simplification, especially on the downstream side. The greenness of the novel process windows has been addressed some years ago [43].

Although there are many advantages for small-scale continuous processing, some of the challenges should also be mentioned. The manufacture of small scale reactors has a certain expenditure, which needs to pay off. In addition, small-scale equipment might have higher demands during changeover, for example, when purging or cleaning, or it may have shorter service lifetimes because of fouling (clogging) in the micro/millichannels [44]. The main benefit – to achieve fine-tuned microfluidics – is also a disadvantage in the sense that the underlying knowledge is not part of the education curriculum of universities and other teaching institutions so that there is an absence of education/training to prepare chemists and engineers to use this kind of equipment. This is a similar situation that other interdisciplinary subjects, such as nanotechnology, face and companies respond to it by creating teams possessing different key skills from the traditional disciplines within chemistry and engineering.

In order to evaluate the environmental pros and cons as well as economic competitiveness, sustainability and greenness of continuous processing production at small scale compared to traditional batch process or large scale continuous processing, the simplified green metrics, LCC, and LCA methodology can be used as the objective and valid judging tools to provide helpful decision support from the beginning of R&D stages until pilot plant processing, and these will be discussed in the following section.

6.2
Environmental Analysis through Green Chemistry Metrics and Life Cycle Assessment

Taking in mind the information given before about the "green" and "sustainable" potentials, now a closer look is given to the measurement of this greenness and sustainability of microflow continuous processing in the fine chemical and pharmaceutical industry. This will include green chemistry metrics such as mass intensity (MI)/process mass intensity (PMI), E-factor, and atom economy, but also comprehensive life cycle assessment and life cycle costing analysis.

6.2.1
Green Chemistry Metrics

Green chemistry metrics were covered in Chapter 1 of this book and the reader is referred to it for the general definitions. Green chemistry metrics, combined with LCC and LCA, can serve to quantify the environmental performance, efficiency or economy of chemical processes, and allow changes in performance to be measured. The motivation for using metrics is the expectation that quantifying technical, environmental, and economic improvements can make the benefits of new technologies more tangible, perceptible, or understandable. This, in turn, is likely to aid in the communication of research and potentially facilitate wider adoption of green chemistry technologies in industry[3].

Some of the typical green chemistry metrics used, as described in Chapter 1 and elsewhere in the literature are mass intensity/process mass intensity (PMI), mass productivity, reaction mass efficiency, carbon efficiency, solvent rate, environmental factor (E-factor), and atom economy. In addition to these metrics, standard environment, health, and safety metrics and assessments are also

---

3) https://en.wikipedia.org/wiki/Green_chemistry_metrics

typically used to evaluate processes. One example of these traditional metrics is the NFPA rating, which is a general assessment metric for the relative safety and environmental hazards of materials, sponsored by the US National Fire Protection Association (NFPA) [45].

6.2.2
Life Cycle Assessment (LCA)

Life cycle assessment is a tool used to evaluate the impacts of an activity or product over their entire life cycle. The general methodology for life cycle assessment is covered in Chapter 4 of this book, and the reader is referred to it for further detail. Given the particular data needs for an LCA, and the degree of development of microflow continuous processes, conducting LCAs on these systems come with specific challenges.

In practice, several sources of uncertainties exist in early design stages. Especially, when a new process is developed, the required extensive data for the assessment are often not available in the commercial software and database, which restricts the development of detailed LCA study. A simplified LCA (SLCA) (also called "streamlined" LCA) is a suitable alternative at this stage [46]. It needs less time and data input to run the assessment, since it allows for the exclusion of certain life cycle stages, system inputs/outputs or impact categories as well as the use of generic data modules to fill data gaps [47]. However, the initial identification of hot spots within the whole life cycle should result in a comprehensive LCA at the end of the process design step in order to allow for a profound analysis.

Kralisch and colleagues have first investigated the environmental impact and ecological potential of microreaction technology at the stage of process development by means of a simplified LCA methodology several years ago [48]. Some more examples of LCA studies of microreactor processing for the fine chemical and pharmaceutical industry were given by Kralisch *et al.*, Hessel *et al.*, and Lapkin *et al.* [15,46,49–55].

6.3
Application of Green Chemistry Metrics and Life Cycle Assessment to Assess Microflow Processing

Green chemistry metrics and life cycle assessment were used for the following aspects of sustainable process design in continuous microflow processing. Each aspect will be discussed as its own section using different examples as follows:

Reaction level

- Use as a benchmarking tool for continuous versus batch; at lab and production scale
- Use as a decision support tool for single innovation drivers such as the choice of a microreactor, catalyst, or solvent

Pharma process level = reaction + separation

- Use as a decision support tool for bundled innovation drivers such as multi-facetted process optimization versus process intensification
- Use as a benchmarking and decision support tool for cascaded multistep flow syntheses, that is, new process integration and simplification schemes

Fine- and bulk chemical process level = reaction + separation + heat integration + preconditioning and more

- Use for process-design guidance as a benchmarking tool compared to conventional scale processes

6.3.1
Use as Benchmarking Tool for Continuous versus Batch; at Lab and Production Scale

In 2006, Kralisch and Kreisel investigated the pros and cons of microreaction technology using LCA for the first time [48]. First thing to do was to show the greenness of the new approach as compared to classical batch technology.

The comparison was based on a two-step synthesis of m-anisaldehyde from m-bromoanisole (Scheme 6.1) [48]. The main parameters for the batch reactor and the continuous microreactor at laboratory scale and industrial scale are shown in Table 6.1 [48]. A macroscale semicontinuous batch process and a continuous microscale setup at laboratory scale were analyzed. In a second step the impact of the latter was extrapolated to an industrial scale.

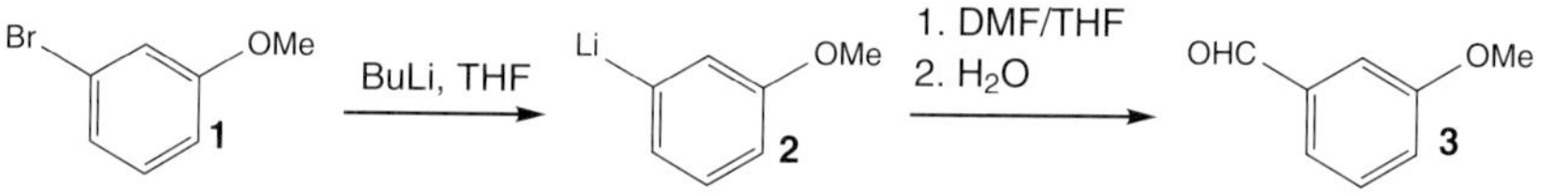

Scheme 6.1 Two-step synthesis of m-anisaldehyde from m-bromoanisole. (Reproduced with permission from Ref. [48]. Copyright 2007, Elsevier).

The relative change in environmental impact categories when switching from macroscale batch to the microflow mode at laboratory scale and industrial scale is given in Figures 6.2 and 6.3, respectively [48]. At laboratory scale, significantly lower environmental impact potentials for many impact categories can be achieved [48]. The decrease within the other environmental impact categories can be related to the savings in energy consumption, the reduction of solvents, and the increase of the reaction yield achieved in the microscale laboratory setup (Figure 6.2). However, the LCA impact categories HTP, FAETP, and TETP are increased as compared to the batch mode. This is due to the emission of chromium and nickel from stainless steel production that is needed for microreactor manufacture (glass being the batch reactor material) (Figure 6.2). The short lifetime of the microscale setup plays a dominant role in this example [48]. In the case of the industrial scale process (Figure 6.3), significant ecological advantages are achieved with continuous synthesis in the microreactor due to the increase

Table 6.1 Batch and continuous microreactor parameters [48].

	Laboratory scale		Industrial scale	
	Batch process	Microreaction process	Batch process	Microreaction process
Reactor	10 l double-walled reactor	Two stainless steel Cytos®-Lab-System modules (100 × 150 × 10) connected in series, product flow of 0.06 kg/h	400 l stainless steel vessel	10 Cytos® microreactors in parallel, product flow of 0.6 kg/h
Temperature (K)	223	273	193	273
Reaction time	Several hours	A few seconds	Several hours	A few seconds
Cooling system	Thermostat	No	Ammonia as primary and nitrogen as secondary cooling media	Electrically tempered
Product yield (%)	60	88	88	88
Product output	10 kg	10 kg	1 ton	1 ton

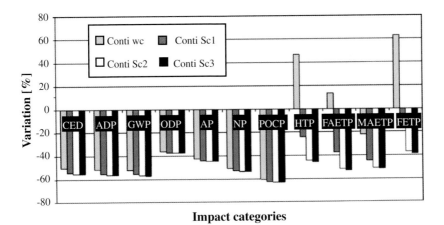

Figure 6.2 Laboratory scale ($Y_{batch} = 60\%$, $Y_{conti} = 88\%$, FU = 10 kg m-anisaldehyde)-variation of environmental impact potentials as a consequence of the change from macroscale batch (0%) to microreaction mode; four scenarios regarding the lifetime of the microstructured devices (conti wc: 1 week, conti Sc1: 3 months, conti Sc2: 3 years, conti Sc3: 10 years). (Reproduced with permission from Ref. [48]. Copyright 2007, Elsevier).

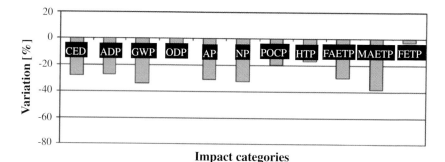

Figure 6.3 Industrial scale ($Y_{batch,conti}$ = 88%, FU = 1 ton *m*-anisaldehyde) – variation (%) within the impact categories as a consequence of the change from macroscale batch (0%) to microreaction mode. (Reproduced with permission from Ref [48]. Copyright 2007, Elsevier).

of the reaction temperature by 80 K (Table 6.1, reaction temperature of 193 K in the batch process and 273 K in the continuous microreaction process.) and thus the avoidance of a cryogenic system. Due to the much longer lifetimes of typical chemical manufacturing facilities, the energy and material demand associated with the fabrication of the microstructured reactors is negligible. Therefore, in this case the use of a microreactor for production results in a better profile for all environmental impact categories.

In order to analyze the influence of single module steps on the overall result, the influence of the representative environmental impact category, cumulative energy demand (CED), associated with the reaction chemicals and the energy demand during synthesis and workup at laboratory scale and industrial scale were presented [48]. At laboratory scale, the electrical supply of the synthesis unit plays a major role, while the supply of the work-up and distillation equipment exhibits a minor effect for both the batch and the microreactor. In the case of the microreactor, the shorter processing times and the higher yield results in a lower CED demand for both types of equipment microreactor. Another major contribution comes from the reaction materials and solvent used. Again, the better yield lowers the CED impact of the microreactor. A third minor contribution is associated with the production of less waste in case of the microreactor that leads to lower respective CED impact.

At an industrial scale, a similar picture is provided for the three contributions mentioned already – the energy needed for synthesis and work-up, for making the starting materials and the solvent, and for the waste disposal. For the reason mentioned, the microreactor has fewer environmental impacts associated with its use. An additional facet is provided by the large CED demand needed for making the liquid nitrogen that is used for the cryogenic operation. Since this is not required for the microreactor, a large environmental benefit is associated with just this one feature.

In the case of the conventional industrial batch production, the supply of the educts and solvents account for 49% of the total impact. The supply of liquid nitrogen for the cryogenic system amounts to 36%, and the electricity consumed during the reaction process comes to 10% of the overall CED. The CED for the microreactor processing is dominated by the impacts associated with the chemicals used in the process (67%) and the electricity demand (27%) [48].

From above, it is evident [48] that

- significant ecological advantages can be gained from the microreaction technology in comparison to the macroscale batch technology. This outcome can be obtained for the laboratory scale syntheses as well as for the industrial scale process,
- on the laboratory scale, the advantages consist in the savings in energy consumption, the reduction of solvents, and the increase of the reaction yield achieved in the microscale setup,
- on the industrial scale, the avoidance of a cryogenic system by increasing the reaction temperature is the most important feature, and the fabrication of the reactors and the peripheral equipment only plays a minor role, and
- microreaction technology can be a promising way to achieve increased ecological sustainability of production processes.

6.3.2
Use as Decision Support Tool for Single Innovation Drivers – Choice of Type of Microreactor and Type of a Catalyst (Including Use/Not Use)

Once it is clear that continuous microflow processing brings environmental benefits, the next question might be how to optimize that through the right choice of single innovation drivers. The most straightforward choice is that of the microreactor itself.

Besides the reactor, the type of catalyst is of crucial importance. The main difference in the choice of catalysts when compared to batch processes is typically found in the order(s) of magnitude higher reaction rate that is demanded of the catalyst, and that might require new catalyst development. This kind of environmentally driven selection is shown here by the example of the phase transfer catalysis of benzoyl chloride reacting with phenol to yield phenyl benzoate.

Phenyl benzoate is an industrially relevant chemical substance, which is used for polymer modification, as an antioxidant [56], and as an intermediate in the production of liquid crystals [57]. A SLCA, together with a cost evaluation, was developed for the phenyl benzoate synthesis by Huebschmann et al. [49].

6.3.2.1 Reaction Conditions of Batch Process and Continuous Microflow Process

The reaction conditions and yields of batch and continuous microflow process are summarized in Table 6.2 [49]. As phase transfer catalysts, the following ionic liquids were used: 1-butyl-3-methylimidazolium chloride [BMIM]Cl, 1-octadecyl-3-methylimidazolium bromide [C_{18}MIM]Br, and 1-butylsulfonate-3-methylimidazolium [MIM]BuSO$_3$. Three different glass micromixers are chosen to

Table 6.2 Reaction conditions and yields used for SLCA investigating the phase transfer catalysis of phenol and benzoyl chloride giving phenyl benzoate.

Entry	Catalyst	T/°C	Mixing structure	Residence time (s)	Yield (%)
1	No catalyst	57	Interdigital mixer	0.83	10
2	No catalyst	57	Herringbone mixer	18.8	26
3	[C_{18}MIM]Br	62	Herringbone mixer	18.8	45
4	[C_{18}MIM]Br	68	Emulsification mixer	1.355	51
5	[C_{18}MIM]Br	75	Interdigital mixer	0.83	62
6	[MIM]BuSO$_3$	30	Interdigital mixer	0.83	70
7	No catalyst	66	Batch synthesis	5400 (90 min)	66
8	[BMIM]Cl	45.5	Batch synthesis	2700 (45 min)	72

Source: Reproduced with permission from Ref. [49]. Copyright 2011, Royal Society of Chemistry.

investigate the influence of the interfacial area between the two phases, and two batch syntheses, without a catalyst and with the ionic liquid in catalytic amounts that are conducted to reveal the phase transfer catalytic potential of ionic liquids.

6.3.2.2 SLCA Results

The toxicity potentials as proposed by CML [58] for the discussed reaction conditions are shown in Figure 6.4. The highest toxicity potentials are caused by the microflow synthesis in an interdigital mixer without the use of a phase-transfer catalyst (entry 1). Under these conditions, yields are low. A herringbone-structured mixer results in a much longer residence time and a higher yield, which in turn affords a better environmental profile (entry 2). By speeding up the reaction through the use of a better phase-transfer catalyst in combination with the high mixing efficiency of the interdigital mixer affords an even better environmental profile that is as good as the batch case (entry 6). However, the residence time for the flow process differs by an order of magnitude, which means that the micromixer has a much higher space-time yield (entry 7).

Then the human toxicity potential (HTP), the global warming potential (GWP), and the cumulative energy demand (CED) are chosen for a more detailed analysis of environmental impacts of the biphasic reaction of benzoyl chloride and phenol giving phenyl benzoate. The results show that the supply of phenol and the energy consumption of the periphery(without reaction) in the continuously running processes give the highest contribution to the HTP, and the latter also exhibits the highest influence on GWP. Further, reactions without the implementation of a phase-transfer catalyst (entries 1 and 2) reveal a higher HTP than the reactions implementing ionic liquids as catalysts. However, without the implementation of a catalyst, the herringbone microstructured mixer is more advantageous than the interdigital mixer. In both HTP and GWP impact categories, the interdigital mixer performed best under phase-transfer catalytic conditions implementing [MIM]BuSO$_3$(entry 6) in the continuous syntheses,

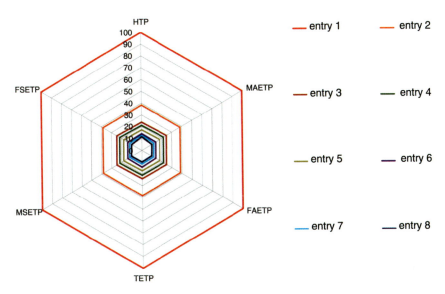

Figure 6.4 Overview of various toxicity potentials of the biphasic reaction of benzoyl chloride and phenol giving phenyl benzoate (HTP – human toxicity potential, MAETP – marine aquatic ecotoxicity potential, FAETP – fresh water aquatic ecotoxicity potential, TETP – terrestrial ecotoxicity potential, MSETP–marine sediment ecotoxicity potential, FSETP – fresh water sediment ecotoxicity potential). (Reproduced with permission from Ref. [49]. Copyright 2011, Royal Society of Chemistry).

which are comparable to the reactions with and without catalyst in batch mode laboratory scale (entries 7 and 8).

The ionic liquids have only a minor influence on the environmental impact. This is because [BMIM]Cl, [C_{18}MIM]Br, and [MIM]BuSO$_3$ were implemented only in catalytic amounts instead of being used more commonly as solvents. The results of overall life cycle impacts show that [BMIM]Cl is the most beneficial ionic liquid used as phase-transfer catalyst due to its plain reactants and the easy synthesis in comparison to [C_{18}MIM]Br and [MIM]BuSO$_3$ [49].

In addition, considering all energy consumed during the whole processing chain from the supply of raw materials and electrical current, the reaction itself, as well as the workup and the disposal of wastes, CED results indicate that under similar conditions, that is, under the same energy demand for pumping and controlling, the continuous syntheses would be two to three times more beneficial from an ecological point of view. It is worth mentioning that the workup procedure considered increases the environment burdens of the process considerably [49]. The result is mainly due to the evaporation of water from the aqueous phase prior to the disposal of the organic residues, including the ionic liquids. Future work is focused on the substitution of the aqueous phase by an ionic liquid (both a solvent and a phase transfer catalyst) allowing for an automated phase separation and recycling of the ionic liquid.

6.3.2.3 Economic Evaluation

As was foreseeable from a consideration of the laboratory-scale performance, the total costs for producing 1 kg of phenyl benzoate by the continuous phase-transfer catalysis process were higher than the selling price of this substance. However, considering a batch process on the same laboratory scale (applying [BMIM]Cl as the phase-transfer catalyst), the total cost would be 4.5 times higher than the selling price due to very high labor costs. The total costs for the continuous synthesis process at the laboratory scale are composed of 43.8% fixed costs, 43.7% labor costs, and 12.5% material costs (Figure 6.5) [49].

By estimating the costs for the continuous synthesis using $[MIM]BuSO_3$ (entry 6) and comparing these to the batch reaction using [BMIM]Cl as a phase-transfer catalyst (entry 8), the following can be concluded [49]:

- The biphasic esterification carried out continuously at laboratory scale is more economically preferable than the batch reaction at laboratory scale.
- The main cost drivers in the base case are fixed and labor costs, which may become negligible by enhancing the production capacities.

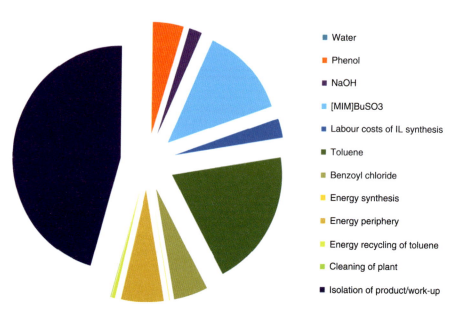

Figure 6.5 Variable costs (without labor costs of the esterification) for the continuously running synthesis of phenyl benzoate at laboratory scale. (Reproduced with permission from Ref. [49]. Copyright 2011, Royal Society of Chemistry).

- A reduction in 99% of total costs seems to be possible by transferring the continuous biphasic esterification from laboratory scale to industrial scale.

6.3.2.4 Conclusions

The environmental burdens of the synthesis of phenyl benzoate were largely influenced by the reaction yields obtained, as higher yields lead to less process waste and to a more efficient use of resources. Thus, the energy consumed by the microreactor plant had the highest contribution to the global warming potential, whereas the human toxicity potential was almost equally dominated by the energy consumption and the presence of phenol. The performance of different micromixers has also been found to be dependent on another process intensification parameter, the phase-transfer catalyst. Processing with the interdigital micromixer in combination with [MIM]BuSO$_3$ was shown to have the lowest environmental impact among all continuous syntheses processes. This result matches the expectation because under the given flow conditions this is supposed to be the best mixing device [49].

Additionally, the economic potential of the processes was demonstrated, although at a screening level. The superficial cost estimation of the laboratory-scale synthesis of phenyl benzoate showed that under laboratory-scale conditions the batch syntheses resulted in higher costs due to higher manpower requirements. Under industrial scale conditions, syntheses with the highest yields will be the most cost-efficient, since the total costs will be influenced mainly by the cost of raw materials. Fixed costs, especially investment costs, will have then a minor share [49].

6.3.3
Use as Decision Support Tool for Single Innovation Drivers – Solvent Choice and Role of Recycling

Once it is clear which reactor and catalyst system has the best environmental performance profile for continuous microflow processing, the next choice for a single innovation driver might be the solvent. The pharmaceutical industry has put enormous focus on the use of solvents as a major issue within their sustainability roadmaps [59].

With this as motivation, Yaseneva *et al.* investigated the synthesis of the API artemether in a microreactor and in a batch reactor process using life cycle assessment [54]. Artemisinin is the WHO-recommended drug for treating uncomplicated *Plasmodium falciparum* malaria, and is obtained by extraction from the plant *Artemisia annua* [60]. The active pharmaceutical ingredients of the artemisinin-based drugs are ethers, such as artesunate or artemether, which are obtained via reduction of artemisinin into dihydroartemisinin (DHA) using sodium borohydride in methanol or ethanol as solvents, followed by etherification to the final APIs artemether (ARM) (see Scheme 6.2) [61,62].

The synthesis of DHA from artemisinin and artemether from DHA were both carried out in batch reactors and in microreactors. Cradle-to-gate life cycle

Artemisinin Dihydroartemisinin β–Artemether

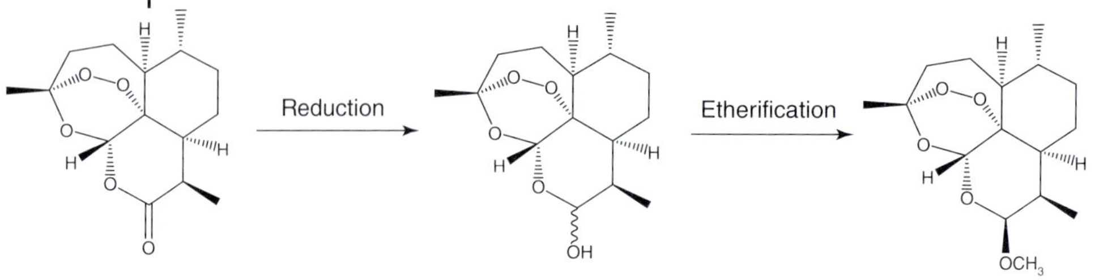

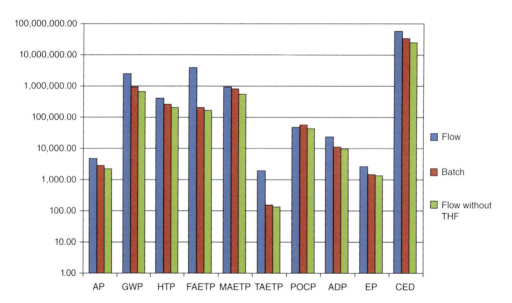

Scheme 6.2 A common route to conversion of artemisinin to artemisinin-based APIs, exemplified by artemether. (Reproduced with permission from Ref. [54]. Copyright 2015, Elsevier).

assessment was performed to reveal the impact of the conversion of artemisinin through a catalytic etherification when the established batch protocols were switched to a new flow process.

In the case of the reduction of artemisinin to DHA, CML impacts and cumulative energy demand (CED) for three alternatives were determined: reduction of artemisinin to DHA using a conventional batch process, reduction carried out in a flow process using Me-THF as the solvent, and a superhydride reducing agent, shown in Figure 6.6 [54]. The introduction of the more complex solvent and the

Figure 6.6 CML impact scores and CED for artemisinin to DHA reaction. Comparison of flow and batch processes and flow process without THF for superhydride. (Reproduced with permission from Ref. [54]. Copyright 2015, Elsevier).

reducing agent resulted in a significant increase in the calculated environmental impacts and CED. However, it is predicted that a significant reduction of CED would be achieved if the reducing agent is recycled.

To illustrate the significance of the solvent on the life cycle environmental impacts in the case of etherification of DHA to ARM, the authors performed a scale-up analysis of the flow process with and without solvent recycle. Results are shown in Figure 6.7 for the CML impacts and CED [54].

The impacts of the flow process appears to be worse than that of batch. However, the solvent is the most dominant impact in the case of the flow process. Therefore, reduction of the solvent inventory through in-process recycling results in a significant reduction in the impacts of the flow process, making it advantageous over batch technology [54].

A detailed LCA study of both reaction steps shows the importance of the choice of solvents in pharmaceutical processes: replacement of THF by Me-THF, as well as introduction of solvent recycle leads to significant decrease in the LCA impacts of the flow processes, making them cleaner than the best literature batch reactions [54].

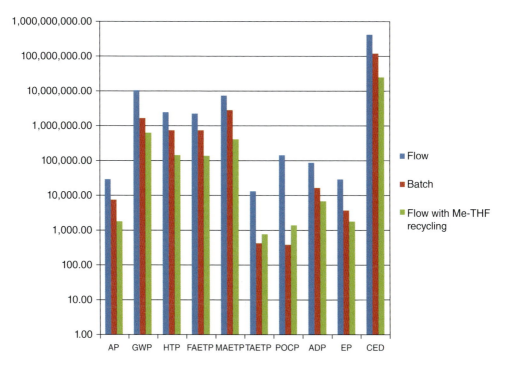

Figure 6.7 CML impact scores and CED for DHA to ARM reaction. Comparison of flow and batch processes. (Reproduced with permission from Ref. [54]. Copyright 2015, Elsevier).

6.3.4
Use as Decision Support Tool for Bundled Innovation Drivers Such as Multifacetted Process Optimization versus Process Intensification

Once a reaction is optimized in microflow through all major innovation drivers, the whole process level, that is, addressing all the other unit operations, needs to be addressed. On a pharma level, this might involve other reactions and will certainly involve separation and purification. In the case of fine chemical manufacturing, other unit operations need to be considered such as preconditioning, heat integration, and more. This will be discussed in Section 6.3.6 using the example of adipic acid manufacture. Coming back to the pharmaceutical API manufacturing case there is a need to consider purification and other reactions, and in all probability not all will be done under continuous microflow conditions. In some cases, batch processing will be best and in other cases, continuous processing options will be best. Thus, normal process optimization and process intensification needs to be combined into an integral process development approach and any environmental analysis needs to consider that.

With this background, Sanofi Company, as is virtually the case for all pharmaceutical companies, is interested in more productive, competitive, and simultaneously environmentally benign process routes through the application of green chemistry, green engineering, and process intensification principles. In this context, Ott *et al.* used LCA to compare two processing options for the synthesis of a Sanofi Company API – by a continuous millireactor-based process and by a batch process [52]. In addition to a consideration of transferring the process from a batch process to continuous processing, alternative catalytic systems and different process options, especially on the downstream side, were included in the environmental impact analysis. The main aim was to identify bottlenecks and potential areas for further process improvement and development activities. This analysis provided a real-life scenario of how existing process optimization approaches (which are done with classical equipment) can be supplemented and strengthened by process intensification and the use of new kinds of equipment (such as microflow reactors).

6.3.4.1 API Production Process at Sanofi
The generic API production process is shown in Figure 6.8 [52]. The main bottlenecks of the process are: (1) low reaction efficiency, (2) high solvent consumption, (3) fresh catalyst is required for each batch, (4) high losses of the desired product (KOM) along the process chain (mainly due to purification and crystallization steps), and, as a result (5) a high waste production rate [52].

To resolve these bottlenecks, some different optimization strategies are as stated in the following section.

6.3.4.2 Process Alternatives for Optimization and Intensification
Tables 6.3 and 6.4 provide an overview of the different batch and continuous process alternatives considered herein [52]. This allows an investigation of the

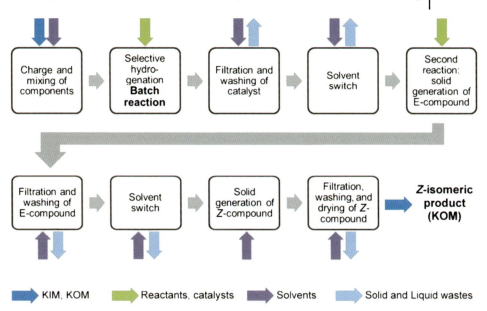

Figure 6.8 Simplified flow sheet of the API batch manufacturing process investigated in this study; Z-isomeric compound as economically valuable key output material (KOM). (Reproduced with permission from Ref. [52]. Copyright 2014, Wiley-VCH Verlag GmbH).

environmental effects associated with single- and coupled-process parameter variations. For simplification, the reference process at Sanofi is called the actual process (AP), whereas the new process options comprise batch process (BP) and continuous process (CP) alternatives.

6.3.4.3 Ecological Profile Comparison of Crude Batch and Continuous Operation

In Figure 6.9, the LCIA for the batch (AP, see Table 6.3) and the current continuous operation processing are compared to a benchmark catalyst (CP1, see Table 6.4) [52].

The execution of the current continuous processing (CP1) with a newly developed catalyst results in a yield enhancement, yet at the price of higher solvent load and higher energy consumption. Thus, it does not bring environmental benefits with regard to the impact categories such as GWP, TETP, NLTP, and FDP. However, significant benefits are revealed within other LCIA impact categories, such as HTP, POFP, MDP, TAP, and FEP. As a whole it was found, despite the higher solvent and energy consumption in the continuous reaction, that changing the catalyst in combination with a yield enhancement outweigh the negative environmental burdens associated with the current state of development.

Starting from the current batch process AP, alternative continuous-mode processes were evaluated comparatively and the results are shown in Figure 6.10. Transferring from a batch to continuous reaction process without any further optimization results in 1 and 41% reduction of GWP and HTP, respectively [52].

Table 6.3 Summary of the main process characteristics of considered batch scenarios.[a]

Process	Operating T (°C)	Solvent	Initial KIM c (g/L)	Reducing agent	Catalyst	Crystallization method	Z-separation + reduction of KOM losses	KOM drying	Overall yield (%)
AP	30	CAN	87	NH_4HCO_2	Pd@C powder, 5 wt% Pd	Indirect	—	Vacuum	47
BP 1	30	CAN	87	NH_4HCO_2	Pd@C powder, 5 wt% Pd	Direct	—	Vacuum	50
BP 2	50	MeOH/H_2O	71	$Na_2S_2O_4$	—	Indirect	—	Vacuum	42
BP 3	50	MeOH/H_2O	71	$Na_2S_2O_4$	—	Direct	—	Vacuum	44
BP 4[b]	30	CAN	87	NH_4HCO_2	Pd@C powder, 5 wt% Pd	Indirect	X	Vacuum	54
BP 5[b]	30	CAN	87	NH_4HCO_2	Pd@C powder, 5 wt% Pd	Indirect	—	Microwave	47
BP 6[b]	30	CAN	30	NH_4HCO_2	Pd@C powder, 5 wt% Pd	Indirect	—	Vacuum	47

a) Common process conditions: $p = 1$ bar; reaction time = 24 h; reactor volume, $V = 90$ l.
b) Hypothetical scenarios for demonstration of effects of parameter variations, based on Aspen simulation and/or estimations.

Source: Reproduced with permission from Ref. [52]. Copyright 2014, Wiley-VCH Verlag GmbH.

Table 6.4 Summary of main process characteristics of continuous process alternatives.[a]

Process	Operation mode	Catalyst	Crystallization method	Z-separation and reduction of KOM losses	Overall yield (%)
CP 1	Continuous reaction	Pt/V@C	Indirect	–	52
CP 2	Continuous reaction	Pt/V@C	Direct	–	55
CP 3[b]	Continuous reaction	Pt@ZnO	Indirect	–	52
CP 4[c]	Continuous reaction	Pt/V@C	Indirect	×	60
CP 5[c]	Continuous reaction	Pt/V@C	Direct	×	64
CP 6[d]	Continuous process	Pt/V@C	Indirect	–	52

[a] Common process conditions: $T = 55\,°C$; $p = 35$ bar; liquid flow rate, $F_l = 41$ mLmin^{-1}; residence time 9 s; estimated reactor volume, $V = 10$ mL (for H-Cube or similar) for assumed bed porosity of 60% (Pt/V@C) (bed porosity of Pt@ZnO still to be determined); reaction solvent: ACN; KIM initial concentration $c_0 = 30$ g L^{-1}; reducing agent: H$_2$; catalyst: fixed bed; Pt/V@C: 1 wt% Pt/2 wt% V @C, Pt@ZnO: 2 wt% Pt @ZnO; KOM drying by means of a vacuum dryer. Liquid flow rate, F_l, calculated according to $P = c_0 F_l$; p: targeted productivity (kg h^{-1}), c_0: KIM initial concentration.
[b] In accordance to CP 1, hypothetical scenario considering newly developed hydrogenation catalyst.
[c] Hypothetical scenarios, forecasted to demonstrate the effect of optimized continuous processing.
[d] Hypothetical scenario, RM demand adapted to known process output, based on Aspen simulation.
Source: Reproduced with permission from Ref. [52]. Copyright 2014, Wiley-VCH Verlag GmbH.

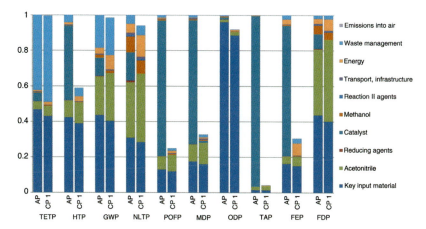

Figure 6.9 Comparison of ecological fingerprints of a conventional batch API manufacturing process (AP) and a novel continuous reaction process (CP1) in the development state. Scaled effects. (Reproduced with permission from Ref. [52]. Copyright 2014, Wiley-VCH Verlag GmbH).

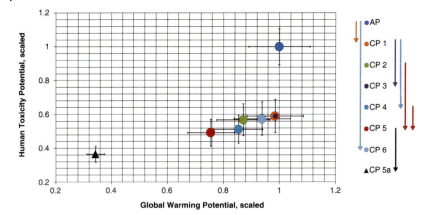

Figure 6.10 Effect of single- and multi-parameter variations within continuous operation mode scenarios compared with the batch benchmark process AP (see Tables 6.3 and 6.4). Scaled effects of GWP and HTP. → Effect of transition from batch to continuous reaction; → effect of transition from batch to full continuous processing; → effect of direct crystallization; → effect of catalyst change; → effect of KOM recovery; → effect of direct crystallization and KOM recovery; → effect of solvent recycling.

Some conclusions can be obtained from Figure 6.10 [52]:

- A process that included a continuous reaction followed by a downstream processing option (direct crystallization) could be readily implemented (scenario CP2, see Table 6.4). An additional reduction in GWP and HTP by 12 and 4%, respectively, can be gained this way (compared to CP1).
- Pt@ZnO can be used as another promising and stable hydrogenation catalyst that is favored, for example, over Pt/V@C
- The share of the GWP and HTP impact associated with the catalyst used in the alternative batch processing options is about 10 and 42%, whereas in the continuous processing alternatives it is < 0.1%. In general, this is due to (1) higher efficiency, (2) reduced consumption, and (3) lower environmental burdens within almost all impact categories compared to using Pd@C in a slurry batch reactor.

Overall, the study showed that existing process optimization options for batch processing can be coupled to new continuous, heterogeneously catalyzed processes, and supplement each other. This, for example, holds for the coupling of direct crystallization and other improved downstream processing. Another challenging task, but highly profitable from an environmental point of view, concerns the recycling and reuse of organic solvents, and represents the current best case already described (CP5, see Table 6.4). This is expected to be implemented at the Sanofi site.

6.3.4.4 Cost Analysis of Batch and Continuous Operation

The LCA analysis was further coupled with a cost analysis to forecast eco-efficiency. Figure 6.11 gives a comprehensive overview compounding the ecological and economic results [52]. The assessment of economic impact was based on Ref. [24]. The higher the value in cost and environmental efficiency, the larger the preference for the appropriate scenario. The following Pareto ranking of

6.3 Application of Green Chemistry Metrics and Life Cycle Assessment to Assess Microflow Processing

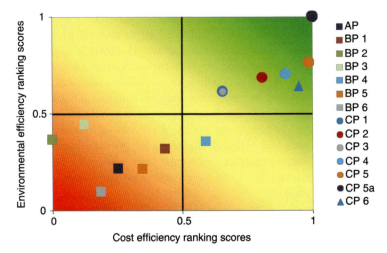

Figure 6.11 Eco-efficiency ranking supported by D-Sight [63]: multicriteria outranking of KOM production alternatives, according to their environmental and cost efficiency. D-Sight parameters: criteria weights – equal, linear minimization of LCIA criteria, and costs. (Reproduced with permission from Ref. [52]. Copyright 2014, Wiley-VCH Verlag GmbH).

process design options was found: CP5a > CP5 > CP4 < > CP6 > CP2 > CP1 ~ CP3 > BP3 < > BP4 > BP1 < > BP2 > BP5 > AP > BP6 ($x > y$: x is preferred to y, $x < > y$: x is noncomparable with y, $x \sim y$: x is indifferent to y). Thus, a clear order of preferences could be concluded for most of the alternative processes considered. For CP4 and CP6, BP3 and BP4, and BP1 and BP2, the order may change based on a weighting priority toward environmental protection or cost savings.

6.3.4.5 Conclusions
The following conclusions can be drawn [52]:

- The LCIA categories of the Sanofi API production process were mostly influenced by the supply of key input material (KIM), solvent, and catalyst. The following strategies show promise for further ecologically related improvements to the process development: yield enhancement, reduced solvent demand or solvent recycling, and fixation/change of catalyst. In particular, multiparameter variations, including improved downstream processing, are recommended.
- The transition from batch to continuously running operation mode by using fixed catalysts improves the overall ecological sustainability, although the concentration of KIM in the solvent needs to be reduced. The positive effect is, however, an indirect cause, since it is mainly due to changing the catalyst.

6.3.5
Cascading Reactions Into a Microreactor Flow Network – Greenness of Multistep Reaction/Separation Integration

Having insight about the greenness of the transformation from batch to continuous microflow synthesis for a single reaction step (as given in Sections 6.3.1–6.3.4

180 | *6 Green Chemistry Metrics and Life Cycle Assessment for Microflow Continuous Processing*

and with or without purification), the next big step is to consider a whole production chain in flow. In this case, it is probably best to start off with a bulk chemical molecule. Eminent questions for this case are (1) if there is an environmental advantage obtained by cascading several flow reactions into a microreactor flow network or is it better to perform those reactions as separate, isolated flow reactions and (2) how does flow multistep synthesis compare to batch multistep synthesis. This means that the greenness of process-design intensification is addressed after evidence has been given for the greenness of transferring the reaction to flow conditions and chemical intensification [38,64]. This is only a first step in mimicking nature's complex bioassembly lines. Among the next steps are a consideration of reaction space compartmentalization and the inclusion of homogeneous and biocatalysts.

In this context, Ott *et al.* [15] investigated the environmental profile of a multistep batch to flow conversion for the rufinamide synthesis, yielding a fully continuous microreactor network. The evaluation was done using simplified metrics (e.g., PMI, solvent rate, CED) and holistic LCA methodology.

Rufinamide is an antiepileptic drug used in combination with other medications to treat the Lennox-Gastaut syndrome. Rufinamide is one of the best-selling five-membered ring heterocyclic pharmaceuticals [65]. The common route to synthesize rufinamide starts from nitrobenzene as the bulk chemical and proceeds through nine steps (see Scheme 6.3) to the final drug. The ecological impacts associated with the first five steps were included, but the scope of the study encompassed specific variations for the last four synthetic steps leading to rufinamide: chloridehydroxylation (step 6), azidation (step 7), [3 + 2] Huisgen

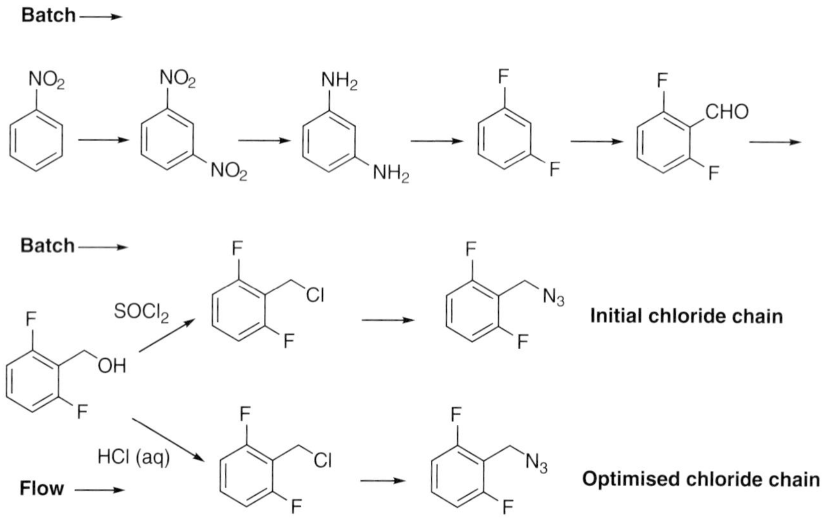

Scheme 6.3 Schematic comparing the literature batch route (via "initial chloride chain") and the author's flow route (via "optimized chloride chain") to 2,6-difluorobenzyl azide. (Reproduced with permission from Ref. [15]. Copyright 2015, Royal Society Chemistry).

cycloaddition (click chemistry, step 8), and amidation (step 9). These are typical fine-chemical/medicinal-chemical batch reactions that may profit from conversion to a microflow system.

In order to optimize these processes, the following two targets were set [15]:

- To investigate the conversion of a batch process to flow conditions for each of these four steps, from benzyl alcohol toward rufinamide, based on experimental data of diverse kinds of flow intensification.
- To bring these steps together to create a multistep microreactor flow network.

6.3.5.1 LCA Study for Single-Step Analyses in Batch and Flow

Based on the first target, three process optimizations were proposed to intensify single steps (before connecting them) [15]:

1) Choice of the chloride route to make 2,6-difluorobenzyl azide, which is the greenest option when compared to using bromide or iodide.
2) The replacement of the very reactive, yet toxic thionyl chloride (batch–batch synthesis) by less reactive aqueous, yet greener HCl (continuous flow–flow synthesis) (see Scheme 6.4) in the steps 6 and 7 of the whole rufinamide synthesis.
3) The choice of (E)-methyl-3 methoxyacrylate (EMMA) as the dipolarophile for Huisgen cycloaddition.

A comparative analysis of the 10 LCIA impact categories for the batch-to-batch processes and the new optimized flow–flow chains is given in Figure 6.12. The abovementioned optimizations significantly reduce the overall environmental impact by approximately 30–90%. This is mainly due to the higher yields (88%) compared to the literature overall yield of 65%. It is also due to the avoidance of materials like dichloromethane and thionyl chloride as well as sulfur dioxide emissions in the alcohol synthesis, which dominate the environmental burdens within ODP and TAP [15].

For the following triazole ring-closure step, the use of (E)-methyl-3-methoxyacrylate (EMMA) as a dipolarophile results in a number of advantages. Some simplified green chemistry metrics along with a holistic LCIA and an economic evaluation are introduced by Ott et al. [15]. Green metrics, including the NFPA rating, PMI, and solvent rate, are used as typical screening tools for the initial selection. The central step in the rufinamide synthesis is the 1,3-dipolar Huisgen cycloaddition. Four choices were considered for the dipolarophile in this reaction (see Table 6.5) [15]. All data presented here reflect batch performance and the aim is to select the most promising dipolarophile for further investigation in microflow processing.

For the hazard potential based on the NFPA rating, (E)-methyl-3-methoxy acrylate EMMA ranks best, taking into consideration the complete package dipolarophile/solvent/catalyst. The solvent used for the EMMA workup, methanol, turned out to be environmentally friendlier compared to the solvents used for the other dipolarophiles that are ethyl acetate and ethyl ether. Focusing on

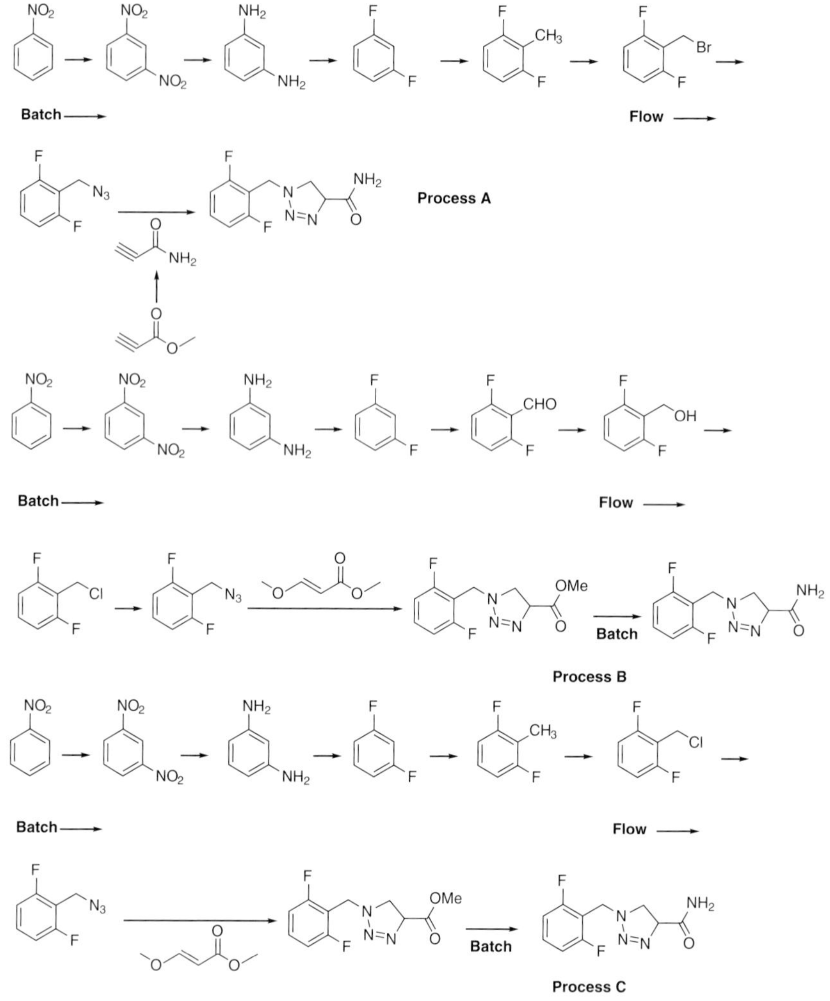

Scheme 6.4 Comparison of rufinamide process routes. Process A developed by Zhang et al. [66], Process B developed by Ott et al. (Reproduced with permission from Ref. [15]. Copyright 2015, Royal Society Chemistry).

the NFPA rating only, the dipolarophile EMMA itself features a lower hazard than the others. Concerning process and economic aspects, though EMMA (entry 3) is the cheapest option among the four, entries 1, 2, and 4 result in a higher overall yield and a shorter reaction time. However, the other three dipolarophiles require either a catalyst or subsequent separation of the undesired regioisomer that leads to a higher PMI and solvent rate values compared to the route via EMMA. Thus, entry 3 using the dipolarophile (E)-methyl 3-methoxyacrylate for the Huisgen cycloaddition was chosen as the most promising

6.3 Application of Green Chemistry Metrics and Life Cycle Assessment to Assess Microflow Processing

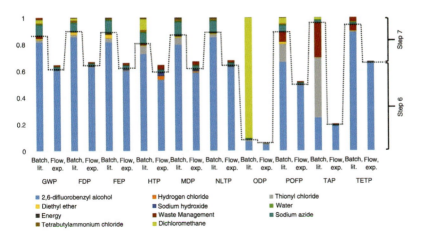

Figure 6.12 Reduction of LCIA categories for 2,6-difluorobenzyl azide supply by changing its upstream process pathways. (Reproduced with permission from Ref. [15]. Copyright 2015, Royal Society Chemistry).

Table 6.5 Overview on common rufinamide batch process routes.

Process options	Entry 1	Entry 2	Entry 3	Entry 4
1,3-dipolar Huisgen cycloaddition				
Dipolarophile assessment	Methyl propiolate	Propiolic acid	(E)-Methyl-3-methoxy-acrylate (EMMA)	2-Chloroacrylonitrile
NPFA rating	2;3;0	3;2;0	2;2;0	4;3;2
Costs ($/mol)	45	13	2	12
CED median value (MJ/kg)]	60	55	271	94
Solvent for synthesis	Ethanol	tBuOH/H_2O	Neat	Water
NFPA rating	0;3;0	1;3;0	—	—
Solvent for work-up	Ethyl acetate	Ethyl ether	Methanol	Cyclohexane
NFPA rating	1;3;0	1;4;1	1;3;0	1;3;0
Catalyst	$CuSO_4 \cdot 5\,H_2O$	$CuSO_4$/ascorbic acid	—	—
Reaction temperature (°C)	25	40	135	80
Reaction time (h)	12	2	28	24
Yield to rufinamide precursor (%)	94	94	85	86
Overall yield to rufinamide (%)	82	86	79	85
Solvent for work-up	Ethyl acetate	Ethyl ether	Methanol	Cyclohexane
Process metrics				
PMI w/o H_2O (kg/kg)	23 ± 6	59 ± 10	18 ± 2	4
PMI incl H_2O (kg/kg)	39 ± 13	76 ± 10	19 ± 3	30
Solvent rate (kg/kg)	16 ± 4	36 ± 10	14 ± 2	2

Source: Reproduced with permission from Ref. [15]. Copyright 2015, Royal Society Chemistry.

candidate for microflow processing and further process optimization, as given above and in the following.

6.3.5.2 LCA Study for "Two-Reactor Network" Process Designs

As an intermediate step to the microreactor flow network, the combination of two steps among the last four was analyzed. This concerned the formation of the azide and the subsequent Huisgen dipolar cycloaddition of EMMA to yield the rufinamide precursor molecule methyl 1-(2,6-difluorobenzyl)-1H-1,2,3-triazole-4-carboxylate. Four scenarios were compared consisting of three experimental scenarios (batch–batch, flow-batch, and one-pot batch) and one hypothetical scenario (batch recycled). Both metrics and LCIA analysis (see Figure 6.13) showed only minor differences for the batch-to-batch and continuous-to-batch processes. This was to be expected as the flow process that is equivalent to the batch process performance in terms of selectivity, but does not exceed it (which is the major key to better environmental performance). A stronger effect is observed by changing from a two-pot batch process to a one-pot batch process. Omitting the intermediate separation avoids the use of a large volume of solvent. The environmental impact of solvent use can be further decreased when recycling is considered for the solvents. The environmental performance of the batch recycled mode was considerably better as well [15].

6.3.5.3 LCA Study for "Three-Reaction Network" Process Designs

The LCIA comparison was expanded to consider three multistep flow scenarios. A literature microflow process from Zhang *et al.* [66], termed Process A, is based on the bromide route. This process does not, therefore, profit from the

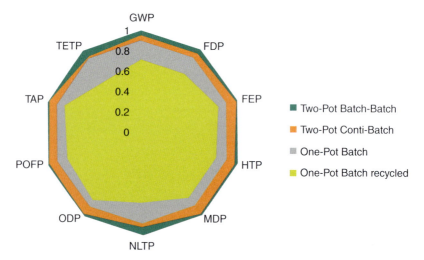

Figure 6.13 Life cycle impact assessment of experimental and hypothetical scenarios according to Mudd and Stevens [78] and Borukhova *et al.* [79] scaled effects of LCIA categories. (Reproduced with permission from Ref. [15]. Copyright 2015, Royal Society Chemistry).

abovementioned chloride intensifications, and also omits one step (making the aldehyde for the alcohol) upstream. Process B is the experimentally validated, intensified microflow route based on the aldehyde–alcohol–chloride conversion, as discussed already. Process C was aimed at combining the best aspects of Processes A and B. It includes the experimentally validated, chloride-related intensification microflow route, as discussed already, but does not use the aldehyde–alcohol conversion [15] (see Scheme 6.4). In all cases, EMMA was considered as the dipolarophile for the Huisgen cycloaddition. For processes B and C, batch data are used for the last step, the amidation, since poor solubility makes a flow step difficult. Please note that for process A this is not needed, as the dipolarophile already contains the amide group. Thus, process A benefits by having one less synthesis step.

For the environmental process option analysis, the global warming potential (GWP) and the human toxicity potential (HTP) were considered (Figure 6.14). These two categories are believed to have the most predictive power and partly represent other categories (e.g., energy), when a true multicriteria decision is not required. The three microflow multistep syntheses have better environmental performance than the multistep batch scenario, but in different aspects. Process A has a lower GWP than process B, which is due to the avoidance of two synthesis steps and respective savings of reactants and solvents. Process B is better in HTP, as a result of using green reactants such as the dipolarophile. As desired, process C can achieve benefits in both directions, with environmental impact reduced to about half for both categories. For process B, options for further improvement include yield increase, the introduction of recycling, and the combined effect. If these options are implemented, a performance similar to process

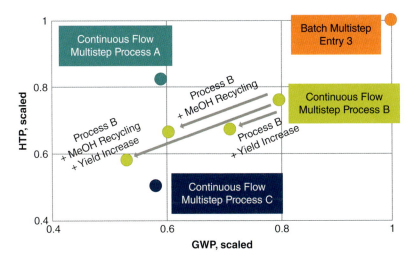

Figure 6.14 Overall life cycle impact assessment comparison of batch and continuous multistep processing of rufinamide by means of GWP and HTP. Scaled effects. (Reproduced with permission from Ref. [15]. Copyright 2015, Royal Society Chemistry).

C can be reached, and if these same options are implemented for process C, it would be improved much further.

It can be concluded that process simplification, that is, omitting reaction steps within the flow routes, has enormous impact, even if only one out of nine steps is omitted. Process integration has similar relevance, as all flow routes are better than the batch one [15], but this is only the case with the assumption of (much) simplified or nonexisting separation steps in between. The improved environmental performance of the integration still largely depends on the types and intensification of the single steps, as the comparison of the routes A and C show. Thus, the improvement in environmental performance for the microreactor flow networks is realized in three different ways – in process improvements within single steps, simplification of the overall process sequence, and improvements through process integration.

Taking into account all improvements considered in the investigations described already, there is a significant reduction in total LCIA impacts by an average of 45% when switching from multistep batch to multistep flow processing and by simultaneously integrating solvent recycling strategies (not shown here as figure) [15].

To access an even higher degree of process simplification, the superficial scenario of telescoping the last flow process steps from the 2,6-difluorobenzyl alcohol to the rufinamide precursor, that is, methyl 1-(2,6-difluorobenzyl)-1H-1,2,3-triazole-4-carboxylate, was investigated. This is referred to as a single-injection, single-flow mode, as opposed to the multistep, multi-injection, single flow considered so far. The last case can be classified as a nontandem reaction, for which all necessary reagents and solvents are added at different stages of the reaction. In contrast, the first case is classified as an orthogonal tandem reaction, for which all reagents and catalysts are present from the start of the reaction.

The LCIA comparison is shown in Figure 6.15 and again HTP and GWP are used as illustrative categories. The single injection, single flow is predicted to have better performance up to about 20% in HTP and GWP. Concerning all impact categories, there is on average a 33% improvement forecasted compared to the multistep flow procedure (not shown here as figure) [15].

6.3.6
Use as Process-Design Guidance and Benchmarking Tool Against Conventional Processes

As mentioned before, when considering fine chemical or even bulk chemical production, a more complex assembly of the process equipment has to be considered as compared to pharmaceutical production. Besides intensification, the effect of introducing an entirely new reactor technology might affect the whole plant itself in terms of process simplification. Thus, the environmental gains might be largely due to changes to plant utilities and auxiliary unit operations and thus an indirect result of the reactor innovation, yet enabled through the implementation of the new reactor. Thus, changes are more far-reaching and

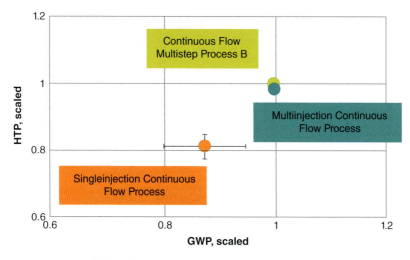

Figure 6.15 Overall life cycle impact assessment comparison of continuous multistep and streamlined processing of methyl 1-(2,6-difluorobenzyl)-1H-1,2,3-triazole-4-carboxylate by means of GWP and HTP scaled effects. (Reproduced with permission from Ref. [15]. Copyright 2015, Royal Society Chemistry).

include the entire process, not just those related to selectivity and solvent load improvements at the reaction level. This means that the scope and data management requirements are consequently larger. Because such plants are large, the extent of the extrapolation for the environmental analysis is high, which means that the uncertainty of the prediction is large. However, the gain in environmental opportunities is also large. Continuous flow, intensified chemistry, and processing can open doors to new chemical routes or even to new business opportunities.

That given, the best way to show the potential that is possible with process intensification is to focus on a production process that is done at close to world-scale volumes, that is, at the bulk chemical level. Adipic acid (ADA) is an important precursor for the production of Nylon-6,6, which is produced in large amounts [69]. There are two conventional ways to produce adipic acid; the two-step catalyzed air/nitric acid oxidation of cyclohexane (see Figure 6.16, left) [70], and the carboxylation/carbomethoxylation of butadiene. As an alternative to the conventional process for adipic acid manufacture, a one-step (direct) batch reaction with high conversion and yield (without recycle) has been reported [71–73], which may provide alternatives to the current two-step industrial process at low conversion and with large recycle ratio (see Figure 6.16).

However, the use of concentrated hydrogen peroxide to oxidize cyclohexene would make the scale-up of the new direct batch technology route difficult. Continuous flow processing can better address safety concerns associated with the large volume use of hydrogen peroxide in a batch reactor. Therefore, a high-temperature flow chemistry protocol was established for the direct adipic synthesis in a millipacked bed reactor with microsized fluid interstices [74]. These

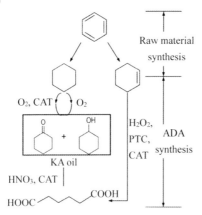

Figure 6.16 Two different production routes of adipic acid, two- and one-step. (Reproduced with permission from Ref. [53]. Copyright 2013, Elsevier).

experimental results were used as input for a green chemistry metrics and LCA study comparing world-scale scenarios for the new direct and intensified flow process and the conventional process by Wang *et al.* [53].

6.3.6.1 Process Simulation and CAPEX Cost Study

A full-chain process simulation was made for the flow-based, intensified direct synthesis of adipic acid using Aspen software (see Figure 6.17). Based on the relevant patents and literatures, the process simulation for the two-step conventional synthesis was built as well (see Figure 6.18). This was done at a world scale of 400 000 t/a [53].

In addition, a CAPEX cost analysis was given for both flow sheets [64]. The major equipment for the processes were determined and rough cost estimates calculated for both routes (see Figure 6.19). The total capital investment is significantly reduced that is simply a consequence of process simplification of the intensified direct route.

Figure 6.19 shows that as a consequence of having fewer reaction steps in the direct route, less equipment is used compared to the conventional route and the

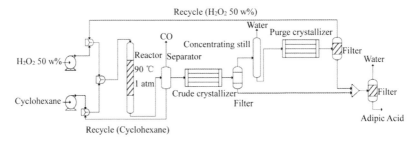

Figure 6.17 Flow sheet for the direct microflow synthesis of ADA. (Reproduced with permission from Ref. [53]. Copyright 2013, Elsevier).

6.3 Application of Green Chemistry Metrics and Life Cycle Assessment to Assess Microflow Processing

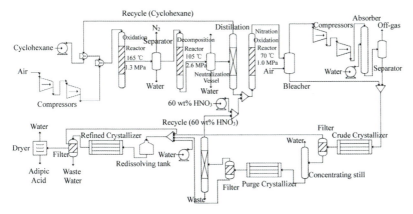

Figure 6.18 Flow sheet for the two-step conventional synthesis of ADA. (Reproduced with permission from Ref. [53]. Copyright 2013, Elsevier).

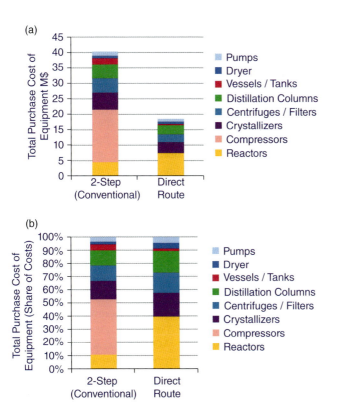

Figure 6.19 Total purchase cost of all equipment for the conventional route and direct route. (Reproduced with permission from Ref. [53]. Copyright 2013, Elsevier).

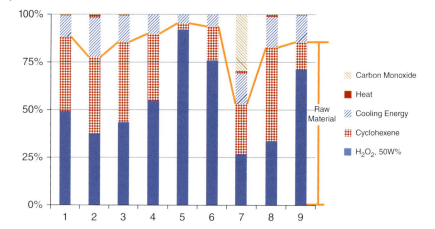

Figure 6.20 Environmental profile of the direct microflow synthesis of ADA. (Functional unit: 1 kg ADA, conversion of cyclohexene is 50%, yield of ADA is 46.8%) 1. AP 2. GWP 20a 3. EP 4. FAETP 20a 5. HTP 20a 6. MAETP 20a 7. POCP 8. Depletion of abiotic resources 9. TAETP 20a. The line drawn separates raw material from process technology contributions. (Reproduced with permission from Ref. [53]. Copyright 2013, Elsevier).

total costs for the direct route are therefore lower. In particular, the direct route does not require the expensive air compressor required for the chemical reaction. On the contrary, it is evident that the new process-intensified, continuous flow reactor cost is about 40% of the total cost of the direct route. It should be noted that while the absolute cost for the continuous flow reactor is higher than any other apparatus in the conventional process (with the exception of the compressor), the overall costs are still lower. There is a reasonable possibility that the costs of continuous-flow reactors will decrease with improvements and development of microscale manufacturing techniques, the numbers of plates/reactors manufactured, and new functionally advanced and simplified designs. It should also be noted that the absence of a compressor and the need to use air as the oxidant leads to additional operational costs in using hydrogen peroxide as the reagent and the purification of its aqueous solutions [53].

6.3.6.2 LCA for Continuous Flow Synthesis of ADA

The environmental profile for the direct microflow route of ADA is detailed for 9 impacts (see Figure 6.20). Even at a glance it is obvious that the production of H_2O_2 dominates most of these impact categories. This is because H_2O_2 production is an energy and waste intensive process. 95% of all H_2O_2 is produced on an industrial scale by the anthraquinone oxidation (AO) process, which was developed by BASF [75].[4] It involves the sequential hydrogenation and oxidation of an anthraquinone precursor dissolved in a mixture of organic solvents followed by liquid–liquid extraction to recover H_2O_2 [67]. The high energy input and

---

4) www.BASF.com (accessed April 25, 2017).

generated waste have large negative environmental impacts and increase production costs [67,75]. The transport, storage, and handling of bulk H_2O_2 involves many hazards and escalating expenses [67]. Moreover, H_2 as the raw material for H_2O_2 is also made in an energy consuming process through steam reforming of methane.

The life cycle environmental impacts associated with H_2O_2 dominate (>50%) all toxicity-related impacts (FAETP 20a, HTP 20a, MAETP 20a, and TAETP 20a). The impacts are, however, somewhat different when referring to AP, GWP 20a, EP, POCP, and depletion of abiotic resources where the life cycle impacts associated with H_2O_2 production are "only" in the 25–50% range.

The second and third dominant factor for most impact categories is associated with cyclohexene production followed by energy used in the process for cooling.

6.3.6.3 LCA for Two-Step Conventional Synthesis of ADA

For the two-step conventional synthesis of ADA, the dominating contribution for most of the impact categories is also associated with raw materials production (see Figure 6.21). Nitrogen oxide is the dominant parameter for AP and EP, while waste treatment is the dominant parameter for POCP.

6.3.6.4 Complete LCA Picture

The quantitative results for the life cycle environmental impact categories are listed in Table 6.6 and give a complete LCA picture. The following two main conclusions may be drawn [53]:

- Several LCA categories are low for the direct microflow process, several for the two-step conventional process. The LCA categories associated with the

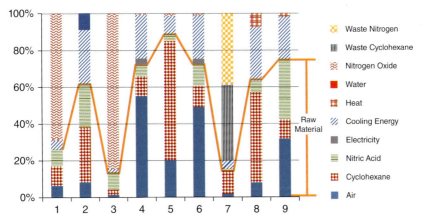

Figure 6.21 Environmental profile of two-step conventional synthesis of ADA. (Functional unit: 1 kg ADA, conversion of cyclohexene = 4.6%, yield of ADA = 4.16%, considering the recycle of unreacted cyclohexane and nitric acid). 1. AP 2. GWP 20a 3. EP 4. FAETP 20a 5. HTP 20a 6. MAETP 20a 7. POCP 8. Depletion of abiotic resources 10. TAETP 20a. (Reproduced with permission from Ref. [53]. Copyright 2013, Elsevier).

Table 6.6 Quantitative results for the ADA synthesis by the direct microflow route and two-step conventional route (functional unit: 1 kg ADA from raw material)

Impact category	DR $X^{a)} = 4.6\%$	DR $X = 40\%$	DR $X = 50\%$	DR $X = 98\%$	CTR $X = 4.6\%$
AP ($\times 10^{-3}$ kg SO_2)	18.7	13.8	13.8	13.9	48.8
POCP ($\times 10^{-3}$ kg ethylene)	2.7	1.8	1.8	1.8	10.1
EP ($\times 10^{-3}$ kg NO_x)	11.0	7.5	7.5	7.4	94.7
FAETP 20a (kg 1,4-DCB)	1.9	1.5	1.5	1.3	1.0
MAETP 20a (kg 1,4-DCB)	1.2	0.9	0.9	0.8	0.6
Depletion of abiotic resources ($\times 10^{-3}$ kg antimony)	83.7	53.2	53.0	51.0	46.7
TAETP 20a ($\times 10^{-3}$ kg 1,4-DCB)	0.3	0.2	0.2	0.2	0.2
HTP	8.7	7.7	7.8	8.6	5.6
GWP 20a	11.0	6.3	6.3	5.9	16.7

a) X denotes the conversion of cyclohexene/cyclohexane.
Source: Reproduced with permission from Ref. [53]. Copyright 2013, Elsevier.

new intensified technology are mixed, and therefore require that a typical multicriteria decision methodology be used.
- Even with radical process improvements, such as increasing conversion efficiency from 4.6% to 98%, the overall environmental profile and advantage/disadvantage of a given process may not be changed. Exceptions may occur when both processes have similar data on one impact. In these cases, the smart changes made possible through process intensification can decisively reduce the overall environmental life cycle impacts.

6.3.6.5 Green Metrics Compared for the Direct Microflow Route and Conventional Two-Step Route

The atom economy (AE) is much higher for the direct microflow route than for the two-step conventional route (see Figure 6.22a). Second, this is an intrinsic effect of the lower reactant load of a one-step synthesis as compared to a two-step route. The direct microflow route has a lower E-factor than the two-step conventional route when considering the recycling of unreacted raw materials in both routes (see Figure 6.22b), due to high selectivity and low reactant entry. The C-efficiency is higher for the direct microflow route when compared on the same basis as the two-step conventional one (see Figure 6.22c) [53].

The use of green metrics, as described earlier in this chapter, is useful for analyzing organic chemistry reactions like those used in batch chemical manufacturing for the pharmaceutical and other specialty chemical industries. Here, the process chemistry (the reactants, reagents, solvent, etc.) stands in focus. No consideration of the environmental impacts associated with the chemicals used in

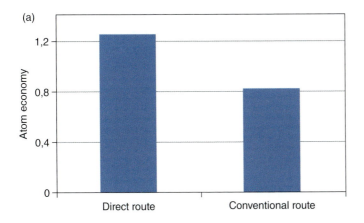
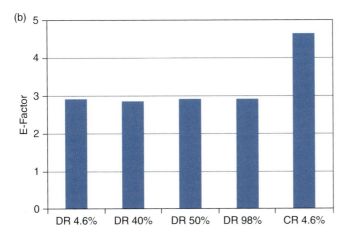
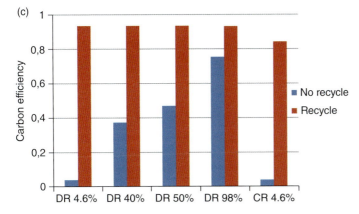

Figure 6.22 Green metrics for the direct microflow route and two-step conventional routes. (a) AE; (b) E-factor; and (c) C-Efficiency. DR: the conversion of cyclohexane by direct route, CR: the conversion of cyclohexane by conventional route. (Reproduced with permission from Ref. [53]. Copyright 2013, Elsevier).

the upstream syntheses, but only the given reaction itself is considered. The process equipment used is often of multipurpose type and thus does not differ too much from process to process nor are different process equipment configurations considered by the metrics except to the extent that they may reduce the mass of reactants or increase the mass efficiency.

In contrast, LCA considers both upstream chemistries and is suited to consider equipment and its heat flows so that different process equipment configurations and options may be evaluated. This is the view needed for a large production volume fine chemical or bulk chemical synthesis with their complex dedicated plants [53]. Changing the chemistry in these manufacturing plants inevitably results in changing the entire process, including up- and downstream operations. Thus, a holistic picture is needed.

Process intensification tools open several opportunities on a large production scale. Since most of these cannot easily be tested, even on a pilot scale, it makes sense to use LCA for option analyses when considering superficial changes and different process scenarios [53].

6.3.6.6 Conclusions

From the above analysis of the direct ADA synthesis as an intensified flow, bulk-scale process using green metrics and LCA, some conclusions can be drawn; also reflecting the use of the tools themselves [53].

- "Green" or "greener" chemicals do not necessarily make green or greener processes. Innovation through flow intensification still suffers from the environmental impacts associated with the process. Thus, only by replacing the process with a more environmentally sustainable process will the H_2O_2-based ADA become "greener" on a process level. Such a process might be the direct H_2O_2 synthesis out of the elements.
- Raw materials consumption largely determines the environmental impact profiles of large-scale chemical production processes. The most desirable processes first of all require smart synthetic paths with "green" chemicals. Process technology then enables one to bring innovation from the reaction level to the process level. This is commonly not included when moving a process to an intensified process. Thus, green chemistry and green engineering need to be aligned.
- The inherent complexity of the processes demands complexity in process analysis. A variety of metrics can be used for an initial analysis of different synthetic pathways when concerning intensified bulk chemical process development. Thereafter, LCA is needed for further and deeper analysis, for example, to allow for process-options selection.
- Greener routes than the anthraquinone process to produce H_2O_2 have to be found, as, for example, by directly obtaining it from H_2 and O_2 (without carrier), so that a double-direct route to ADA may be implemented. The production of H_2 using renewable energy (wind, solar) will also be beneficial.

6.4
Economic Analysis and Snapshot on Applications with Continuous Microflow Processing

6.4.1
Life Cycle Costing (LCC)

From an industrial viewpoint, the major drivers for pilot or scale-up research into new processes are to lower costs through greater efficiency and to implement more environmentally benign and safe processes. Any novel pathway to a current production process should be more economical than the one it replaces. For the implementation of new developments in industrial practice, it is thus important to determine the associated life cycle costs as part of any basic decision-making tool, at the early stages of process R&D and to compare with the costs of different processing alternatives to be able to respond flexibly to future market needs [68].

LCC is a methodology for calculating the total cost of a system from inception to disposal, also commonly referred to as "cradle-to-grave" or "womb-to-tomb" costs[5]. The purpose of an LCC is to estimate the overall costs of project alternatives and to select the design that ensures the facility will provide the lowest overall cost of ownership consistent with its quality and function. It can be used as a suitable tool for efficient decision support within cost reducing development target [68,76].

Within the integrated technology and product life cycle, whole life cycle costs consist of the individual costs during development, usage, and disposal phase, shown in Figure 6.23 [68].

The integration of different cost figures in the analysis depends on the goal and scope of the study. Lowest life cycle cost is the most straightforward and easy-to-interpret measure of economic evaluation. In the context of investment

Figure 6.23 Composition of the life cycle costs. (Reproduced with permission from Ref. [68]. Copyright 2014, Wiley-VCH Verlag GmbH).

---

5) https://en.wikipedia.org/wiki/Whole-life_cost (accessed April 25, 2017).

decisions, several alternatives can be compared holistically based on life cycle costs. Therefore, the LCC should be performed early in the design process while there is still a chance to refine the design to ensure a reduction in life cycle costs.

6.4.2
Snapshot on LCC Applications with Continuous Microflow Processing

LCC analyses have been reported in the literature for the use of continuous processing at small scale in the fine chemical and pharmaceutical industry [24,68,76]. There are many factors contributing to the costs and, owing to the environmental focus of this book and chapter, this will not be given in its details. Rather some relevant literature will be quickly discussed for further information.

Denčić *et al.* [24] compared the process and cost analysis results for a high-value, low-volume, active pharmaceutical ingredient (API) using a millireactor-based process and a batch process developed at Sanofi as a reference case. Five optimized and intensified scenarios, including an intensified reaction, continuous processing, alternative catalysts, a change of solvent, a change in the purification sequence, recovery of the key product, and an intensified drying option, were compared to the reference case. The results indicated that the intensified processes had the highest net present values (NPV), mainly due to reductions in operating costs. Moreover, compared to the reference batch process, total product cost was reduced by 31–35% (different ratio with different raw materials price) under continuous operation due to the much lower equipment size and capital costs and consequently less labor that was needed, as shown in Figure 6.24 [24]. The cost of the main reagents as the key input material (KIM)

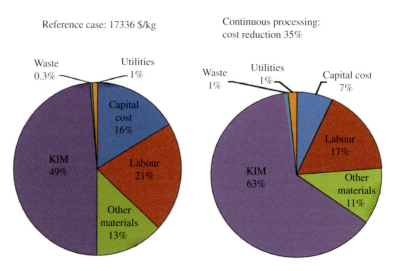

Figure 6.24 Production cost breakdown: reference case and continuous operation scenario. (Reproduced with permission from Ref. [24]. Copyright 2014, American Chemical Society).

6.4 Economic Analysis and Snapshot on Applications with Continuous Microflow Processing

makes the largest contribution to the process cost. All in all, through the cost analysis of the reaction mode, operating mode, purification, and the drying method, in combination with other metrics, the authors believe that these indicator-based analysis help to systematically recognize process bottlenecks. For the complex processes with many more unit operations, streams, and components involved, process-design needs have to be considered in a holistic manner to achieve the maximal gain. Preliminary cost analysis can be used to support process selection decisions and as a starting point for the next step in decision-making [24].

Krtschil *et al.* conducted a cost analysis of a microchemical process in the production of 4-cyanophenylboronic acid using the economic fine chemical process of the custom chemical producer AzurChem GmbH for the basic assessment [76]. In order to get a sufficiently large capacity in microdevices, three other scenarios were considered assuming 5- and 10-fold increase in capacity and external numbering-up of 10 microdevices in parallel. In addition, a virtual batch process for achieving 20% higher selectivity according to the reported selectivity achievement was also investigated using comparable parameters with respect to the microchemical process. The results show that operational (variable) costs are the main drivers for developing a business case for microflow process while the equipment costs, including microstructured reactors and balance-of-plant equipment, have a lower share of the costs. Comparison results of total costs and earnings for different scenarios are show in Figures 6.25 and 6.26, respectively. These results highlight possible cost advantages for the synthesis of high-valued chemical products from expensive raw materials using microstructured reactors with higher throughput. The authors believe that plants with microstructured reactors are more profitable when operated at larger scale than their conventional counterparts, and micro-process engineering is, for some cases, a commercially viable processing alternative [76].

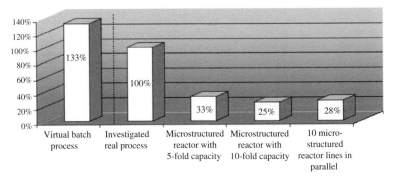

Figure 6.25 Comparison of total cost for different case scenarios. (Reproduced with permission from Ref. [76]. Copyright 2006, Swiss Chemical Society).

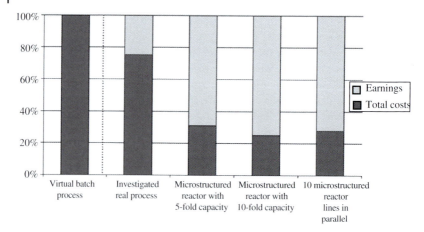

Figure 6.26 Influence of the plant scenario to costs and earnings. (Reproduced with permission from Ref. [76]. Copyright 2006, Swiss Chemical Society).

Sell *et al.* [68] evaluated the LCC coupled with LCA differences between an existing discontinuous production process (batch plant) and a continuously running microreactor-based production process. The representative results for LCC and LCA using global warming potential as the assessment category are shown in Figures 6.27 and 6.28 [68]. Whereas in the case of microreactor processing the

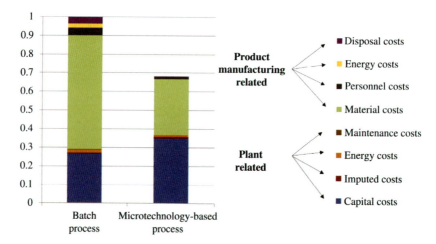

Figure 6.27 Comparison of life cycle costs of chemical production processes in batch and continuous processing mode during utilization phase. (Reproduced with permission from Ref. [68]. Copyright 2014, Wiley-VCH Verlag GmbH).

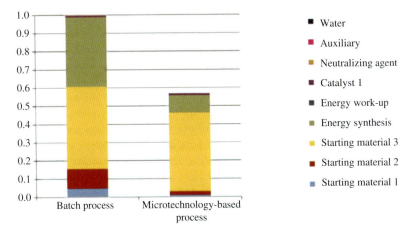

Figure 6.28 Comparison of environmental effects concerning global warming potential for batch and microreaction technology-based processing. (Reproduced with permission from Ref. [68]. Copyright 2014, Wiley-VCH Verlag GmbH).

plant-related costs are dominant, product manufacturing related costs are lower compared to batch processing due to the higher yield and selectivity obtained. Despite partially higher capital costs, microreaction technology-based processes can represent an economically more favorable alternative due to lower operating costs and a faster market entry [68]. Moreover, a reduction of greenhouse gas emissions during the product life cycle can be achieved through shortened reaction times that result in a reduced demand for energy, or less starting materials and solvents are needed as a result of obtaining higher yields and selectivity, and less effort for product workup, solvent recycling, and waste disposal. The authors believe that in addition to possible cost superiority, a considerable reduction in environmental impact is also important in the implementation of microflow processes into industrial processes.

Typically, compared to batch procedures, microflow processing can result in lower life cycle costs. This is because material costs comprise the major share of chemical processes, most often being around two thirds of the total costs. There are other contributions to costs and the reader is referred to the cited papers for further insight.

6.5
Conclusions and Outlook

As outlined with a few selected examples, a number of different green metrics, and a few tools like LCA and LCC analysis, it is possible to make a comprehensive assessment of a continuous microflow route in the fine chemical and pharmaceutical industry. The green metrics presented in this chapter and elsewhere are especially well suited to address single targets that have been set as strategic goals, for example, by the pharmaceutical industry and associated industries. Thus, they are to be used when a microflow process comes close to production. So far, this has not been published and it looks like while parts of the processes have indeed been made continuous, the product did not make it to the market. Here and there, cost issues are addressed in a global manner; however, the main drivers of industrial use of microflow technology are the clear potential in the interplay of operating and capital costs of existing equipment and respective costs on the microflow processing side [77]. However, information on the use of microflow production is limited and, the only information that is generally found is in press releases or conference presentations.

Since microflow benefits can be huge and much larger than what is commonly obtained through routine process optimization, it might be possible that the wrong choice of key metrics may predict an environmentally unfavorable process, while the overall profile might be positive. An example of this might be the high-temperature microflow synthesis, which consumes more energy, but gains orders of magnitude higher reaction speed, possibly even better selectivity and possibly reduction of solvent load.

The main task of environmental analysis concerning continuous microflow synthesis was and still is to confirm the greenness of the new approach relative to batch reactor synthesis and to provide guidance for further improvements in continuous flow technology. As this is done from a generic and global ambition, focusing on only one kind or one set of green metrics may result in missing other opportunities to reduce the overall environmental burden associated with a process, or it may result in incorrect decisions. Therefore, the integrated use of green metrics, LCA, and LCC analysis is essential to provide a more accurate and comprehensive assessment of the chemical process.

In this way, it was shown that continuous microflow processing can contribute to finding more environmental benefits, gaining increased economic benefits, and thereby resulting in advanced eco-efficiency. This shows that microflow approaches are promising alternatives/supplements to conventional batch processing in the fine chemical and pharmaceutical industry. After such a global environmental optimization has been done for one microflow process in the early stages of process development, the use of specific green metrics shall guide the late-stage process development.

It is expected that the environmental issues and the concerns raised here will help to promote the development of continuous flow processing in the chemical industry. The sustainability and environmental push increased in 2016, as evidenced by the United Nations delivering "17 Sustainable Development Goals" that promote the continued application of green chemistry principles while also highlighting the vital need to apply the concepts of green engineering to radically improve process intensification for the betterment and sustainability of the planet.

In addition, the push from legislative authorities continues to increase. In May 2015, the FDA opened a new era in pharmaceutical manufacturing and called on pharma manufacturers and CMOs to begin making a switch from batch to continuous production with a call to complete the transition by 2025. In April 2016, the FDA approved Johnson & Johnson's Janssen Pharmaceuticals application for the continuous production of HIV drug Prezista in its plant in Gurabo, Puerto Rico. Beyond this legislative authority push, the ACS Green Chemistry Pharmaceutical Roundtable, and thus the pharmaceutical industry, has endorsed the same view and declared continuous manufacturing as its top priority. The vision of a complete continuous end-to-end pharmaceutical manufacturing has also recently been demonstrated by Novartis and MIT while Bosch Company and the Formulation Competence Center RCPE in Graz/Austria have announced their collaboration on continuous tablet production.

References

1 Anastas, P.T. and Warner, J.C. (1998) *Green Chemistry: Theory and Practice*, Oxford University Press, New York, p. 30.

2 Anastas, P.T. and Zimmerman, J.B. (2003) Design through the twelve principles of green engineering. *Environmental Science & Technology*, **37** (5), 94–101.

3 Hessel, V., Kralisch, D., and Krtschil, U. (2008) Sustainability through green processing – novel process windows intensify micro and milli process technologies. *Energy & Environmental Science*, **1** (4), 467–478.

4 Abraham, M.A. and Nguyen, N. (2003) Green engineering: defining the principles – results from the Sandestin conference. *Environmental Progress*, **22** (4), 233–236.

5 Winterton, N. (2001) Twelve more green chemistry principles. *Green Chemistry*, **3** (6), G73–G75.

6 Jiménez-González, C., Poechlauer, P., Broxteman, Q.B., Yang, B.S., Ende, D., Baird, J., Bertsch, C., Hannah, R.E., Orco, P.D., Noorman, H., Yee, S., Reintjens, R., Wells, A.S., Massonneau, V., and Manley, J. (2011) Key green engineering research areas for sustainable manufacturing: a perspective from pharmaceutical and fine chemicals manufacturers. *Organic Process Research and Development*, **15**, 900–911.

7 Jimenez-Gonzalez, C., Ponder, C.S., Broxterman, Q.B., and Manley, J.B. (2011) Using the right green yardstick: why process mass intensity is used in the pharmaceutical industry to drive more sustainable processes. *Organic Process Research and Development*, **15** (4), 912–917.

8 Dunn, P.J., Wells, A.S., and Williams, M.T. (2010) *Green Chemistry in the Pharmaceutical Industry*, Wiley-VCH Verlag GmbH, Weinhein, p. 21.

9 Jiménez-González, C., Constable, D.J.C., and Ponder, C.S. (2012) Evaluating the "Greenness" of chemical processes and products in the pharmaceutical industry – a green metrics primer. *Chemical Society Reviews*, **41**, 1485–1498.

10 Sheldon, R.A. (2007) The E factor: fifteen years on. *Green Chemistry*, **9** (12), 1273.

11 Sheldon, R.A. (1994) Consider the environmental quotient. *ChemTech*, **24** (3), 38–47.

12 Trost, B. (1991) The atom economy – a search for synthetic efficiency. *Science*, **254** (5037), 1471–1477.

13 Curzons, A.D., Constable, D.J.C., Mortimer, D.N., and Cunningham, V.L. (2001) So you think your process is green, how do you know? – using principles of sustainability to determine what is green – a corporate perspective. *Green Chemistry*, **3**, 1–6.

14 Constable, D.J.C., Curzons, A.D., and Cunningham, V.L. (2002) Metrics to green chemistry which are the best? *Green Chemistry*, **4**, 521–527.

15 Ott, D., Borukhova, S., and Hessel, V. (2015) Life cycle assessment of multi-step rufinamide synthesis – from isolated

reactions in batch to continuous microreactor networks. *Green Chemistry*, **18**, 1096–1116.

16 Fleischer, G. and Schmidt, W.P. (1997) Iterative screening LCA in an eco-design tool. *International Journal of Life Cycle Assessment*, **2** (1), 20–24.

17 Azapagic, A. (1999) Life cycle assessment and its application to process selection, design and optimization. *Chemical Engineering Journal*, **73** (1), 1–21.

18 Guinée, J.B., Heijungs, R., Huppes, G., Zamagni, A., Masoni, P., Buonamici, R., Ekvall, T., and Rydberg, T. (2011) Life cycle assessment: past, present, and future. *Environmental Science and Technology*, **45** (1), 90–96.

19 Tufvesson, L.M., Tufvesson, P., and Woodley, J.M. (2013) Life cycle assessment in green chemistry: overview of key parameters and methodological concerns. *International Journal of Life Cycle Assessment*, **18**, 431–444.

20 Jiménez-González, C., Curzons, A.D., Constable, D.J.C., and Cunningham, V.L. (2004) Cradle-to-gate life cycle inventory and assessment of pharmaceutical compounds. *International Journal of Life Cycle Assessment*, **9** (2), 114–121.

21 ISO (2006) ISO 14040, International Organization for Standardization, Geneva.

22 ISO, (2006) ISO 14044, International Organization for Standardization, Geneva.

23 Freissmuth, J. (2013) Facing the pharmaceutical future. Pharma Manufacturing. Available at http://www.pharmamanufacturing.com/articles/2013/021/ (accessed June 3, 2017).

24 Denčić, I., Ott, D., Kralisch, D., Noël, T., Meuldijk, J., Croon, M., Hessel, V., Laribi, Y., and Perrichon, P. (2014) Eco-efficiency analysis for intensified production of an active pharmaceutical ingredient: a case study. *Organic Process Research and Development*, **18**, 1326–1338.

25 Plumb, K. (2005) Continuous processing in the pharmaceutical industry: changing the mind set. *Chemical Engineering Research and Design*, **83**, 730–738.

26 Srai, J.S., Harrington, T., Alinaghian, L., and Phillips, M. (2015) Evaluating the potential for the continuous processing of pharmaceutical products – a supply network perspective. *Chemical Engineering and Processing: Process Intensification*, **97**, 248–258.

27 Anderson, N.G. (2012) Using continuous processes to increase production. *Organic Process Research and Development*, **16**, 852–869.

28 Calabrese, G. and Pissavini, S. (2011) From batch to continuous flow processing in chemicals manufacturing. *AIChE Journal*, **57**, 828–834.

29 Cervera-Padrell, A.E., Skovby, T., Kiil, S., Gani, R., and Gernaey, K.V. (2012) Active pharmaceutical ingredient (API) production involving continuous processes – a process system engineering (PSE)-assisted design framework. *European Journal of Pharmaceutics and Biopharmaceutics*, **82**, 437–456.

30 Kralisch, D., Streckmann, I., Ott, D., Krtschil, U., Santacesaria, E., Serio, M.D., Russo, V., Carlo, L.D., Linhart, W., Christian, E., Cortese, B., Croon, M.H.J., and Hessel, V. (2012) Transfer of the epoxidation of soybean oil from batch to flow chemistry guided by cost and environmental issues. *ChemSusChem*, **5** (2), 300–311.

31 Baraldi, P.T., Noël, T., Wang, Q., and Hessel, V. (2014) The accelerated preparation of 1,4-dihydropyridines using microflow reactors. *Tetahedron Letters*, **55**, 2090–2092.

32 Lapkin, A.A. and Plucinski, P.K. (2009) Engineering factors for efficient flow processes in chemical industries, in *Chemical Reactions and Processes under Flow Conditions* (eds S.V. Luis and E. Garcia-Verdugo), Royal Society of Chemistry, pp. 1–43.

33 Hessel, V., Vural-Gursel, I., Wang, Q., Noël, T., and Lang, J. (2012) Potential analysis of smart flow processing and micro process technology for fastening process development: use of chemistry and process design as intensification fields. *Chemical Engineering and Technology*, **35** (7), 1184–1204.

34 Hessel, V., Hardt, S., and Löwe, H. (2004) *Chemical Micro Process Engineering*, Wiley-VCH Verlag GmbH, Weinheim, pp. 97–105.

35 Borukhova, S., Seeger, A.D., Noël, T., Wang, Q., Busch, M., and Hessel, V. (2015) Pressure-accelerated azide–alkyne cycloaddition: micro capillary versus autoclave reactor performance. *ChemSusChem*, **8**, 504–512.

36 Kockmann, N., Gottsponer, M., and Roberge, D.M. (2011) Scale-up concept of single-channel microreactors from process development to industrial production. *Chemical Engineering Journal*, **167**, 718–726.

37 Hessel, V., Kralisch, D., and Kockmann, N. (2014) *Novel Process Windows*, Wiley-VCH Verlag GmbH, Weinheim.

38 Hessel, V., Kralisch, D., Kockmann, N., Noël, T., and Wang, Q. (2013) Novel process windows for enabling, accelerating, and uplifting flow chemistry. *ChemSusChem*, **6**, 746–789.

39 Hessel, V., Cortese, B., and de Croon, M.H.J.M. (2011) Novel process windows – concept, proposition and evaluation methodology, and intensified superheated processing. *Chemical Engineering Science*, **66**, 1426–1448.

40 Illg, T., Hessel, V., Löb, P., and Schouten, J.C. (2011) Novel process window for the safe and continuous synthesis of tert.-butyl peroxy pivalate in a micro-reactor. *Chemical Engineering Journal*, **167**, 504–509.

41 Hessel, V. (2009) Novel process windows – gate to maximizing process intensification via flow chemistry. *Chemical Engineering and Technology*, **32**, 1655–1681.

42 Hessel, V., Vural-Guersel, I., Wang, Q., Noel, T., and Lang, J. (2012) Potential analysis of smart flow processing and micro process technology for fastening process development: use of chemistry and process design as intensification fields. *Chemical Engineering and Technology*, **35**, 1184–1204.

43 Hessel, V., Kralisch, D., and Krtschil, U. (2008) Sustainability through green processing – novel process windows intensify micro and milli process technologies. *Energy & Environmental Science*, **1**, 467–478.

44 Hessel, V., Hardt, S., and Löwe, H. (2004) *Chemical Micro Process Engineering*, Wiley-VCH Verlag GmbH, Weinheim, p. 28.

45 National Fire Protection Association (2012) NFPA 704.

46 Huebschmann, S., Kralisch, D., Hessel, V., Krtschil, U., and Kompter, C. (2009) Environmentally benign microreaction process design by accompanying (simplified) life cycle assessment. *Chemical Engineering and Technology*, **32**, 1757–1765.

47 SETAC (1997) Simplifying LCA: just a cut? Final report of the SETAC-Europe Screening and Streamlining Working-Group, Society of Environmental Chemistry and Toxicology (SETAC), Brussel.

48 Kralisch, D. and Kreisel, G. (2007) Assessment of the ecological potential of microreaction technology. *Chemical Engineering Science*, **62**, 1094–1100.

49 Huebschmann, S., Kralisch, D., Loewe, H., Breuch, D., Petersen, J.H., Dietrich, T., and Scholz, R. (2011) Decision support towards agile eco-design of microreaction processes by accompanying (simplified) life cycle assessment. *Green Chemistry*, **13**, 1694–1707.

50 Kralisch, D., Staffel, C., Ott, D., Bensaid, S., Saracco, G., Bellantoni, P., and Loeb, P. (2013) Process design accompanying life cycle management and risk analysis as a decision support tool for sustainable biodiesel production. *Green Chemistry*, **15**, 463–477.

51 Reinhardt, D., Ilgen, F., Kralisch, D., König, B., and Kreisel, G. (2008) Evaluating the greenness of alternative reaction media. *Green Chemistry*, **10**, 1170–1181.

52 Ott, D., Kralisch, D., Denč, I., Hessel, V., Laribi, Y., Perrichon, P.D., Berguerand, C., Kiwi-Minsker, L., and Loeb, P. (2014) Life cycle analysis within pharmaceutical process optimization and intensification: case study of an API production. *ChemSusChem*, **7** (12), 3521–3533.

53 Wang, Q., Gürsel, I.V., Shang, M., and Hessel, V. (2013) Life cycle assessment for the direct synthesis of adipic acid in microreactors and benchmarking to the commercial process. *Chemical Engineering Journal*, **234**, 300–311.

54 Yaseneva, P., Plaza, D., Fan, X., Loponov, K., and Lapkin, A. (2015) Synthesis of the antimalarial API artemether in a flow reactor. *Catalysis Today*, **239**, 90–96.

55 Shang, M., Noël, T., Wang, Q., and Hessel, V. (2013) Packed-bed microreactor for continuous-flow adipic acid synthesis from cyclohexene and hydrogen peroxide. *Chemical Engineering and Technology*, **36** (6), 1001–1009.

56 Lowrance, W. (1971) Process for the synthesis of phenyl esters, US Patent 3,772,389, filed Jun. 24, 1971 and issued Nov. 13, 1973.

57 Ikeda, T., Horiuchi, S., Karanjit, D.B., Kurihara, S., and Tazuke, S. (1990) Photochemically induced isothermal phase transition in polymer liquid crystals with mesogenic phenyl benzoate side chains. 2. Photochemically induced isothermal phase transition behaviors. *Macromolecules*, **23** (1), 42–48.

58 Guinee, J.B. (2001) *Life Cycle Assessment – An Operational Guideline to the ISO Standards* (eds M. Gorree and R. Heijungs), Leiden University, The Netherlands.

59 Constable, D.J.C., Jimenez-Gonzalez, C., and Henderson, R.K. (2007) Perspective on solvent use in the pharmaceutical industry. *Organic Process Research and Development*, **11** (1), 133–137.

60 Lapkin, A.A., Plucinski, P.K., and Cutler, J. (2006) Comparative assessment of technologies for extraction of artemisinin. *Journal of Natural Products*, **69**, 1653–1664.

61 Klayman, D.L., Lin, A.J., Acton, N., Scovill, J.P., Hoch, J.M., Milhous, W.K., Theoharides, A.D., and Dobek, A.S. (1984) Isolation of artemisinin (qinghaosu) from Artemisia annua growing in the United States. *Journal of Natural Products*, **47** (4), 715–717.

62 Brossi, A., Venugopalan, B., Gerpe, L.D., Yeh, H.J.C., Flippenanderson, J.L., and Buchs, P. (1988) Arteether, a new antimalarial drug: synthesis and antimalarial properties. *Journal of Medicinal Chemistry*, **31**, 645–650.

63 Blaser, H.U. (2010) Developing catalysts and catalytic processes with industrial relevance. *CHIMIA International Journal for Chemistry*, **64**, 65–68.

64 Gürsel, I.V., Wang, Q., Noël, T., and Hessel, V. (2012) Process-design intensification: direct synthesis of adipic acid in flow. *Chemical Engineering Transactions*, **29**, 565–570.

65 Baumann, M., Baxendale, I.R., Ley, S.V., and Nikbin, N. (2011) An overview of the key routes to the best selling 5-membered ring heterocyclic pharmaceuticals. *Beilstein Journal of Organic Chemistry*, **7**, 442.

66 Zhang, P., Russell, M.G., and Jamison, T.F. (2014) Continuous flow total synthesis of rufinamide. *Organic Process Research and Development*, **18**, 1567.

67 Campos-Martin, J.M., Blanco-Brieva, G., and Fierro, J.L.G. (2006) Hydrogen peroxide synthesis: an outlook beyond the anthraquinone process. *Angewandte Chemie, International Edition*, **42**, 6962–6984.

68 Sell, I., Ott, D., and Kralisch, D. (2014) Life cycle cost analysis as decision support tool in chemical process development. *ChemBioEng Reviews*, **1** (1), 50–56.

69 Herzog, B., Kohan, M.I., Mestemacher, S.A., Pagilagan, R.U., and Redmond, K. (2013) Polyamides, in *Ullmann's Encyclopedia of Industrial Chemistry*, Wiley-VCH Verlag GmbH, Weinheim.

70 Steeman, J.W.M., Kaarsemaker, S., and Hoftyzer, P.J. (1961) A pilot plant study of the oxidation of cyclohexane with air under pressure. *Chemical Engineering Science*, **14**, 139–149.

71 Sato, K., Aoki, M., and Noyori, R. (1998) A "Green" route to adipic acid: direct oxidation of cyclohexenes with 30 percent hydrogen peroxide. *Science*, **281**, 1646–1647.

72 Buonomenna, M.G., Golemme, G., De Santo, M.P., and Drioli, E. (2010) Direct oxidation of cyclohexene with inert polymeric membrane reactor. *Organic Process Research and Development*, **14**, 252–258.

73 Deng, Y., Ma, Z., Wang, K., and Chen, J. (1999) Clean synthesis of adipic acid by direct oxidation of cyclohexene with H_2O_2 over peroxytungstate–organic complex catalysts. *Green Chemistry*, **1**, 275–276.

74 Shang, M.J., Noël, T., Wang, Q., Su, Y.H., Miyabayashi, K., Hessel, V., and Hasebe, S. (2015) 2- and 3-Stage temperature ramping for the direct synthesis of adipic acid in micro-flow packed-bed reactors. *Chemical Engineering Journal*, **260**, 454–462.

75 Centi, G. and Perathoner, S. (2009) One-step H2O2 and phenol syntheses: examples of challenges for new sustainable selective oxidation processes. *Catalysis Today*, **143**, 145–150.

76 Krtschil, U., Hessel, V., Kralisch, D., Kreisel, G., and Küpper, M. (2006) Cost analysis of a commercial manufacturing process of a fine chemical compound using micro process engineering. *Chimia*, **60** (9), 611–617.

77 Dencic, I. and Hessel, V. (2013) Industrial micro-reactor process development up to production, in *Micro-reactors. Organic Chemistry and Catalysis*, 2nd edn, (ed T. Wirth), Wiley-VCH Verlag GmbH, p. 442.

78 Mudd, W.H. and Stevens, E.P. (2010) An efficient synthesis of rufinamide, an antiepileptic drug. *Tetrahedron Letters*, **51**, 3229–3231.

79 Borukhova, S., Noël, T., Metten, B., Vos, E., and Hessel V. (2013) Solvent- and catalyst-free Huisgen cycloaddition to rufinamide in flow with a greener, less expensive dipolarophile. *Chem Sus Chem*, **6**, 2220–2225.

7
Benchmarking the Sustainability of Biocatalytic Processes
John M. Woodley

7.1
Introduction

There can be little doubt about the importance of ensuring that chemical production processes of the future are designed in such a way that they are sustainable with respect not only to the economy, but also the environment [1–3]. Indeed, it is not only good engineering practice to design processes with such a "sustainability" objective, but it also results in almost all cases in a better economy as well [4]. There are two basic process design approaches to meet the sustainability objective. In the first approach, chemical production processes are designed based on the use of more sustainable feedstocks and reagents. A good example of this is the implementation of processes based on renewable bio-based feedstocks (such as those based on biomass) [3]. In the second approach, processes use chemistry with improved selectivity and more benign operating conditions, which also helps reduce energy demand and minimizes waste. The use of catalytic (rather than stoichiometric) reactions in chemical production processes is an excellent example of this second approach [5]. There are plenty of examples in this book illustrating these two approaches. In this chapter, the focus will be on bioprocesses, which are frequently highlighted in the scientific literature as one of the most effective ways of achieving sustainability in production processes.

7.2
Biocatalytic Processes

The application of biological catalysts, using either fermentation or biocatalysis, to produce value-added chemical products has attracted enormous interest in recent years as one of the key routes toward more sustainable production [6–8]. Both fermentation and biocatalysis involve the use of one or more enzymes to catalyze highly selective reactions under mild conditions. Fermentation processes, which use growing cells, consume some of the carbon substrate for this

Handbook of Green Chemistry Volume 11: Green Metrics, First Edition. Edited by David J. Constable and Concepción Jiménez-González.
© 2018 Wiley-VCH Verlag GmbH & Co. KGaA. Published 2018 by Wiley-VCH Verlag GmbH & Co. KGaA.

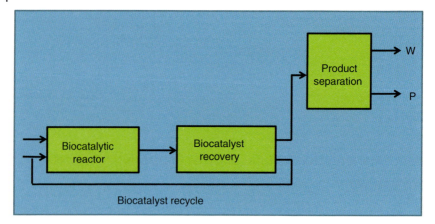

Figure 7.1 Typical biocatalytic process structure. (W, waste; P, product).

growth, consequently lowering the reaction (carbon) yield. Likewise, the productivity in such processes is limited by the growth rate of the cells. In contrast, in biocatalytic processes, the biocatalyst is not growing (and may even be used outside the confines of a cell). Hence, the biocatalyst remains at a constant mass throughout the reaction – a true catalyst. A number of possible options exist for the format of the biocatalyst, such as "resting" whole cells, isolated soluble enzyme(s), or "immobilized" enzyme(s) (with enzyme(s) adsorbed or covalently bound to an inert support material). Since the biocatalyst may still retain activity at the end of the reaction, it may be advantageous to recycle it in order to reduce the cost contribution of the biocatalyst to the final product. This, in particular, is the driver for implementation of immobilized enzyme processes [9] to simplify biocatalyst separation from the reaction mixture and enable biocatalyst recycle (Figure 7.1).

Likewise in reactions where reaction rate is important (e.g., to minimize reactor volume), the space-time yield (or productivity) of the reaction can be increased by adding more enzyme, provided a matching increase in product formation results (so as not to increase the cost contribution of the biocatalyst to the final product). In principle, these features result in potentially highly effective bioprocesses, overcoming some of the limitations often observed in fermentations (Table 7.1), and such processes will be discussed further in this chapter.

Although enormous progress has been made in recent years in the application of biocatalysis in industry [6,10–12], process development still requires significant effort. Among other attributes, processes based on biocatalysis can potentially use both approaches already outlined to fulfill the basic requirements of sustainability. Indeed today, the chemical scientific literature is full of examples where biocatalytic reactions complement conventional chemical steps in new routes to existing or novel molecules [13]. Increasingly, such examples are multi-step, involving several enzymes in simultaneous one-pot cascades or sequential one-pot cascades [14–18]. More recently, chemoenzymatic conversions have

Table 7.1 Biocatalytic process efficiency and potential limitations.

Performance	Potential limitation
Selective	Enzyme
Reaction rate	Amount of catalyst
	Mass transfer
Reaction yield	Thermodynamics
	Stoichiometry

also been explored, where both enzymatic and chemical catalysts are present in one pot either simultaneously or, more commonly, sequentially [19,20]. As the availability of enzymes increases, alongside improved enzyme expression levels, increasing numbers of potential synthetic routes have been proposed, involving one or more biocatalytic steps [21]. Although today there are several hundred industrial examples reported of biocatalytic processes (mostly in the high-value pharmaceutical sector), many examples remain in the laboratory. There are several reasons for this including the fact that industrially relevant synthetic problems are rarely found in nature. One consequence of this is that the available enzymes are usually unsuitable, without some (protein) modification, for the types of molecules and concentrations of reagents used in industry. It is clear, therefore, that the translation of such exciting proof of concept chemistry into scalable processes still requires considerable work [22,23]. Indeed, evaluation of many laboratory-based biocatalytic reactions rapidly reveals that while the selectivity of most of these reactions is without precedent, scale-up requires significant improvements in reaction yield, product concentration, enzyme turnover, and space-time yield. These so-called costing (or economic) metrics [22,24–26] provide the basis for benchmarking processes at different stages of development against existing processes, or alternatively against previously calculated targets based on economic requirements. Both enzyme modification (by protein engineering [27]) and process development (by reaction and reactor engineering [22]) can contribute to such improvements, and the benchmarking methodology provides a systematic way of identifying the extent of improvement required. This is particularly valuable for protein engineering where method development takes time and setting targets is a useful tool for effective screening [28,29].

Although new processes always need to meet economic thresholds to achieve commercial viability, a major motivation for implementation of many biocatalytic processes remains the claim that such processes are environmentally improved versions of chemical processes [30–32]. There are of course many potential environmental benefits of operating with biocatalysis, and many scientists and engineers perceive biocatalysis as an important route to the development of green chemical manufacturing processes for the future [33]. Nevertheless such claims are usually qualitative, and therefore carry limited value. With the objective of quantifying such claims several attempts to use full life cycle assessment (LCA) on biocatalytic processes have been reported [34,35]. However, such methods are fraught with difficulty, as will be discussed later. An alternative approach might be to simplify the quantitation of eco-efficiency of a process to a single metric as a means of quantitating the sustainability of a biocatalytic process [36]. Other authors suggest that the use of more than one (easily calculated) metric would be better to give a more accurate picture, while still maintaining simplicity [3,37]. In order to benchmark one process against another (or alternatively a predefined target), in much the way as we have advocated for economic benchmarking, we also reasoned that a better approach might be to use a combination of metrics, but even more importantly in our opinion, to use them at the appropriate stage of process development [25]. In this chapter the role of benchmarking based on green chemistry metrics and its role in the development of biocatalytic processes will be discussed.

7.3
Biocatalytic Process Design and Development

Biocatalytic reactions are first discovered by biologists and chemists in the laboratory and, if they are found of sufficient commercial interest will later be scaled-up for industrial implementation, at scales varying from kilograms to thousands of tons per year, dependent upon the final market. Conventionally, such scale-up has been carried out by chemical engineers applying established design and scale-up principles to bioprocesses, in much the same way as has been done for many petrochemical processes. However, in the last few years it has become clear that such an approach brings a number of risks. One of the major risks is not the increase in scale itself, but rather ensuring the reaction of interest is carried out at sufficient process intensity (i.e., that the previously mentioned costing metrics are satisfied). This requires that the reactor operates in such a way that the starting materials are efficiently used and the product concentration is high enough and production rate fast enough for a given economic and market scenario. It seems likely, therefore, that future processes might follow a two-part development paradigm of first increasing intensity, prior to a subsequent increase in scale (schematically represented in Figure 7.2).

7.4
Sustainability of Biocatalytic Processes

Many excellent reviews describe the enormous potential of biocatalytic processes, with respect to selective chemistry and the opportunity for new product and process innovation. Such potential has developed a growing industrial interest in a wide range of applications. Today, many processes have already been implemented, especially in the pharmaceutical industry, which benefit from high selectivity and mild reaction conditions to avoid multiple protection (and

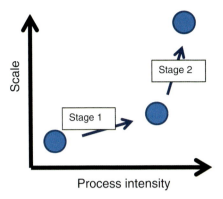

Figure 7.2 Two-stage development paradigm proposed for biocatalytic processes. Stage 1: Lab scale increase in intensity, with focus on enzyme improvement (by protein engineering) and process development (by reaction and reactor engineering). Stage 2: Scale-up.

deprotection) steps [38]. In itself, this leads to significant environmental improvements. The high selectivity of such processes also brings the potential benefit of simplified processes and reduced recovery operations, meaning that the waste from such processes is also much reduced. Combined with the mild reactions conditions (around neutral pH, ambient temperature, and atmospheric pressure), savings can also be made in utilities and energy consumption. Table 7.2 lists some of the claimed environmental benefits of implementing biocatalytic processes.

The use of water as a reaction medium is often cited as a great advantage from an environmental perspective. While it is true that water is benign, as will be discussed later, its contribution in improving the environmental profile of a biocatalytic process is more limited than appears at first, and for this reason it has been excluded from Table 7.2.

For fermentation processes it is obvious that the microbial cells are renewable, since they are grown on sugar, water, and air, with some nitrogen. This is also the case for those microbial cells and enzymes (originating from microbial cells) used as biocatalysts. The focus of this chapter is on chemical products, but in reality, today enzymes have a role in the production of many industrial products (not just chemicals). An interesting quantitative environmental assessment (using LCA) of the role enzymes play in improving the sustainability of

Table 7.2 Claimed environmental benefits of implementing biocatalytic processes.

Claim	Rationale
Reduced utilities	Mild reaction conditions (temperature)
Safe operation	Mild reaction conditions (pressure, pH)
Reduced waste	High selectivity
Renewable catalyst	Biocatalyst produced by fermentation based on sugar, air, and water

production processes, clearly documents biocatalysis to be one of the best examples [34]. The use of a renewable catalyst (rather than the use of metal catalysts (such as Rh, Ru, Pd, or Pt) is also a strong argument for implementation of biocatalytic processes.

Several of these arguments have also been the basis for claims that biocatalytic processes are greener than conventional chemical processes. As argued in several scientific reviews on the topic, such claims need to be substantiated and methods need to be implemented to quantitatively measure the sustainability of biocatalytic processes [33,36,39,40].

7.5
Quantitative Measuring of the Sustainability of Biocatalytic Processes

One of the most important issues in applying a methodology to assess processes for their sustainability (eco-efficiency) is perhaps the lack of clarity in reported examples. Indeed, in the scientific literature it would appear at first sight to be somewhat confusing. For example, there are many different methods proposed to carry out such an assessment [41–46]. There are also several excellent examples [34,35,47–51] attempting to quantify the green chemistry credentials of biocatalysis, but, here too, different objectives and methodologies have been used.

From the perspective of the green chemistry profile of a process, efforts toward systematic design methods, taking into account waste minimization were already developed over 20 years ago [52], and have today developed into sophisticated software tools such as LCA [53,54] (or its alternative, LCIA). However, such sophistication comes at a price. Such methods require significant input data, much of which is often lacking for biocatalytic processes, and in any case such data are often not available at an early stage of process development. This is a particular concern for biocatalytic processes since evaluation of alternative routes and processes needs to happen early in process development, at the point where biocatalytic processes are often in need of significant improvement. A consequence of the proposed two-stage approach to design ((1) intensification and (2) scale-up) is the need to assess processes very early in the development cycle. This is also necessary to take advantage of the unique ability to alter biocatalyst properties by protein engineering. Clearly, when such changes are required it would be beneficial to carry them out ahead of the major process design effort, meaning an early identification is required. All these arguments indicate that there is value in early stage process evaluation. For conventional chemical processes there are several methods available for evaluation of processes, including metric-based approaches for early stage (or rapid) evaluation [55–57] and detailed full LCA later in development [53,54]. For biocatalytic processes, the first evaluation frequently needs to take place at the initial stages of development. At this early stage, while some quantitation is both desirable and possible, a full analysis of the process is out of scope. Although a rough flow sheet may be drawn, the full details are unknown and information about utilities and process conditions are undecided. Second, and of equal importance, is that at an early stage the risks of process development are too great to justify extensive time spent on analysis. Instead it is appropriate to assess processes rather quickly with the primary objective of helping decision-making in process development.

Later when a process is already "intensified," and ready for scale-up (Figure 7.2), a more detailed analysis can be undertaken, potentially including LCA. But even for late-stage processes, the problem is exacerbated in the case of bioprocesses, by the lack of basic information on relevant compounds available in databases for LCA and a lack of suitable process models. It has long been argued that accurate comparisons between processes require the use of full LCA for environmental benchmarking. Indeed these tools have an important role to play in comparing implemented processes, but are not helpful for processes still in the development phase. A related argument for delaying benchmarking using these tools is that some inputs, such as energy efficiency and material costs, are scale dependent.

These are the drivers for the implementation of simple, readily measurable process metrics at an early development stage of a process. Benchmarking such metrics can be used both to assess the feasibility of a process, as well as to drive process development.

7.6
Early Stage Sustainability Assessment

Evaluating potential biocatalytic processes can initially be done *in silico*, prior to any experimentation. Metrics can be used at this early stage to eliminate routes that should not go to the laboratory, based on environmental or economic concerns. In the laboratory, the development toward full-scale production requires several rounds of iterative enzyme improvement and reaction engineering. In many cases the reactor configuration is predetermined but how to operate it remains an open question. At this point there are two basic tasks that need to be done. The first is to improve the intensity of the process such that a minimum threshold performance is achieved in the laboratory. As suggested previously, it would appear preferable to achieve this prior to scale-up, where the volume, rather than intensity of the process is increased. The primary task, therefore, of those charged with process development of new biocatalytic processes is to achieve the necessary intensity. Both environmental and economic metrics can be used to assess the process at the different experimental stages, but the focus in this text is on environmental metrics.

The methodology proposed divides the process into two stages: Evaluation of (1) route feasibility and (2) biocatalyst and reaction development. The objective at each stage is different and the type of environmental assessment at each stage is, therefore, also different (see Table 7.3).

Table 7.3 Potential staged approach to benchmark biocatalytic processes on an environmental basis.

Stage	Benchmarking objective	Methodology
1	Route feasibility	*In silico* metrics: AE; CME
2	Biocatalyst and reaction development	*In vitro* metrics: PMI; SI; WI; E-factor

7.6.1
Evaluation of Route Feasibility

Routes for the synthesis and production of industrially useful chemical products usually involve multiple reaction and separation steps [58]. Ideally, using retrosynthetic tools, alternative routes can be evaluated [59]. Recently, it has been emphasized that the number of enzymes commercially available now means that retrosynthesis can also incorporate biocatalysis [21] for key chemical transformations, especially those involving chiral centers. This opens the way to a broader selection of alternative routes. There are several considerations here, including the availability of enzymes and substrates, but aside from this, each route should be evaluated in terms of two simple green chemistry metrics, atom economy (AE) and carbon mass efficiency (CME).

7.6.1.1 Atom Economy

Abbreviation: AE

Definition: MW product/MW substrates

Rationale: Atom economy measures how much of the starting material ends in the desired product [60]. It is dependent upon the reaction stoichiometry, so different reactions and routes can be expected to give different values for atom economy. A high atom economy will ensure a route with effective use of the substrate and minimal waste, with a higher proportion of substrate atoms found in the product molecule [61].

7.6.1.2 Carbon Mass Efficiency

Abbreviation: CME

Definition: Mass of carbon in product/mass of carbon in reagents

Rationale: Measures how much of the carbon in the substrate ends in the final product. It is very useful, therefore, as a means of assessing if carbon is wasted by being produced as by-product (although this does not include water). It is normally expressed as a percentage.

The objective behind using these green chemistry metrics [56] is to avoid the need for full evaluation of the environmental impact of the route of interest at such an early stage. It is a shortcut method and must, therefore, be treated with some caution. The metrics have been chosen to reflect the most important environmental effects. Unlike costing metrics, the green chemistry metrics to be used in a given case are independent of the value of the product.

From an economic perspective, the value added and theoretical yield (allowing for thermodynamic limitations) should also be assessed at this point. The green chemistry metrics can be evaluated *in silico*, ahead of experimental work, since the values obtained will be independent of the way the reaction is operated. Already at this first stage some routes may be eliminated on an economic or environmental basis. With respect to biocatalysis, this also places the biocatalytic steps in the context of the surrounding chemical steps.

7.6.2
Evaluation of Biocatalyst and Reaction Development

At this second stage, each biocatalytic reaction from the route is tested in the laboratory. The aim of such laboratory tests is first to establish proof of concept and second to provide an initial set of (economic) metrics, which can subsequently be used as the basis for defining the necessary improvements in performance to achieve the required process intensity such that the economic constraints on the process are satisfied. With more data, sensitivity analyses can be used to assess *in silico* the effects of potential changes on improving metrics. Changes can be made at the level of the biocatalyst or the reaction. Interestingly, the economic metrics measured in the laboratory can also be used to calculate environmental metrics, such as the process mass intensity, solvent intensity, water intensity, and E-factor.

7.6.2.1 Process Mass Intensity

Abbreviation: PMI

Definition: Total mass used in a process/mass of product

Rationale: Measures at a high level the environmental sustainability of a process. The value should be as low as possible, but gives an excellent idea about how far a process is from being sustainable. Interestingly, it has been selected by the ACS Green Chemistry Institute Roundtable as the single most useful metric to compare processes [62,63]. As argued previously, the use of a metric like PMI avoids the detailed information necessary for a full LCA.

7.6.2.2 Solvent Intensity

Abbreviation: SI

Definition: Total mass of solvent used/mass of product

Rationale: Measures at a high level the amount of solvent used in a process. The value should be as low as possible and reflects the downstream process, as well as the reaction stage.

7.6.2.3 Water Intensity

Abbreviation: WI

Definitions: Total mass of water used/mass of product

Rationale: Water intensity measures at a high level the amount of water used in a process. The value should be as low as possible.

7.6.2.4 E-factor

Abbreviation: E–factor

Definition: Mass of waste/mass of product

Rationale: Measures at a high level the amount of waste generated in a process [64]. It does not take into account the type of waste.

It is important to recognize that these values will be far from ideal since the process is not defined yet for optimal use of the enzyme, but such values may be compared (or "benchmarked") against standards, or known processes, as the basis for setting targets for improving the process. This is the premise then for defining a development plan on a systematic basis.

7.7 Benchmarking

Economic benchmarking provides a plan for development of a process, as well as ultimately an assessment of feasibility. In contrast, sustainability (environmental) benchmarking is different in that it allows an estimate of the eco-efficiency profile of a given process at a given stage of development. It may provide extra information of use in decision-making, but it can also provide guidance for process improvement and development.

Application of the reaction intensity metrics (PMI, SI, WI, and E-factor) enables a quick benchmarking of a reaction. However, they should be applied at the correct stage of process development. For example, comparing a single liquid phase with a fed-batch or biphasic system will make significant differences to the values obtained. Likewise, comparison of environmental metrics at an early stage of development with another process at a late stage will lead to the wrong value being quoted. Most important is to benchmark processes at exactly the same point in the process development cycle. Most chemical processes do not see the same degree of process improvement as bioprocesses, where order of magnitude, or more, improvements are common.

7.7.1 Route Selection

Routes to advanced chemicals involve multiple steps, and while achieving step economy is a major objective in retrosynthetic design of such routes, so also is incorporation of biocatalytic steps. Indeed, recent developments in biocatalysis now mean that many conventional reaction steps can be replaced by enzymatic ones, opening the possibilities for multistep biocatalysis. The use of multistep biocatalytic cascades is now well established, meaning that innovative new routes to products can also be considered. This also implies that alternative routes will need to be evaluated using suitable costing metrics (e.g., value added by the route and theoretical thermodynamic yield of the route) and green chemistry metrics (e.g., atom economy and carbon mass efficiency). It is perhaps not unreasonable to select routes in which all the atoms of the substrate make their way to the product molecule, but this also depends on the type of reaction. Cleavage reactions and transferase-type reactions will inevitably have lower values for atom economy. Likewise, CME values close to unity could be expected, but again are dependent on the type of reaction.

7.7.2
Biocatalyst and Reaction Development

Benchmarks in the biocatalyst and reaction development stage are more complicated perhaps, but a PMI of around 10 would be expected for an individual process step. This correlates with the E-factor (= PMI−1). E-factors for biocatalytic conversions are correlated with the initial concentration of the substrate and reaction yield, such that high initial concentrations and reaction yields lead to lower E-factors [36]. Low-substrate concentrations are typically used in laboratory reactions, placing emphasis on the need for biphasic and fed-batch systems in processes. Likewise, E-factors measured in the reaction stage are somewhat higher than would be expected in an industrial setting when operating a commercial process. However evaluated, the dominant part of the E-factor comes from the water, when used in dilute aqueous phase reactions. Values of 10 kg/kg would not seem unreasonable for a typical reaction. It is important to be aware that the benchmarks very much depend on the number of steps in a process. More steps will lead to higher values for the E-factor.

Water intensity (WI) in general will be high, since this is the solvent of choice for many bioprocesses. However, it comes with its own problems, but can be expected to be over 10 kg/kg even for optimized systems. On the other hand, solvent intensity (SI) might be predicted to be zero, but in reality many processes use solvent as a partial replacement for some of the water. Values around 5 kg/kg would seem to be reasonable.

7.8
Examples

It is clear that during development enormous improvements in a process are possible. Unlike chemical processes, where optimization and clever process configurations alone lead to improvement, in biocatalysis there is the enormously powerful option of protein engineering. Protein engineering can be used to improve multiple aspects, which might otherwise limit a process. In the two examples which follow, the improvements due to the combined efforts of protein engineering and process engineering are documented. These two industrial examples serve well to illustrate the importance of using the right metrics at the right point of development.

7.8.1
Biocatalytic Route to Atorvastatin

Atorvastatin is the active ingredient in the drug Lipitor®, which is a cholesterol-lowering drug, and is one of the biggest selling drugs of all time with annual sales that exceeded USD 10 billion in 2010. The molecule contains two chiral centers and, therefore, biocatalytic routes have always been of interest to help synthesize the product in an effective way. This, coupled with the high demand for the drug, makes it a particularly interesting example. Early work to synthesize atorvastatin by biocatalytic methods focused on kinetic resolutions using whole cells, as well as chemoenzymatic routes using nitrilases and lipases. Nevertheless, these routes failed to meet the requirements for a scalable, commercial process, and were, therefore, not implemented. At Codexis Inc. (Redwood City, CA, USA), scientists began work on another route, using a ketoreductase (KRED) linked with glucose dehydrogenase (GDH) to recycle the NADPH required by the KRED. This step is followed by a second enzymatic step, using halohydrin dehalogenase (HHDH). The new route was found to have an atom economy of 0.45. Although this is rather low, the reductant was glucose which is a renewable feedstock. In principle, the aim should be an atom economy of 1.00, but judgment is also necessary in interpreting the values. Likewise, the carbon mass efficiency is 50%. Again a value of 100% would be desirable, although inspection shows the loss is due to NADPH recycle, which is essential from an economic perspective. Having identified a suitable route (see Figure 7.3), initial experiments were carried out with the two reactions independently.

In the first reaction while the selectivity was found to be suitably high (ee greater than 99.5%), the reaction rate was very low, implying a large amount of enzyme was required in order to obtain a sufficient space-time yield. Both the KRED and GDH protein were improved using several rounds of DNA shuffling (protein engineering), 7-fold and 13-fold, respectively. This meant that less enzyme was used, resulting not only in a better biocatalyst yield (g product/g biocatalyst), but also in improvements for the downstream product recovery. The GDH recycle reaction produced gluconate, implying the pH would drop unless neutralized and this, therefore, demanded the use of a stirred tank reactor with pH measurement and control.

In the subsequent step HHDH was used. In initial experiments, the activity and stability were found to be low, and this was also improved via DNA shuffling. Finally, the kinetics of the HHDH reaction was found to exhibit strong product inhibition. Although attempts were made to overcome this by *in situ* product removal (ISPR), ultimately a better solution in this case was found by

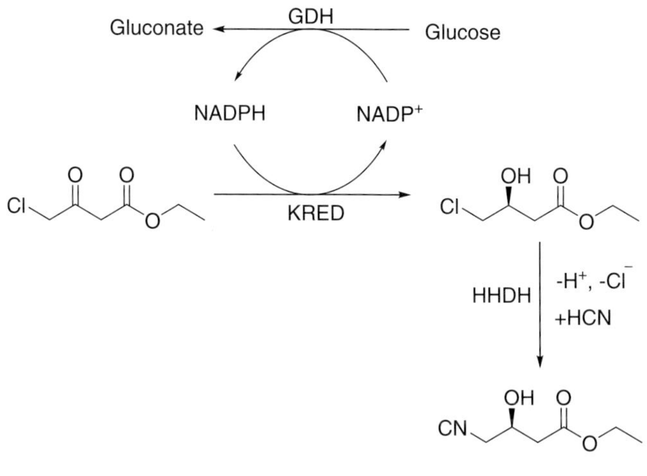

Figure 7.3 Biocatalytic synthesis of atorvastatin.

improving the enzyme. The activity was increased 2500-fold over the wild type. Such improvements are without precedent in chemical processes but indicate the extent of improvement possible in bioprocesses. The increase in concentration possible, as a result of the protein engineering efforts, has a direct impact on the sustainability of the process. Interestingly, metrics for cascades or routes will of course be higher than for individual steps. For example, in this synthesis it was found that the individual E-factors (excluding water) for the first and second enzymatic step were 2.3 and 3.5, respectively, making a route total of 5.8. Interestingly, including water in the calculation gives E-factor values of 6.6 and 11.4 for the two enzymatic steps, respectively, and a route value total of 18.0. E-factors for multistep routes and cascades are calculated additively (as long as each stage is normalized to the production of 1 kg of final product) and will of course be higher than for individual steps.

Today, the process is run at large scale with a space-time yield of 20 g/(l h), biocatalyst yield of 178 g/g, and product concentration of 152 g/l [47]. These are typical values for industrial commercialized biocatalytic process, when benchmarked against other such processes [24,25] (although much higher than reported for typical laboratory reactions). The enormous improvement in the enzymes in this example enabled the implementation of an economic, but also sustainable process. The huge improvement over the development cycle of the process also indicates the importance of benchmarking processes at similar development stages.

7.8.2
Biocatalytic Route to Sitagliptin

Sitagliptin is an antidiabetic compound and its phosphate salt is the active ingredient in a pharmaceutical product marketed as Januvia® by Merck and Co

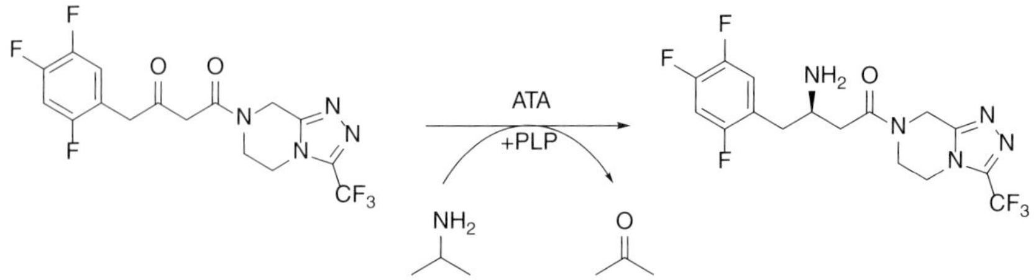

Figure 7.4 Biocatalytic synthesis of sitagliptin.

(Rahway, NJ, USA). It is a highly complicated molecule and for sometime was manufactured using asymmetric hydrogenation of an enamine. The reaction takes place at high pressure (17 bar) using a rhodium-based chiral catalyst [65]. However, the application of pressure and use of the metal-based catalyst, potentially contaminating the product, are not ideal from a sustainability perspective. Likewise, the stereoselectivity of the rhodium-based catalyst is limited, making the process inefficient for the production of the optically pure chiral amine product. For these reasons, scientists at Merck and Co (Rahway, NJ, USA) considered using an aminotransferase (specifically an ω-transaminase) to asymmetrically synthesize the chiral center with a suitable amine donor (see Figure 7.4).

Although in the last decade there has been significant research focus on the application of ω-transaminases [66–71] and other biocatalytic approaches to chiral amine synthesis [72–74], nevertheless the molecule to be converted to synthesize sitagliptin was without precedent. Initially, *in silico* design techniques were applied to provide a suitable starting point to engineer the enzyme to accept the large substituent adjacent to the ketone. Using substrate walking (in which the substrate is altered in steps, so as to allow the enzyme to be altered and screened in stages) the large binding pocket of the enzyme was altered and subsequently evolved for activity. This enabled a 75-fold improvement of activity over the first mutated enzyme. This provided the starting point for fine tuning the enzyme.

Although an active enzyme had now been created and it could only work at concentrations of 2 g/l, clearly too low for feasible commercial operation of the reactor, failing targets for both economic and sustainability metrics. The problem was made worse since the water solubility of the compound is low, and therefore, in the next stage of development, various solvents were assessed. Methanol and DMSO both proved effective solvents. In addition, the equilibrium of the reaction needed to be pushed toward the product, meaning an excess of the amine donor was required [75]. Indeed despite the reasonable thermodynamic favorability of this particular reaction, a 1M concentration (a fourfold excess) of the amine donor was still required to achieve satisfactory conversion. Since IPA (amine donor) is converted to acetone (by-product) during the reaction, clearly the enzyme must also be tolerant to high concentrations of acetone. In an extensive enzyme evolution (protein engineering) program, the enzyme tolerance was improved to be able to effectively convert substrate concentrations from 2 to 100 g/l, IPA from 0.5 to 1 M, DMSO from 5 to 50%, pH from 7.5 to 8.5, and temperature from 22 to 45 °C [76]. A selectivity of more than 99.99% was achieved. Simultaneous optimization included improving stability and expression levels in *Escherichia coli*, so as to reduce the biocatalyst cost. The optimal process at this point was able to convert 200 g/l with greater than 99.5% conversion at 92% yield with 6 g/l enzyme in 50% DMSO. Compared to the rhodium-based process this gave a 10–13% increase in yield, 53% increase in space-time yield (g/(l h)), and 19% reduction in total waste. Additionally, the process was able to operate in nonspecialized plant at ambient pressure.

In a further development [77], a second-generation biocatalytic process was developed to operate in neat organic solvent using immobilized ω-transaminase, to afford still further benefits for the product recovery and potentially allow continuous operation. Indeed operation could be sustained for 200 h, which meant that the biocatalyst yield was improved to manageable levels. This was important for the final implemented industrial process.

Today, the process is implemented at large scale, with excellent metrics. A direct comparison of the metrics for the laboratory and process cannot easily be made since a different form of biocatalyst was used, and for a true comparison to be made, immobilized enzyme loading data would also be required. This again emphasizes the importance of comparing metrics at similar points in the development cycle. Additionally, benchmarking is also dependent upon similar biocatalyst format. The entire development procedure is summarized in a seminal paper by the joint Codexis and Merck team [76].

7.9
Future Perspectives

7.9.1
Process Development

The metrics that reflect reaction and process intensity (PMI, SI, WI, and E-factor) are very important as a tool in development. Interestingly, the real target is PMI, since the balance of SI and WI is dependent upon the balance of water and solvent in the process. Nevertheless, it is clear that recycling reagents (where economically feasible) can also have a "green" benefit. For example, in the synthesis of atorvastatin [47], the solvent intensity (which was reported as contributing 50% to the E-factor, excluding water) was assumed with a recycle of 85%. It is obvious that without recycle the contribution would be even greater. Clearly, evaluating recycle options is an essential part of process development, both to satisfy economic, but also sustainability, metrics. It is important to consider the ease of recycle early enough when selecting reagents, rather than as an optimization task later in the design.

The use of aqueous solutions for reactions looks highly attractive at first glance, but in reality it hides a further problem downstream. The aqueous

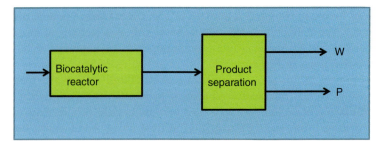

Figure 7.5 Standard separation in aqueous phase biocatalytic process. (W, waste; P, product).

material contains the product at a relatively low concentration, meaning removal of water is required ahead of separation and purification (Figure 7.5).

Given the high boiling point of water, this makes evaporation expensive and potentially more attractive to extract the product into a lower boiling solvent (at high concentration), which can then be evaporated more easily (Figure 7.6). Other options could include membrane-based unit operations. Nevertheless, the large amount of water makes this step difficult.

Water replacement using solvents can be back integrated into the reactor itself. For example, two-liquid phase biocatalysis using biphasic reaction media has proven very attractive for many biocatalytic reactions (Figure 7.7).

Furthermore, the concentration of substrate (and therefore product) is of key importance in the metrics. As expected, both PMI and E-factor (since they are related) are strongly correlated with substrate concentration. Thus, a low aqueous phase concentration (high aqueous intensity), leads to a high PMI and hence also a high E-factor. Two obvious implications result from this. The first is that to decrease the PMI, it is beneficial to replace some of the water by organic

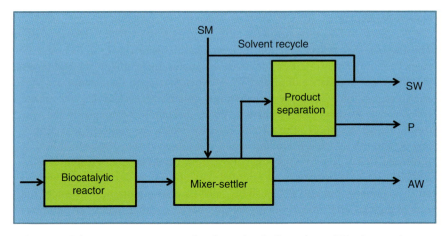

Figure 7.6 Solvent swap to recover product from a low boiling solvent. (SM, solvent makeup; SW, solvent waste; P, product; AW, aqueous waste).

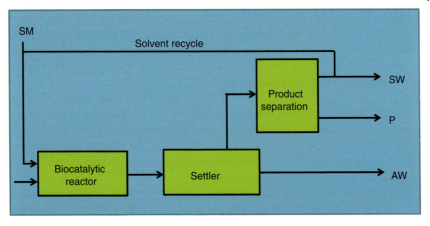

Figure 7.7 Two-liquid phase biocatalysis. (SM, solvent makeup; SW, solvent waste; P, product; AW, aqueous waste).

solvent (as indicated previously) and the second is to increase the concentration as much as possible (even in organic solvent this will decrease SI). This also has a direct beneficial impact on the process economics, since the product concentration is the costing metric which is used to reflect the operating and capital cost of the downstream process.

Finally, improvement can also come through the integration of reactions together. Cascade biocatalysis enables reactions to be carried out without the need to isolate intermediates and, therefore, results in an improvement in the overall sustainability performance of the process. Matching the reaction conditions (i.e., with one enzyme reaction leading to the next) also minimizes the need for changes in conditions.

7.9.2
Methodology

One of the most confusing issues in the scientific literature documenting sustainability concerns the nomenclature and use of different metrics. For instance, E-factor is sometimes quoted including water, and sometimes without water. This makes a big difference for biocatalytic processes, which require at least some water for operation. For example, as highlighted previously, in the synthesis of atorvastatin (Section 7.8.1), the E-factor without water was 5.8, and with water was 18. It is a substantial difference. To confuse this further, PMI is also sometimes quoted with, and sometimes quoted without water (even though in principle there should be a direct relationship between the PMI and the E-factor: E-factor (kg/kg) = (PMI (kg/kg) −1). This is very well illustrated in the case of the synthesis of 7-aminocephalosporic acid (7-ACA) [35] where the PMI for a shift from a chemical to an enzymatic process was reduced from 81 to 44 kg/kg, while at the same time the E-factor increased from 93 to 172 kg/kg. The reason for this would appear to be that while water is excluded in the PMI definition, it is included in the E-factor definition used here. It is important the definitions are consistent, even if they exclude water. More accurate though is probably to include water. In fact, in the pharmaceutical industry the high purity of water required for manufacturing will in itself have an impact. Furthermore, an important reason to include water in the E-factor is that when water comes into contact with solvent, the water needs to be treated, and therefore, is an important part of E-factor calculations, since it is no longer benign. However, the nature of the solvent (in contact with the water) determines the subsequent treatment. Hence, using the E-factor alone as a metric is not really sufficient, and a more multivariate approach is probably better.

Likewise, the metrics to be used to benchmark processes are equally varied. For good reasons the PMI is the single metric proposed by the ACS Roundtable [63,78], while other authors propose E-factor on the basis of simplicity [36]. Use of PMI places particular emphasis on the contribution of the extra reagents to the sustainability profile, rather than emphasizing the consequential downstream processing implications of waste removal. Other authors again propose several metrics for evaluation [3]. It seems likely that more than one metric is indeed required, but which metrics should be used at a given stage is also dependent upon the stage of development of the process. It seems that standardization of definitions and application of the metrics should be an essential objective of methodological research in the future.

In several recent publications, metrics for costing bioprocesses have been proposed as a shortcut route to benchmark a new biocatalytic reaction alongside alternative processes. A major driver behind this approach has been to enable informed targets for these metrics to provide a strategy for process development. In this way the development procedure can be targeted on the areas of the process in greatest need of improvement. Likewise, this approach enables the elimination of some routes on the basis that the development required is infeasible. It makes sense now to integrate the green chemistry metrics discussed here together with the costing metrics to provide a systematic, staged methodology to the development, design, and scale-up of biocatalytic processes [25].

7.10
Concluding Remarks

Several conclusions can be drawn from the brief discussion of benchmarking tools available and reported in the scientific literature to help quantify the varied claims for environmental improvement through the introduction of biocatalytic processes. They are best summarized with the following five suggestions for future developments in the field:

1) Standardize the green chemistry metrics (e.g., PMI or E-factor) used for benchmarking and the definitions to be used (e.g., E-factor with or without water).

2) Benchmark processes at similar stages of development (e.g., early stage process with early stage process).
3) Integrate costing metrics (e.g., space-time yield, product concentration, biocatalyst yield, reaction yield) with green chemistry metrics to guide process development.
4) Extend the application of green chemistry metrics to complete route benchmarking, rather than solely single steps.
5) Delay full LCA and implementation of costing software to later in the development cycle.

Used correctly, such tools can not only help validate the environmental claims for biocatalysis, but also provide a useful complement to the costing tools currently available to guide protein and process engineering, to ensure economically feasible implementation of sustainable bioprocesses at an industrial scale.

References

1 Gwehenberger, G. and Narodoslawsky, M. (2008) Sustainable processes – the challenge of the 21st century for chemical engineering. *Process Safety and Environmental Protection*, **86**, 321–327.
2 Anastas, P. and Eghbali, N. (2010) Green chemistry: principles and practice. *Chemical Society Reviews*, **39**, 301–312.
3 Sheldon, R.A. (2014) Green and sustainable manufacture of chemical from biomass: state of the art. *Greem Chemistry*, **16**, 950–963.
4 Tucker, J.L. (2006) Green chemistry, a pharmaceutical perspective. *Organic Process Research & Development*, **10**, 315–319.
5 Sheldon, R.A. (1997) Catalysis: the key to waste minimization. *Journal of Chemical Technology and Biotechnology*, **68**, 381–388.
6 Woodley, J.M., Breuer, M., and Mink, D. (2013) A future perspective on the role of bioprocesses for chemical production. *Chemical Engineering Research and Design*, **91**, 2029–2036.
7 Van Dien, S. (2013) From the first drop to the first truckload: commercialization of microbial processes for renewable chemicals. *Current Opinion in Biotechnology*, **24**, 1061–1068.
8 Sanford, K., Chotani, G., Danielson, N., and Zahn, J.A. (2016) Scaling up of renewable chemicals. *Current Opinion in Biotechnology*, **38**, 112–122.
9 DiCosimo, R., McAuliffe, J., Poulose, A.J., and Bohlmann, G. (2013) Industrial use of immobilized enzymes. *Chemical Society Reviews*, **42**, 6437–6474.
10 De Wildeman, S.M.A., Sonke, T., Schoemaker, H.E., and May, O. (2007) Biocatalytic reductions: from lab curiosity to 'first choice'. *Accounts of Chemical Research*, **40**, 1260–1266.
11 Moore, J.C., Pollard, D.J., Kosjek, B., and Devine, P.N. (2007) Advances in the enzymatic reduction of ketones. *Accounts of Chemical Research*, **40**, 1412–1419.
12 Bommarius, A.S. (2015) Biocatalysis: a status report. *Annual Review of Chemical and Biomolecular Engineering*, **6**, 319–345.
13 Lalonde, J. (2016) Highly engineered biocatalysts for efficient small molecule pharmaceutical synthesis. *Current Opinion in Biotechnology*, **42**, 152–158.
14 Swartz, J. (2006) Developing cell-free biology for industrial applications. *Journal of Industrial Microbiology and Biotechnology*, **33**, 476–485.
15 Swartz, J. (2012) Transforming biochemical engineering with cell-free biology. *A.I.Ch.E. Journal*, **58**, 5–13.
16 Santacoloma, P.A., Sin, G., Gernaey, K.V., and Woodley, J.M. (2011) Multi-enzyme catalyzed processes: next generation

biocatalysis. *Organic Process Research & Development*, **15**, 203–212.

17 Xue, R. and Woodley, J.M. (2012) Process technology for multi-enzymatic reaction systems. *Bioresource Technology*, **115**, 183–195.

18 Wang, J.-B. and Reetz, M.T. (2015) Chiral cascades. *Nature Chemistry*, **7**, 948–949.

19 Vennestrøm, P.N.R., Christensen, C.H., Pedersen, S., Grunwaldt, J.-D., and Woodley, J.M. (2010) Next generation catalysis for renewables: combining enzymatic with inorganic heterogeneous catalysis for bulk chemical production. *ChemCatChem*, **2**, 249–258.

20 Schwartz, T.J., Shanks, B.H., and Dumesic, J.A. (2016) Coupling chemical and biological catalysis: a flexible paradigm for producing biobased chemicals. *Current Opinion in Biotechnology*, **38**, 54–62.

21 Turner, N.J. and O'Reilly, E. (2013) Biocatalytic retrosynthesis. *Nature Biotechnology*, **9**, 285–288.

22 Lima-Ramos, J., Neto, W., and Woodley, J.M. (2014) Engineering biocatalysts and processes for future biocatalytic processes. *Topics in Catalysis*, **57**, 301–320.

23 Ringborg, R.H. and Woodley, J.M. (2016) The application of reaction engineering to biocatalysis. *Reaction Chemistry & Engineering*, **1**, 10–22.

24 Tufvesson, P., Lima-Ramos, J., Nordblad, M., and Woodley, J.M. (2011) Guidelines and cost analysis for catalyst production in biocatalytic processes. *Organic Process Research & Development*, **15**, 266–274.

25 Lima-Ramos, J., Tufvesson, P., and Woodley, J.M. (2014) Application of environmental and techno-economic metrics to biocatalytic process development. *Green Process Synthesis*, **3**, 195–213.

26 Lundemo, M.T. and Woodley, J.M. (2015) Guidelines for development and implementation of biocatalytic P450 processes. *Applied Microbiology and Biotechnology*, **99**, 2465–2483.

27 Bornscheuer, U.T. (2012) Engineering the third wave of biocatalysis. *Nature*, **485**, 185–194.

28 Woodley, J.M. (2013) Protein engineering of enzymes for process applications. *Current Opinion in Chemical Biology*, **17**, 310–316.

29 Dör, M., Fibinger, M.P.C., Last, D., Schmidt, S., Santos-Aberturas, J., Böttcher, D., Hummel, A., Vickers, C., Voss, M., and Bornscheuer, U.T. (2016) Fully automated high-throughput enzyme library screening using a robotic platform. *Biotechnology and Bioengineering*, **113**, 1421–1432.

30 Bull, A.T., Bunch, A.W., and Robinson, G.K. (1999) Biocatalysts for clean industrial products and processes. *Current Opinion in Microbiology*, **2**, 246–251.

31 Wohlgemuth, R. (2010) Biocatalysis – key to sustainable chemistry. *Current Opinion in Biotechnology*, **21**, 713–724.

32 Wenda, S., Illner, S., Mell, A., and Kragl, U. (2011) Industrial biotechnology – the future of green chemistry? *Greem Chemistry*, **13**, 3007–3047.

33 Jiménez-González, C., Poechlauer, P., Broxterman, Q.B., Yang, B.-S., am Ende, D., Baird, J., Bertsch, C., Hannah, R.E., Dell'Orco, P., Noorman, H., Yee, S., Reintjens, R., Wells, A., Massonneau, V., and Manley, J. (2011) Key green engineering research areas for sustainable manufacturing: a perspective from pharmaceutical and fine chemical manufacturers. *Organic Process Research & Development*, **15**, 900–911.

34 Nielsen, P.H., Oxenbøll, K.M., and Wenzel, H. (2007) Cradle-to-gate environmental assessment of enzyme products produced industrially in Denmark by Novozymes A/S. *The International Journal of Life Cycle Assessment*, **12**, 432–438.

35 Henderson, R.K., Jiménez-González, C., Preston, C., Constable, D.J.C., and Woodley, J.M. (2008) EHS & LCA assessment for 7-ACA synthesis: a case study for comparing biocatalytic & chemical synthesis. *Industrial Biotechnology*, **40**, 180–192.

36 Ni, Y., Holtmann, D., and Hollmann, F. (2014) How green is biocatalysis? *To calculate is to know. ChemCatChem*, **6**, 930–943.

37 Sheldon, R.A. (2011) Utilisation of biomass for sustainable fuels and

chemicals: molecules, methods and metrics. *Catalysis Today*, **167**, 3–13.
38. Pollard, D.J. and Woodley, J.M. (2007) Biocatalysis for pharmaceutical intermediates: the future is now. *Trends in Biotechnology*, **25**, 66–73.
39. Woodley, J.M. (2008) New opportunities for biocatalysis: making pharmaceutical processes greener. *Trends in Biotechnology*, **26**, 321–327.
40. Kim, S., Jiménez-González, C., and Dale, B.E. (2009) Enzymes for pharmaceutical applications – a cradle-to-gate life cycle assessment. *The International Journal of Life Cycle Assessment*, **14**, 392–400.
41. Saling, P., Kircherer, A., Dittrich-Krämer, B., Wittlinger, R., Zombik, W., Schmidt, I., Schrott, W., and Schmidt, S. (2002) Eco-efficiency analysis at BASF: the method. *The International Journal of Life Cycle Assessment*, **7**, 203–218.
42. Shonnard, D.R., Kirchere, A., and Saling, P. (2003) Industrial applications using BASF eco-efficiency analysis: perspectives on green engineering principles. *Environmental Science and Technology*, **37**, 5340–5348.
43. Voss, B., Andersen, S.I., Taarning, E., and Christensen, C.H. (2009) C-Factors pinpoint resource utilization in chemical industrial processes. *ChemSusChem*, **2**, 1152–1162.
44. Andraos, J. (2013) Safety/hazard indices: completion of a unified suite of metrics for the assessment of 'greenness' for chemical reactions and synthesis plans. *Organic Process Research & Development*, **17**, 175–192.
45. Xiao, T., Shirvani, T., Inderwildi, O., Gonzales-Cortes, S., Al Megren, H., King, D., and Edwards, P.P. (2015) The catalyst selectivity index (CSI): a framework and metric to assess the impact of catalyst efficiency enhancements upon energy and CO_2 footprints. *Topics in Catalysis*, **58**, 682–695.
46. Smith, R.L., Ruiz-Mercado, G.R., and Gonzalez, M.A. (2015) Using GREENSCOPE indicators for sustainable computer-aided process evaluation and design. *Computers & Chemical Engineering*, **81**, 272–277.
47. Ma, S.K., Gruber, J., Davis, C., Newman, L., Gray, D., Wang, A., Grate, J., Huisman, G.W., and Sheldon, R.A. (2010) A green-by-design biocatalytic process for atorvastatin intermediate. *Greem Chemistry*, **12**, 81–86.
48. Kuhn, D., Kholiq, M.A., Heinzle, E., Bühler, B., and Schmid, A. (2010) Intensification and economic and ecological assessment of a biocatalytic oxyfunctionalization process. *Greem Chemistry*, **12**, 815–827.
49. Brinkmann, T., Rubbeling, H., Fröhlich, P., Katzberg, M., and Bertau, M. (2010) Application of material and energy flow analysis in the early stages of biotechnical process development – a case study. *Chemical Engineering and Technology*, **33**, 618–628.
50. Jegannathan, K.R. and Henning, P.H. (2013) Environmental assessment of enzyme use in industrial production – a literature review. *The Journal of Cleaner Production*, **42**, 228–240.
51. Illner, S., Plagemann, R., Saling, P., and Kragl, U. (2014) Eco-efficiency analysis as a reaction–engineering tool – case study of a laccase-initiated oxidative C-N coupling. *Journal of Molecular Catalysis B: Enzymatic*, **102**, 106–114.
52. Douglas, J.M. (1992) Process synthesis for waste minimization. *Industrial and Engineering Chemistry Research*, **31**, 238–243.
53. Azapagic, A. (1999) Life cycle assessment and its application to process selection, design and optimization. *Chemical Engineering Journal*, **73**, 1–21.
54. Azapagic, A. and Clift, R. (1999) Life cycle assessment and multiobjective optimization. *The Journal of Cleaner Production*, **7**, 135–143.
55. Curzons, A.D., Constable, D.J.C., Mortimer, D.N., and Cunningham, V.L. (2001) So you think your process is green, how do you know? – Using principles of sustainability to determine what is green – a corporate perspective. *Greem Chemistry*, **3**, 1–6.
56. Constable, D.J.C., Curzons, A.D., and Cunningham, V.L. (2002) Metrics to 'green' chemistry – which are the best? *Greem Chemistry*, **4**, 521–527.

57 Marteel, A.E., Davies, J.A., Olson, W.W., and Abraham, M.A. (2003) Green chemistry and engineering: drivers, metrics, and reduction to practice. *The Annual Review of Environment and Resources*, **28**, 401–428.

58 Butters, M., Catterick, D., Craig, A., Curzons, A., Dale, D., Gillmore, A., Green, S.P., Marziano, I., Sherlock, J.-P., and White, W. (2006) Critical assessment of pharmaceutical processes – a rationale for changing the synthetic route. *Chemical Reviews*, **106**, 3002–3027.

59 Wender, P.A., Verma, V.A., Paxton, T.J., and Pillow, T.H. (2008) Function-oriented synthesis, step economy, and drug design. *Accounts of Chemical Research*, **41**, 40–49.

60 Trost, B.M. (1991) The atom economy – a search for synthetic efficiency. *Science*, **254**, 1471–1477.

61 Sheldon, R.A. (2000) Atom efficiency and catalysis in organic synthesis. *Pure and Applied Chemistry*, **72**, 1233–1246.

62 Jiménez-González, C., Ollech, C., Pyrz, W., Hughes, D., Broxterman, Q.B., and Bhathela, N. (2013) Expanding the boundaries: developing a streamlined tool for eco-footprinting of pharmaceuticals. *Organic Process Research & Development*, **17**, 239–246.

63 Jiménez-González, C., Ponder, C.S., Broxterman, Q.B., and Manley, J.B. (2011) Using the right green yardstick: why process mass intensity is used in the pharmaceutical industry to drive more sustainable processes. *Organic Process Research & Development*, **15**, 912–917.

64 Sheldon, R.A. (2007) The E-Factor: 15 years on. *Greem Chemistry*, **9**, 1273–1283.

65 Hansen, K.B., Hsiao, Y., Xu, F., Rivera, N., Clausen, A., Kubryk, M., Krska, S., Rosner, T., Simmons, B., Balsells, J., Ikemoto, N., Sun, Y., Spindler, F., Malan, C., Grabowski, E.J.J., and Armstrong, J.D. III (2009) Highly efficient asymmetric synthesis of Sitagliptin. *Journal of the American Chemical Society*, **131**, 8798–8804.

66 Koszelewski, D., Tauber, K., Faber, K., and Kroutil, W. (2010) ω-transaminases for the synthesis of nonracemic α-chiral primary amines. *Trends in Biotechnology*, **28**, 324–332.

67 Truppo, M., Rozzell, J.D., Moore, J.C., and Turner, N.J. (2009) Rapid screening and scale-up of transaminase catalyzed reactions. *Organic & Biomolecular Chemistry*, **7**, 395–398.

68 Truppo, M.D., Rozzell, J.D., and Turner, N.J. (2010) Efficient production of enantiomerically pure chiral amines at concentrations of 50 g/L using transaminases. *Organic Process Research & Development*, **14**, 234–237.

69 Seo, J.-H., Kyung, D., Joo, K., Lee, J., and Kim, B.-G. (2011) Necessary and sufficient conditions for the asymmetric synthesis of chiral amines using ω-aminotransferases. *Biotechnology and Bioengineering*, **108**, 253–263.

70 Tufvesson, P., Lima-Ramos, J., Jensen, J.S., Al-Haque, N., Neto, W., and Woodley, J.M. (2011) Process considerations for the asymmetric synthesis of chiral amines using transaminases. *Biotechnology and Bioengineering*, **108**, 1479–1493.

71 Malik, M.S., Park, E.-S., and Shin, J.-S. (2012) Features and technical applications of ω-transaminases. *Applied Microbiology and Biotechnology*, **94**, 1163–1171.

72 Ghislieri, D. and Turner, N.J. (2014) Biocatalytic approaches to the synthesis of enantiomerically pure chiral amines. *Topics in Catalysis*, **57**, 284–300.

73 Kohls, H., Steffen-Munsberg, F., and Höhne, M. (2014) Recent achievements in developing the biocatalytic toolbox for chiral amine synthesis. *Current Opinion in Chemical Biology*, **19**, 180–182.

74 Höhne, M. and Bornscheuer, U.T. (2009) Biocatalytic routes to optically active amines. *ChemCatChem*, **1**, 42–51.

75 Cassimjee, K.E., Branneby, C., Abedi, V., Wells, A., and Berglund, P. (2010) Transaminations with isopropyl amine: equilibrium displacement with yeast alcohol dehydrogenase coupled to *in situ* cofactor regeneration. *Chemical Communications*, **46**, 5569–5571.

76 Savile, C.K., Janey, J.M., Mundorff, E.C., Moore, J.C., Tam, S., Jarvis, W.R., Colbeck, J.C., Krebber, A., Fleitz, F.J., Brands, J., Devine, P.L., Huisman, G.W., and Hughes, G.J. (2010) Biocatalytic asymmetric synthesis of chiral amines from ketones

applied to sitagliptin manufacture. *Science*, **329**, 305–309.

77 Truppo, M.D., Strotman, H., and Hughes, G. (2012) Development of an immobilized transaminase capable of operating in organic solvent. *ChemCatChem*, **4**, 1071–1074.

78 Kjell, D.P., Watson, I.A., Wolfe, C.N., and Spitler, J.T. (2013) Complexity-based metric for process mass intensity in the pharmaceutical industry. *Organic Process Research & Development*, **17**, 169–174.

8
How Chemical Hazard Assessment in Consumer Products Drives Green Chemistry

Lauren Heine and Margaret H. Whittaker

8.1
Introduction

Emerging market opportunities driven by consumer demand and changing regulations in the United States and abroad are compelling businesses and organizations up and down the supply chain to go beyond regulatory compliance to provide greater transparency by disclosing the chemical composition and associated hazards of ingredients in products. This in turn is creating greater awareness among consumers, companies, and regulators, and is moving companies to implement overarching strategies built on transparency, disclosure, collaboration, chemical hazard assessment (CHA), principles of green chemistry, and engineering, and that results in the use of safer chemical alternatives in the design of products.

A commonly used definition of green chemistry is "the design of products and processes that reduce or eliminate the use or generation of hazardous substances" [1]. A key strategy underlying the practice of green chemistry is to reduce risk by reducing the inherent hazard of chemicals or materials used in consumer products and the processes used to make them. This chapter will identify and compare chemical hazard assessment methods and tools used by consumer products companies and related organizations to reduce hazard throughout the supply chain. It is generally hoped that the increased consumer demand for safer chemical alternatives should lead to the development, use, and market adoption of inherently safer chemicals.

Applying green chemistry and engineering principles, along with an over-arching toxics reduction strategy can lead to a systematic transition to consumer products that are designed to be safer for people and the environment. This transition can help meet consumer demand and should result in a more transparent supply chain across the product life cycle without compromising product performance while creating shared values. There are effective and proven tools available for assessing hazards of individual chemicals and assessing life cycle impacts at the product

---

Handbook of Green Chemistry Volume 11: Green Metrics, First Edition. Edited by David J. Constable and Concepción Jiménez-González.
© 2018 Wiley-VCH Verlag GmbH & Co. KGaA. Published 2018 by Wiley-VCH Verlag GmbH & Co. KGaA.

level. Chemical hazard assessment can range from simply compiling lists of chemical ingredients in a consumer product in order to identify known health or environmental hazards (e.g., cancer, skin sensitization, aquatic toxicity), to comprehensive comparative hazard assessment reports that include literature reviews along with advanced toxicological modeling that pinpoints hazard "hot spots" in a product's life cycle (such as occupational hazards) or end-of-life hazards (as for example, the production of dangerous transformation products such as dioxins). Life cycle assessment (LCA) ranges from "life cycle thinking" to data intensive LCAs. But many challenges still exist. Manufacturers are challenged with how and when to apply the various tools individually and in combination without making undesirable trade-offs. While there appears to be a general convergence toward aspirational principles of sustainable product design based on green chemistry and engineering, explicit consensus has not yet occurred, and there is a need to measure pragmatic progress in the process of driving innovation. Incremental improvements that replace toxic chemicals with those that are inherently safer may or may not lead to innovation. Truly disruptive technologies that are inherently safer and more sustainable over the full life cycle may move beyond reformulation into the realm of new materials and product systems requiring metrics at broader temporal and spatial scales.

8.2
What Drives Consumer Product Companies to Look for Less Hazardous Chemical Ingredients

Globally, a chemical of concern can be defined as a chemical that displays one of the following characteristics [2]:

- Carcinogenicity, mutagenicity, reproductive or developmental toxicity
- Potential concern for children's health (for example, because of potential adverse effects from endocrine disruption or known to cause asthma)
- Used in children's products or in products to which children may be highly exposed
- Neurotoxicity
- Persistent, bioaccumulative, and toxic (PBT)
- Very persistent or very bioaccumulative in the environment (vPvB)
- Ozone depleting
- Detected in biomonitoring programs

Chemicals of concern can be found in almost every type of consumer product. Example chemicals of concern include polyhalogenated aromatic or organophosphate chemicals used to provide flame retardancy, bisphenol A used in sealants and found in polycarbonate containers, ortho-phthalate plasticizers used to soften polyvinyl chloride (PVC) plastics, alkylphenol ethoxylate surfactants used in cleaning products, and perfluorinated coatings used to provide stain and water resistance in textiles and oil resistance in food packaging [3,4].

Many people in industry argue that small amounts of toxic chemicals, such as those listed, do not pose an unacceptable risk to human health or the environment. But, it can be difficult to explain to the average consumer why it is acceptable for a product to contain even small amounts of chemicals such as phthalates or Bisphenol A that have been experimentally shown in various mammalian models to disrupt the endocrine system in ways that may cause adverse effects to children. Arguments about acceptable levels of cancer or developmental toxicity risks can fall on deaf ears for those whose primary concern is the health and well-being of their families.

> "Chemical hazard is like a bad reputation – the best strategy is to avoid it in the first place."
>
> – *Lauren Heine and Margaret Whittaker*

Many chemicals are hazardous and their hazards are inherent to their performance or function. For example, fuels are hazardous and we exploit their flammability to rapidly release chemical energy (heat and pressure) to drive a piston in a car engine. Fuels also have a variety of human health and environmental impacts if they are used outside of their particular intended use. Acids and bases are hazardous because they are corrosive, and we use them because of their ability to dissolve metals and a variety of materials or they promote a variety of useful chemical reactions. However, many chemicals, while they have a useful function in a product have been found to have unintended impacts in human or environmental organisms. For example, ortho-phthalate plasticizers are intended to soften rigid PVC plastics, but they have been shown to readily leach out of the plastic into food or out of baby's toys, and these plasticizers have been shown to cause developmental toxicity. In the case of solvents, we use them to dissolve other chemicals and we would like them not to cause cancer.

In our opinion, some manufacturers seem to fear public awareness of chemical hazards and argue that individual chemical hazards should not be revealed to the public for fear that the public will deselect certain chemicals leading to some greater harm, or that consumers will misunderstand the true risk. However, the public is accustomed to disclosure about flammability or the corrosivity potential of a product, and accepts that there are risks associated with consuming chemicals like ethanol. Individuals make informed decisions accordingly. Risk perception does not necessarily align with results from risk assessment. It is possible that even small risks from the presence of toxic chemicals in consumer products are simply less acceptable to some consumers when those risks are presumed to be unnecessary and safer alternatives are available.

8.2.1
Chemical Substitution and Regrettable Substitution

Eliminating or substituting a chemical of concern in a product and throughout a supply chain can be an expensive proposition. When faced with a chemical restriction, the least proactive position a company can take is to eliminate the restricted chemical of concern and replace it with a structurally similar chemical that is currently unrestricted. This approach can be risky because it is possible that the substitute will have the same (or worse) hazard characteristics as the original chemical of concern. If so, then it may just be a matter of time before the chemical substitute also becomes restricted and/or regulated. This phenomenon is also known as *regrettable substitution*, which is defined as the replacement of a toxic chemical with one that has unknown – if not greater – toxic effects [5].

An example of a regrettable substitution is demonstrated by the substitution of one problematic plasticizer for another equally problematic. Di(2-ethylhexyl) phthalate (DEHP) is an ortho-phthalate plasticizer used to soften vinyl plastics such as polyvinyl chloride (PVC). DEHP is toxic to reproduction and is known to interfere with the endocrine system (particularly in males) [6]. DEHP was added to California's Proposition 65 list because of DEHP's potential carcinogenicity and reproductive toxicity. Proposition 65 (also known as the Safe Drinking Water and Toxic Enforcement Act of 1986) is a California statute that requires the California Office of Environmental Health Hazard Assessment to maintain a list of chemicals that are known to cause cancer or birth defects or other reproductive harm. Since being balloted in 1986, the "Prop 65" list has grown to more than 950 naturally occurring or synthetic chemicals that are thought to be carcinogenic and/or reproductively/developmentally toxic [7]. Without a safe use determination by California's Office of Environmental Health Hazard Assessment (OEHHA) or a manufacturer's safe harbor determination, a consumer product marketed in California that contains a listed Proposition 65 chemical must be labeled with a warning so that California consumers are aware that a product contains a Proposition 65 chemical. Due to DEHP's Prop 65 listing, numerous companies, such as the vinyl glove industry, ceased using DEHP to avoid labeling their products with a requisite Proposition 65 warning. Instead, many of these companies substituted DEHP with the ortho-phthalate plasticizer diisononyl phthalate (DINP). In late 2013, OEHHA added DINP to California's Proposition 65 list due to DINP's potential carcinogenicity. OEHHA based DINP's Prop 65 listing on positive cancer findings in multiple cancer studies in animals, along with data and cancer classification of DEHP and another phthalate (butyl benzyl phthalate) [8].

The move from DEHP to DINP is a regrettable substitution because DINP has its own hazards such as potential carcinogenicity. The cancer data that served as the basis of DINP's Prop 65 listing existed well before DINP's Prop 65 listing, in fact, all 12 cancer studies in animals had been published prior to DINP's Proposition 65 listing [8], and a careful chemical hazard assessment would have identified DINP's cancer hazard. As a result of the DINP's Prop 65 listing, entire industries that moved from DEHP to DINP are now moving to other plasticizers. Ideally, such industries will apply CHA methods when vetting safer substitutes to DINP.

Hewlett Packard estimated that the cost of eliminating a chemical from their product supply chain could cost between $5 and 50 million. It is an understatement to

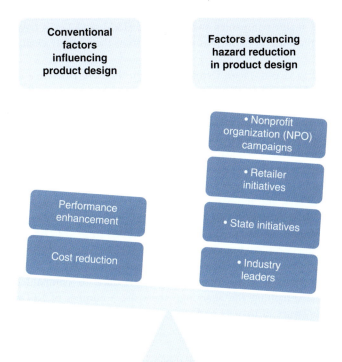

Figure 8.1 Consumer product design and development considerations.

say that replacing a regulated/restricted chemical with another chemical with similar hazard characteristics is a business risk that has the potential to be extremely wasteful. Historically, companies designed and engineered consumer products with two primary considerations in mind: cost reduction and performance enhancement. As shown in Figure 8.1, consumer products companies now face four additional factors that drive hazard reduction as part of the product design and development process. The influence that each factor has upon the design and development of consumer products is discussed further.

8.2.2
Nonprofit Organization (NPO) Campaigns

A major driver for increased awareness about chemicals in products can be credited to nonprofit organizations (NPOs) like Greenpeace that effectively "name and shame" companies and even product sectors whose products and practices are less than exemplary. Greenpeace's Detox Campaign, begun in 2011, has been indisputably effective in motivating individual companies and industry sectors, particularly in the fashion, textile, and electronics sectors, to take action toward the elimination of chemicals of concern. They help to focus action such as prioritizing the elimination of 11 hazardous chemicals from the textile supply chain [9]. To date, 18 major companies such as Nike, H&M, and Levi's have committed to reduce their use of hazardous chemicals from their supply chains and products [10].

Formed in response to Greenpeace's Detox Campaign, the *Zero Discharge of Toxic Chemicals* (ZDHC) Program comprises about 50 organizations, including brands, value chain affiliates, and associates, who are committed for implementing safer chemical management practices in the textile and footwear industries [11]. The Program has four focus areas and two cross-cutting areas that are critical to eliminate hazardous chemicals from the global supply chain. ZDHC has created a roadmap to implement its mission to ensure "widespread implementation of sustainable chemistry and best practices in the textile industry to protect consumers, workers, and the natural environment." Focus areas include a manufacturing restricted substances list and associated guidance, wastewater quality, audit protocol, and research. The two cross-cutting areas are Data & Disclosure and Training. Familiar brands include but are not limited to Levi Strauss & Co., Adidas, Puma, Gap Inc., Marks & Spencer, and Burberry.

Some NPOs apply positive rather than negative pressure to companies and sectors. Rather than "naming and shaming," there are a number of NPOs that effectively address issues of toxicity and/or waste and entice private sector entities to engage in collaborative initiatives.

- The Ellen MacArthur Foundation promotes the circular economy through a number of programs, including the CE 100 Programme [12]. The circular economy 100 was established to enable organizations to work on precompetitive projects, develop new opportunities, and realize their circular economy ambitions faster. It brings together corporations, governments and cities, academic institutions, emerging innovators, and affiliates in a unique multistakeholder platform. Specially developed activities help members learn, build capacity, network, and collaborate with key organizations around the circular economy. The Ellen MacArthur Foundation works on a strategic level with influential businesses across key sectors of the economy to demonstrate circular innovation at scale and to explore the potential of the circular economy as a vehicle for value creation.
- The Green Chemistry and Commerce Council (GC3) is a cross-sectoral, business-to-business network of companies and other organizations working collaboratively to accelerate green chemistry across sectors and supply chains [13]. Started in 2005, the GC3 has over 100 members from leading companies, consulting firms, academia, and other organizations. One GC3 initiative, the Green & Bio-based Chemistry Startup Network, is dedicated to accelerating the development and market adoption of green chemistry technologies by supporting the growth of green and bio-based startup companies. Leveraging the diversity of the GC3 membership, they strive to connect startups and small companies with established chemical suppliers, brands, retailers, and investors who can serve as strategic partners to accelerate the development and growth of promising green and bio-based chemistry

technologies. Established companies can tap into the startup network to identify new, strategic technologies, potential partners, and investments.

Some NPOs work positively to provide product certification and public recognition of market leadership, another driver for using inherently safer ingredients.

- The Cradle to Cradle Products Innovation Institute (C2CPII) builds on the work of Bill McDonough and Michael Braungart to promote Cradle to Cradle design and to certify products to the Cradle to Cradle Certified™ Product Standard [14]. A large part of the multiattribute certification program is material health assessment and optimization. The C2C standard includes banned lists of inputs that are not allowed as intentional inputs in certified products (generally, above 1000 ppm or 0.1% by weight in a homogenous material) such as heavy metals (e.g., mercury), flame retardants (e.g., deca-BDE), halogenated polymers (e.g., PVC) [14]. Chemical constituents in assessed products are evaluated for hazard and risk, and products containing chemicals with inherently low hazard achieve increasing levels of certification (Basic, Bronze, Silver, Gold, Platinum) to demonstrate progress toward use of chemicals and materials that are inherently benign during manufacture, use, and end of life.
- The International Living Future Institute (ILFI) developed the Living Product Challenge (LPC) to certify products that are made with chemically optimized materials and have a net positive benefit to people and the environment [15]. ILFI's LPC Standard and the ILFI Red List are described in greater detail later in this chapter.

8.2.3
Retailer Initiatives

"We didn't get where we are today by being like everyone else and driving the middle of the road."
— *Walmart President and CEO Lee Scott (retired)*

Retailers are on the front lines of consumer concerns about the health and environmental impacts of chemicals in products and they are in a unique and important position to make significant changes in the supply chain. Increasingly, brick and mortar retailers and e-retailers around the world are influencing the chemical composition of consumer products by setting requirements. This phenomenon is known as *Retailer Regulation*. Most trace the start of Retailer Regulation to an October 23, 2005 speech titled "Twenty First Century Leadership" by Walmart President and CEO Lee Scott (retired) [16]. In his speech, delivered in the aftermath of Hurricane Katrina, Mr Scott communicated the following three environmental goals for Walmart:

1) To be supplied with 100% renewable energy.
2) To create zero waste.
3) To sell products that sustain our resources and environment.

Although Retailer Regulation does not carry the rule of law, companies who wish to sell their products through a retailer have little choice but to comply. The ability of retailers to effect change quickly across many different consumer product lines in numerous countries around the world is demonstrated by their immense revenue and global reach. The National Retail Federation reported that revenue generated by the top 250 retailers around the world totaled more than US$4.3 trillion in revenue throughout FY2015, resulting in an average size of US $17.2 billion per retailer [17]. These retailers operate in dozens of countries.

Through the Green Chemistry and Commerce Council's Retailer Initiatives, leaders from seven retailers and five major chemical manufacturers joined together to create the *Joint Statement on Using Green Chemistry and Safer Alternatives to Advance Sustainable Products* [18]. This statement aligned the organizations toward the overarching goal of exploring and accelerating the development and use of more sustainable products through innovation and sourcing of green chemistry solutions. While each company may be at a different stage in its product sustainability development, all share a commitment to having open dialog in five key areas, as illustrated in Figure 8.2.

Other GC3-inspired retailer initiatives include preparation of a GC3 report titled Best Practices in Product Chemicals Management in the Retail Industry and a GC3 database that includes summary descriptions of tools that retailers can use to evaluate chemical ingredients in products [19,20]. These tools fall into three major categories: Restricted Substances Lists (e.g., American Apparel & Footwear Association (AAFA) Restricted Substances List); Standards, Certifications, and Labels (e.g., Cradle to Cradle Product Certification, Health Product Declaration, Green Seal, US EPA Safer Choice); and Chemicals Management Software (e.g., Pharos database, SciVera Lens™). Some of the tools are applicable to all product sectors. Some tools are sector specific (i.e., apparel/footwear/outdoor industry; automotive; building materials and products; chemicals and plastics; cleaning and janitorial products; electronics, food, and beverage;

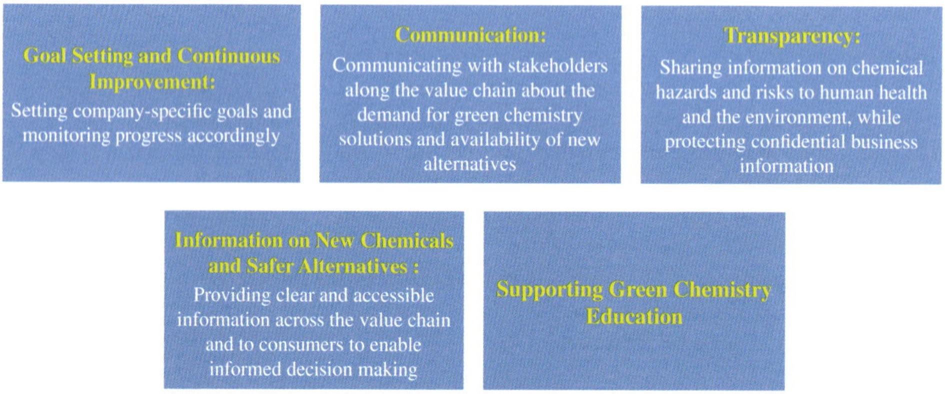

Goal Setting and Continuous Improvement:
Setting company-specific goals and monitoring progress accordingly

Communication:
Communicating with stakeholders along the value chain about the demand for green chemistry solutions and availability of new alternatives

Transparency:
Sharing information on chemical hazards and risks to human health and the environment, while protecting confidential business information

Information on New Chemicals and Safer Alternatives:
Providing clear and accessible information across the value chain and to consumers to enable informed decision making

Supporting Green Chemistry Education

Figure 8.2 GC3 joint statement on using green chemistry and safer alternatives to advance sustainable products.

Table 8.1 Retailer initiatives to advance hazard reduction and green chemistry.

Initiative	Example
Ingredient disclosure	Target publicly committed to disclosing ingredients in beauty, baby care, personal care, and household cleaning products by 2020 [21]
Phase-out and elimination of hazardous chemicals	As part of fulfilling its 2013 Sustainable Chemistry Policy, Walmart achieved a 95% reduction by weight of 16 "high priority" chemicals in formulated (chemical-based) consumer products by the end of 2016 [22,23]. Examples of Walmart HPC chemicals include diethyl phthalate (DEP), nonyl phenol ethoxylates, butyl paraben, and triclosan. Target has flagged more than 2000 high-priority chemicals or chemicals of concern for ultimate reduction and/or elimination, including Proposition 65 chemicals, EU Substances of Very High Concern (SVHCs), US EPA PBT chemicals, among others [24].
Promoting third-party certification/recognition programs	Best Buy features EPEAT-certified electronics on its Web site [25] reports that Best Buy customers purchased more than 3 million EPEAT-registered products, which collectively helped prevent the generation of hazardous materials equivalent to the weight of 35,000 refrigerators. Walmart, Target, and The Home Depot encourage suppliers to obtain third-party standard certification of their products, such as US EPA Safer Choice or Cradle to Cradle.

furniture; hard goods and appliances; health and beauty, cosmetics, pharmacy; lawn and garden; paints and coatings, pool and spa; textiles; tools, hardware, and plumbing; and toys).

Examples of retailer initiatives are detailed in Table 8.1, and are designed to promote ingredient transparency, alert consumers to the presence of hazardous chemicals, and/or identify and promote inherently more sustainable products. Through Retailer Regulation, a retailer may identify specific chemicals for phase-out and elimination, requiring increased transparency and cooperation by suppliers. Some retailers develop policies for sustainable products, including overarching standards for procurement to screen out certain toxic chemicals from products that they sell. Some retailers are farther along the spectrum of advancing chemical hazard reduction and green chemistry.

The US-based NGO Safer Chemicals, Healthy Families published a 2016 report titled "Who's Minding the Store? – A Report Card on Retailer Actions to Eliminate Toxic Chemicals" that ranked the 11 biggest retailers in North America on a 130-point scale, with letter grades calculated from A^+ to F. Retailer scores were based on 13 metrics, including ingredient disclosure, credible safety screening, and efforts to find safer alternatives [25]. Major US retailers earned letter grades ranging from "B" for good progress to "F" for failing to develop and make public even basic safer chemical policies. Walmart, Target, and CVS Health received the highest grades (grades of B, B, and C, respectively) because they each developed and made public the most robust safer chemical management programs during the past 3 years.

8.2.4 State Initiatives

State initiatives and regulations can drive mandatory disclosure and/or restrictions of chemicals in consumer products that may lead to substitution with safer alternatives. As described earlier, one of the earliest such state initiatives is California Proposition 65 ("Prop 65," formally titled "The Safe Drinking Water and Toxic Enforcement Act of 1986") [26]. Prop 65 was passed by direct voter initiative in 1986 to protect drinking water sources from chemicals that may cause cancer and birth defects. It is intended to reduce or eliminate exposures to those toxic chemicals by requiring warning labels, including on consumer products.

Prop 65 regulates substances that have a 1 in 100,000 chance of causing cancer over a 70-year lifetime, or that have developmental or reproductive toxicity. Businesses may not discharge Prop 65-listed substances into drinking water sources, or onto land where the substances can pass into drinking water sources. Businesses must also provide a clear and reasonable warning when products contain listed chemicals in order to help consumers avoid exposure. The list of chemicals on Prop 65 is maintained and updated regularly by Cal/EPA's California Office of Environmental Health Hazard Assessment and made publicly available, and indexes approximately 950 chemicals [7]. Prop 65 forces businesses to be transparent about listed chemicals in products and increases consumer awareness about exposures to hazardous chemicals. Such transparency also provides incentives for manufacturers to remove those chemicals from their products. The law is in part self-policing and can be expensive to businesses when they are subjected to lawsuits for not reporting listed chemicals in products.

More recent state initiatives include regulations that require disclosure and/or restriction of toxic chemicals used in children's products. Such laws have been enacted in Washington, Minnesota, Maine, Connecticut, and Oregon.

The State of Washington enacted the Children's Safe Products Act (CSPA) in 2008 [27]. The law requires manufacturers of children's products sold in Washington to report if their product contains one or more of the 66 listed Chemicals of High Concern to Children [28]. The CSPA also limits the amount of lead, cadmium, and phthalates allowed in children's products and authorizes the State of Washington's Department of Ecology to test. Beginning July 2017, CSPA also limits certain flame retardants in children's products and residential upholstered furniture.

Minnesota passed the Minnesota Toxic Free Kids Act in 2009 [29]. It established a framework by which the Minnesota Department of Health (MDH), in consultation with the Minnesota Pollution Control Agency (MPCA), was required to create and maintain list of chemicals of high concern and publish a list of priority chemicals in children's products. The statute required three reports to the Legislature, including recommendations for how to reduce and phase out the use of priority chemicals in children's products and to promote consumer product design that uses green chemistry principles and that considers a product's impact over its life cycle. Minnesota's first list of chemicals of concern was published in 2010, and the second update to the chemicals of high concern list was published in 2016 [30].

The Oregon Toxic-Free Kids Act was passed in 2015 [31]. This law goes further than Washington's CSPA because it requires manufacturers of children's products sold in Oregon to not only report products that contain one or more of the 66 high priority chemicals of concern for children's health (the list is the same as that in the State of Washington) at levels above 100 ppm, but also to remove these chemicals or to seek a waiver. Products that fall under this law are those that are marketed to or intended for children.

Additional state initiatives are emerging that are designed to compel greater ingredient transparency that is necessary for identifying chemicals of concerns and prioritizing them for replacement. In 2017, New York State Governor Andrew Cuomo announced that he intended to make information about ingredients in cleaning products public by enforcing an underutilized Environmental Conservation law passed in 1970 [32]. As described in the 2017 New York State of the State report, manufacturers must identify chemicals of concern and disclose impurities in their cleaning products, as well as content by weight in ranges. In addition to this information appearing on company websites, the New York Department of Environmental Conservation and the Interstate Chemicals Clearinghouse (IC2), which is partly funded by New York State, will maintain a centralized searchable database [32]. California Senate Bill 258, the Cleaning Product Right to Know Act of 2017, was introduced (February, 2017) in California that would require full disclosure of ingredients in institutional and household cleaning products, and specifically mentions phthalates and Bisphenol A [33].

Sustainable procurement is another approach that can drive the use of greener chemicals in consumer products [34]. Sustainable procurement is a term that describes government-initiated spending that is intended to promote a region's social, environmental, and/or economic policies. Public spending generally accounts for 15–25% of a country's GDP and wields enormous purchasing power [35]. According to the 2017 HEC/EcoVadis Sustainable Procurement Barometer Report [34], sustainable procurement policies can increase revenue growth and help suppliers save money. Looking specifically at sustainable procurement programs, the three most important business drivers that influence sustainable procurement programs are as follows:

- Brand reputation – identified as a critical factor by 63% of organization
- Risk mitigation – identified as a critical factor by 61% of organization
- Compliance – identified as a critical factor by 57% of organizations

8.2.5
Consumer Product Sector Leaders: Setting the Example for Others

Not all organizations striving for sustainable products and use of greener chemicals in their products and processes are driven by NGO pressure or retailer or state initiatives. Some companies build their brands on a commitment to the health and well-being of consumers and the environment and institutionalize their values through corporate policies and practices. Market leaders include companies such as Seventh Generation, Method Home, and Patagonia. These companies have become known for proactive and precautionary approaches to product development linked to health and environmental protection through elimination of toxic chemicals and reduction of waste. Seventh Generation's fragrance- and dye-free cleaning formulations, unbleached, and subsequently dioxin-free diapers and paper products [36]; Method Home's cleaning products that have earned Cradle to Cradle Gold Certification™ and have been packaged in plastic derived from ocean plastic waste [37]; and Patagonia's high-performance clothing made from recycled soda bottles [38] are just three examples of each company's desire to put green chemistry to practice.

Social and environmental entrepreneurs are creating companies to make products designed to solve problems caused by conventional products that use or generate toxic or unsustainable materials. This may be part of an overall rise in benefit corporations [39]. Benefit corporations are a type of for-profit corporate entity that includes positive impacts on society, workers, community, and then environment in addition to profit as part of its legally defined goals. Currently, 30 US states and the District of Columbia authorize benefit corporations.

Emerging examples of social and environmental entrepreneurs include TidalVision USA that makes products "from the ocean and for the ocean" by upcycling crab shell waste and processing it into chitin and chitosan and salmon waste into salmon leather products. Floral Soil, Inc. is displacing petroleum-based floral foam that is contaminated with formaldehyde and other toxic chemicals with a coconut fiber-based medium that can be used for plant starts and that will fully biodegrade in a backyard compost. These companies "push the envelope" by demonstrating what is possible. Often, such companies represent the values of their entrepreneurial founders who envisioned transforming industry for the better while creating products of value. They may compete in the overall market or appeal directly to a market sector, such as the sector that at one time was coined LOHAS, an acronym for Lifestyles of Health and Sustainability.

Awareness of impacts associated with products can go beyond toxic chemicals to address how materials in the products are sourced and manufactured. With food, many consumers care whether or not plants or animals grown for food were genetically modified, treated with pesticides, raised under inhumane conditions, or fed diets filled with antibiotics. These concerns are linked to expectations of nutritional quality and conscience. Likewise, consumers are extending their concern to the supply chain and the manufacture of nonfood consumer products. There is desire for greater transparency about labor practices, manufacturing practices, and whether or not products were manufactured with and/or contain chemicals of concern – whether intentionally or unintentionally added. Some people are willing to pay a premium for products that are certified by labels that give them confidence that the product is what it says it is, and represents the consumer's values.

Whether or not companies are responding proactively or reactively to NPO, retailer, regulatory, or internal or competitive pressures, there has been growing trend toward eliminating chemicals of high concern in consumer products. Leading companies are motivated to establish public chemical policies, to eliminate chemical hazards from their supply chains, and to identify and use safer alternatives to help meet demands for safer consumer products. This not only helps businesses meet a growing demand from consumers, but it also helps to manage risk, verify compliance, inform decision-making, protect workers, identify innovation opportunities, and offer more sustainable choices for customers. This can lead to increased access to markets, increased market shares, development of new products, cheaper access to information by sharing costs through sector initiatives, meeting corporate policies, meeting regulations, improving/ protecting brand image, and better knowledge of what is in products, including awareness of unwanted impurities, innovation, and cost savings.

8.3
What is Chemical Hazard Assessment?

With the growing list of chemicals of concern comes a growing awareness that simply banning or eliminating a chemical of concern may be insufficient to lead to overall risk reduction. Simple bans can lead to substitution with the nearest replacement. This can result in essentially switching from a known chemical of concern to a chemical with unknown or similarly hazardous characteristics. It is important to ensure that the substitutes are inherently safer to avoid unintended negative consequences. In order to make more informed and inherently safer substitutions, one must understand the hazard characteristics of the chemical of concern and any viable alternatives. In a world where formulations are typically considered proprietary, knowing the identity of substitutes, never mind its hazard properties, can be a challenge.

Chemical substitution is one strategy for moving away from the use of chemicals of concern. Sometimes alternative chemicals can be used that are inherently safer and can meet the performance requirements associated with the chemical of concern. These are called "drop-in" replacements. For example, a fragrance containing musk xylene can be replaced with muscone (CAS#: *[541-91-3]*. 3-methyle-cyclopentadecanone) without having to reformulate the product

But, inherently safer drop-in alternatives may not always exist. Replacing one chemical can lead to the need for complete reformulation of the product, not an insignificant hurdle. For example, phosphates, such as the now phased-out detergent ingredient sodium tripolyphosphate (STPP), can perform multiple functions in cleaning products such as a laundry or dishwashing detergents. Phosphates function as builders in detergents, and are one of the major ingredients along with surfactants. Phosphates reduce water hardness because they remove calcium and magnesium ions in wash water (making the surfactant work better), along with buffering pH and suspending certain kinds of dirt [40]. To replace all of these functions may require the use of multiple chemicals with different solubility and chemical compatibilities. Sometimes, it is possible to redesign products so that the function served by the chemical of concern is no longer needed. For example, one can eliminate the need for flame retardants in plastic laptop enclosures by switching to aluminum enclosures that are inherently flame retardant. Phthalate plasticizers used in PVC can be obviated by using a different polymer that does not require addition of a plasticizer to be flexible (such as a thermoplastic copolyester). But, regardless of the substitution or redesign strategy, chemical hazard assessment is emerging as a commonsense way to inform decision-making wherever chemicals are used in product and process design.

Chemical hazard assessment is a systematic process of assessing and classifying hazards across a spectrum of endpoints and severity [41]. By endpoint or hazard endpoint, we mean a specific adverse effect that is linked to mortality (death), morbidity (illness), or physical properties. Hazard endpoints are measured by changes in *in vivo*, *in vitro*, or *in silico* test systems, and/or changes in physiochemical parameters. The Globally Harmonized System of Classification and Labeling of Chemicals (GHS) defines an endpoint as a physical, health, and/or environmental hazard [42].

8.3.1
Globally Harmonized System of Classification and Labelling of Chemicals (GHS)

In order to understand chemical hazard assessment, it is important to understand the Globally Harmonized System of Classification and Labelling of Chemicals (GHS). GHS is a system for harmonizing hazard classification criteria and chemical hazard communication elements worldwide. GHS is not a regulation. Rather, it is a framework with guidance for classifying and labeling hazardous chemicals. The purpose of hazard classification under the GHS is to provide harmonized information to users of chemicals with the goal of enhancing protection of human health and the environment. As shown in Table 8.2, GHS defines criteria for the classification of health, physical, and environmental hazards.

GHS was born out of the United Nation's 1992 Conference on Environment and Development (The Earth Summit) and was written up as Chapter 19 of the Earth Summit report: Agenda 21 UN 1992 [43]. The United Nations developed GHS to promote a worldwide standard for hazard classification and communication (launched 2002). At one time, each country had its own hazard classification and communication system, including different pictograms, labels, and safety data sheets. Needless to say, this created complexity and confusion (and lots of

Table 8.2 GHS hazard classifications by endpoint.

GHS health hazards	High Hazard ⟷ Low Hazard
Acute toxicity	Categories 1, 2, 3, 4, 5
Skin corrosion/irritation	Categories 1 (1A, 1B, 1C), 2, 3
Serious eye damage/eye irritation	Categories 1, 2 (2A, 2B)
Respiratory or skin sensitization	Category 1 (1A, 1B)
Germ cell mutagenicity	Categories 1 (1A, 1B), 2
Carcinogenicity	Categories 1 (1A, 1B), 2
Reproductive toxicity	Categories 1 (1A, 1B), 2
Specific target organ toxicity single exposure	Categories 1, 2, 3
Specific target organ toxicity repeated exposure	Categories 1, 2
Aspiration hazard	Categories 1, 2
GHS physical hazards	**High Hazard ⟷ Low Hazard**
Explosives	Divisions 1.1, 1.2, 1.3, 1.4, 1.5, 1.6
Flammable gases and chemically unstable gases	Categories 1, 2 and A, B
Aerosols	Categories 1, 2, 3
Oxidizing gases	Category 1
Gases under pressure	Compressed/Liquefied/Refrigerated liquefied/Dissolved Gas
Flammable liquids	Categories 1, 2, 3, 4
Flammable solids	Categories 1, 2
Self-reactive substances and mixtures	Types A, B, C and D, E and F, G
Pyrophoric liquids/solids	Category 1
Self-heating substances and mixtures	Categories 1, 2
Substances and mixtures which, in contact with water, emit flammable gases	Categories 1, 2, 3
Oxidizing liquids/solids	Categories 1, 2, 3
Organic peroxides	Types A, B, C, D, E, F, G
Corrosive to metals	Category 1
GHS environmental hazards	**High Hazard ⟷ Low Hazard**
Acute aquatic hazard	Acute categories 1, 2, 3
Chronic aquatic hazard	Chronic categories 1, 2, (3), 4

paperwork) and so GHS was developed to harmonize and simplify hazard classification and labeling of chemical products. GHS specifies information to include on labels of hazardous chemicals and also on safety data sheets. GHS adoption is voluntary and countries may adopt portions of it. To date, 72 countries have adopted GHS as of March 2017 [44]. As shown in Figure 8.3, this includes the United States, the 28 member states in the European Union (who have adapted

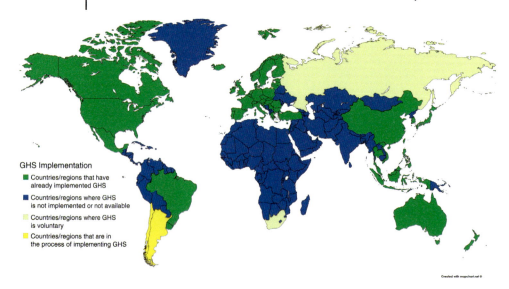

Figure 8.3 Worldwide GHS implementation.

GHS into their Classification, Labelling and Packaging of Substances and Mixtures Regulation), China, Australia, Canada, Mexico, among others.

In addition to GHS, there are publicly accessible, transparent, nonregulatory, stand-alone chemical hazard assessment schemes that build on GHS but are designed to serve slightly different purposes. Best known systems in the United States include US EPA's Design for the Environment (DfE) Program Alternatives Assessment Criteria for Hazard Evaluation [45], GreenScreen® for Safer Chemicals [46], and the Cradle to Cradle Products Innovation Institute's Material Health hazard assessment protocol [14]. The US EPA DfE AA Criteria were developed to support US EPA's Alternatives Assessment Partnerships.

GreenScreen was developed to identify both chemicals of concern as well as safer alternatives, and goes a step further than the DfE AA criteria by incorporating a numeric benchmarking system that facilitates high-level hazard comparison among chemicals and easy identification of chemicals to avoid. The C2C hazard assessment criteria are part of the overall Cradle to Cradle Product Certification™ Standard.

Other CHA methods exist that are not freely and publicly accessible. These methods are used to prioritize chemicals of concern for replacement or for internal product development and to screen for risks. These approaches are typically part of commercially available software and database solutions and are available by subscription. Examples include SciVeraLENS [47], Chemical Compliance Systems, Inc.'s GreenSuite® [48], and the green product scoring module in 3E Company's Ariel® WebInsight [49].

Hazard-based assessment paradigms are powerful because they encompass multiple hazard endpoints. The chemical hazard assessment methods already indicated typically share common hazard endpoints relating to human toxicity, environmental toxicity, and environmental fate. Most of the endpoints are based in GHS. Endpoints include human health hazards such as carcinogenicity or skin sensitization, ecotoxicity hazards such as acute or chronic aquatic toxicity, physical hazards such as flammability and reactivity, or indicators of chemical fate such as persistence and potential for bioaccumulation. Some hazard endpoints are subcategories of other hazard endpoints and may need to be called out or specified. For example, neurotoxicity can be included as a subset of systemic toxicity. To test for neurotoxicity requires an additional battery of tests. To ensure that neurotoxicity was considered as part of systemic toxicity testing, one would need to look to see if the neurotoxicity test battery was run.

GHS has some limitations for use in product design and development and alternatives assessment. It does not currently include criteria for all endpoints that may be needed to assess both hazardous chemicals and safer alternatives. For example, GHS does not include explicit criteria for endocrine disruption/activity. Consensus is still emerging on testing protocols for assessing adverse effects from endocrine activity. However, decisions must still be made about the use of chemicals that may be endocrine active. In addition, GHS does not include criteria for persistence and bioaccumulation potential as stand-alone endpoints. Persistence and bioaccumulation potential are not necessarily hazards. Rather, they augment hazard associated with chemicals that are toxic by increasing the likelihood of exposure in certain environments. While GHS includes classification of chemicals that are both aquatically toxic and persistent in the aquatic environment and/or bioaccumulating, it is useful to have stand-alone criteria for persistence and bioaccumulation potential because this is helpful information when designing products and evaluating alternatives. To fill these gaps, CHA systems, such as US EPA's DfE AA Criteria, GreenScreen, and C2C, have derived stand-alone criteria for persistence and bioaccumulation from national and international precedents. But because there is no global harmonization for these endpoints, the criteria and thresholds for classification vary between the different systems.

Table 8.3 provides a summary of hazard endpoints included in the DfE AA, GreenScreen, and C2C Material Health hazard assessment frameworks. The endpoints in italics are those that are not included in GHS. A number of additional physical hazard endpoints are included in GHS (identified in Table 8.2) that are not typically included in CHA systems that focus on sustainable product design. They are not included in Table 8.3.

8.3.2
Comprehensive and Abbreviated Forms of CHA

The ideal CHA utilizes the best available scientific information and data along with expert judgment to evaluate a broad suite of hazards associated with chemicals and is freely and easily accessible. But this is only ideal. Often data are incomplete, especially when CHAs are based only on publicly accessible data.

Table 8.3 Example hazard endpoints used in chemical hazard assessment.

Human health	Human health cont'd	Environmental toxicity and fate	Physical hazards/ climate hazards
Carcinogenicity	Acute toxicity (oral, dermal, inhalation)	Acute aquatic toxicity (fish, Daphnia, algae)	Reactivity
Mutagenicity and genotoxicity	Systemic toxicity and organ effects (oral, dermal, inhalation)	Chronic aquatic toxicity (fish, Daphnia, algae)	Flammability
Reproductive toxicity	Neurotoxicity	*Other ecotoxicity studies when available∗ (earthworm, bird, bee (from DfE))*	
Developmental toxicity	Skin sensitization	*Persistence (air, water, sediment, soil)*	*Ozone depletion potential*
	Respiratory sensitization		
Endocrine activity	Skin irritation/ corrosivity	*Bioaccumulation*	*Global warming potential*
	Eye irritation/ corrosivity		

The availability of comprehensive CHAs is limited by cost, time, and technical resources. Because of these constraints, a number of approaches to CHA have been developed. They vary based on the scope and depth of assessment, including the number of hazard endpoints evaluated and the variety of information sources and tools used to generate information. This results in a range of confidence. They can also vary based on the degree to which they can be automated.

8.3.2.1 GreenScreen for Safer Chemicals

GreenScreen for Safer Chemicals (v. 1.3) is a chemical hazard assessment framework designed to identify both chemicals of concern and less hazardous chemicals using a standardized approach that incorporates both human health, environmental fate and toxicity, and physical hazard endpoints. GreenScreen was developed by Clean Production Action and was launched in 2007 (*Disclosure*: GreenScreen development was led by coauthor Heine) [46,50]. A GreenScreen chemical hazard assessment communicates what is known and what is not known about the hazards associated with a chemical. It is intended to help manage the risk of exposure to hazardous chemicals by reducing the use of hazardous chemicals.

In a GreenScreen assessment (Figure 8.4), chemicals used in a formulation (present at threshold levels of either 100 or 1000 ppm) are evaluated against hazard endpoints relating to human health, environmental toxicity, and physical hazards. Each endpoint is given a score of low hazard (L), moderate hazard (M), high hazard (H), or data gap (DG). For some hazard endpoints, very low (vL) and very high hazard (H) can also be assigned. Bold font indicates higher confidence

GreenScreen® Hazard Ratings for Tralopyril

Group I Human					Group II and II* Human								Ecotox		Fate		Physical	
C	M	R	D	E	AT	ST		N		SnS*	SnR*	IrS	IrE	AA	CA	P	B	Rx	F
						Single	Repeated*	Single	Repeated*										
L	L	*L*	L	L	vH	DG	H	DG	H	L	DG	*M*	*M*	vH	vH	*H*	vL	L	L

Note: Hazard levels (very high (vH), high (H), moderate (M), low (L), very low (vL)) in *italics* reflect estimated values, authoritative B lists, screening lists, weak analogues, and lower confidence. Hazard levels in **bold** font are used with good quality data, authoritative A lists, or strong analogues. Group II Human Health endpoints differ from Group II* Human Health endpoints in that they have four hazard classifications (i.e., vH, H, M, and L) instead of three (i.e., H, M, and L), and are based on single exposures instead of repeated exposures.

Figure 8.4 Hazard ratings for the biocide tralopyril. (Reproduced with permission from Ref. [51]. Copyright 2015, Tox Services.)

and italic font indicates lower confidence, a convention borrowed from the DfE AA approach.

Figure 8.4 is an example hazard table for a GreenScreen performed on the antifoulant paint additive tralopyril.

Hazard classifications for the screened endpoints are then collectively evaluated according to an algorithm to assign one of four different benchmark scores, which incorporate the hazard scores for each of the individual endpoints and serves as an overall indicator of hazard, as illustrated in Figure 8.5:

- Benchmark 1: Avoid (chemical of high concern)
- Benchmark 2: Use (but search for safer substitutes)
- Benchmark 3: Use (but still opportunity for improvement)
- Benchmark 4: Prefer (safer chemical)
- In addition, chemicals that have insufficient data or data gaps for specific hazard endpoints will be assigned a benchmark score of unspecified ("U").

A GreenScreen can be performed on each chemical ingredient present in a mixture. An overall mixture score is currently under development, as is a more robust polymer hazard assessment paradigm. Although the GreenScreen for Safer Chemicals is a publicly accessible framework, training and licensing is required to generate a GreenScreen that is eligible for use in standards such as the US Green Building Council's Leadership in Energy and Environmental Design (LEED) v4 certification or for making GreenScreen benchmark claims. In lieu of training to become a certified practitioner to perform a GreenScreen, a consumer products company or supplier can commission a CPA-authorized third-party GreenScreen Profiler to perform a GreenScreen.

A GreenScreen report can be 40 pages or longer because it contains summaries of test data, literature, and QSAR modeling results for 18 different human health, environmental fate and toxicity, and physical hazard endpoints. It also

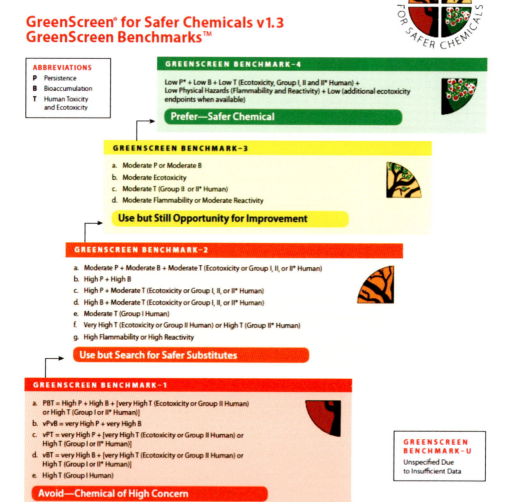

Figure 8.5 GreenScreen® for safer chemicals benchmarks. (Reproduced with permission from Clean Production Action.)

contains justification for hazard classifications presented in the hazard table (shown in Figure 8.4), predictions of environmental transformation products, and an overall benchmark score. A great deal of toxicological expertise is needed to evaluate the quality of existing data and the strength of any models used. Despite this extensive process, a GreenScreen assessment is still a screen and more information may be needed in order to conclude that a chemical is indeed a safer chemical and appropriate for a specific application. One helpful modification is to present hazard classifications in a stratified hazard table based on exposure routes. GreenScreen benchmarks can occasionally be misleading and it is always recommended to look at the hazard table and the summary sections for each individual hazard endpoint. The benchmarking system is not nuanced for specific product applications and the benchmarks are broad tiers. For example, a chemical can be a benchmark 2 based only on very high ecotoxicity or based on very high ecotoxicity plus moderate toxicity for *all* of the Group 1 human health endpoints. The benchmarking system does not discriminate between human health and environmental fate and toxicity effects. CHA is limited only by the state of the science and the ability of the assessor to access and evaluate chemical hazard using the toxicology scientific literature and tools. This includes evaluating standard toxicological test results and emerging science, using quantitative and other structure activity relationship (Q/SAR) models as well as other relevant *in vitro* and *in silico* test methods, applying read-across strategies and using chemical hazard surrogates appropriately, analyzing modes of action based on high throughput screening and more. The more data types and information generated, the more important it is for the assessor to have expertise and to be able to weigh the various lines of evidence to make an expert judgment. Most GreenScreen assessments require use of a minimum set of freely and publicly accessible models. Some proprietary models are commercial and can be very expensive. Some users have gone beyond the scope of GreenScreen to include additional ecotoxicity endpoints. Criteria for toxicity to earthworms, birds, and bees have been established as part of the DfE AA criteria. These endpoints and criteria can be easily incorporated into GreenScreen reports.

Comprehensive CHAs provide a high level of credibility and confidence in results and they can be used to document chemicals of concern, prompt the acquisition of safer alternatives, and as part of a chemicals management program. The downside of a comprehensive CHA is the cost and level of expertise needed to generate a CHA report. Table 8.5 provides an estimate of cost ranges of various tools discussed in this chapter. A comprehensive GreenScreen assessment of a single chemical can start at ~$900 and go to more than $5000 depending on the amount of data and modeling requirements. Often the cost is at the lower end because GreenScreen Profilers have created business models that allow them to reuse assessments.

GreenScreen and DfE AA criteria are examples of comprehensive CHA methods. One major different between the DfE AA method and that of GreenScreen is that GreenScreen includes both a hazard assessment table and an overall benchmark score for a chemical. The DfE AA criteria are used to generate a hazard table but the US EPA does not produce an overall benchmark score, making it more difficult to easily identify a less hazardous (or conversely, safer) chemical when following their criteria.

8.3.2.2 Quick Chemical Assessment Tool (QCAT)

Because of the high level of technical and financial resources required for performing a full CHA, a simpler assessment method called the Quick Chemical Assessment Tool (QCAT) was developed [52]. The State of Washington Department of Ecology (WA DOE) developed QCAT (currently, v. 2.0) to assist small and medium-sized businesses in screening chemicals in their products and processes to help prioritize chemicals of concern and to identify potentially safer alternatives. QCAT uses a subset of the hazard endpoints and proscribes a limited set of data sources, all of which are publicly available.

The trade-off for using a QCAT assessment with fewer endpoints and prescribed data sources is decreased comprehensiveness and confidence in the results. Some hazards could be missed during the evaluation process. QCAT includes nine hazard endpoints, which include six priority human health effects along with persistence, bioaccumulation, and acute aquatic toxicity. As with GreenScreen and the DfE AA Criteria, these nine endpoints are evaluated and assigned a level of concern for each chemical. But the QCAT grading system is intended to be distinctive to avoid confusion. QCAT places chemicals along a continuum of concern and assigns each chemical one of four possible letter grades (Table 8.4).

Evaluating chemicals using the QCAT provides several advantages. The QCAT focuses on important hazard endpoints, lowers data and expertise requirements, and provides a significant amount of information with a relatively low investment of resources in comparison to a full CHA. QCAT provides a quick and easy method to identify chemicals that are equally or more toxic than the chemical being reviewed. QCAT assessments can be performed by individuals with backgrounds in science, but they do not need to be expert toxicologists. Using QCAT, limited resources can quickly identify chemicals that are not viable safer alternatives to the chemical of concern. QCAT also helps users prioritize chemicals on the basis of their hazard. The QCAT assists users in identifying chemicals of concern and that information may be used, for example, to prioritize chemicals at a particular manufacturing facility for substitution or for a more comprehensive CHA.

Using QCAT, it is possible that an endpoint of concern could be overlooked either because certain endpoints are not included in the assessment or because newly published studies were not included in the proscribed public information

Table 8.4 Grade levels from the QCAT assessment process.

Grade A	Few concerns, that is, safer chemical	Preferable
Grade B	Slight concern	Improvement possible
Grade C	Moderate concern	Use but search for safer
Grade F	High concern	Avoid

sources. For example, new carcinogenicity data may be available on a chemical that has not yet been reviewed by the International Agency for Research on Cancer (IARC) or EPA. A comprehensive CHA would use new and emerging evidence as well as modeled results that would not be included in a QCAT assessment and could, therefore, result in assigning a different level of concern.

Because of the reduced amount of information assessed, a QCAT does not provide sufficient confidence that a chemical is indeed safer. To do so, a comprehensive CHA should be performed. QCAT can be used to eliminate nonviable alternatives, a useful initial step in assessing chemical alternatives. For some chemicals, it only takes one high hazard for a priority endpoint to characterize a chemical as a chemical of high concern. The bar is higher to identify an inherently safer chemical because one needs to rule out an entire suite of hazards. For example, if a chemical is listed as a carcinogen, then it can be identified as a chemical of concern. Assessing the other hazard endpoints will not change that. But the absence of carcinogenicity does not prove that the chemical has low hazard for the rest of the hazard endpoints. Proving that something has an inherently lower hazard requires more data and higher confidence based on available data.

8.3.2.3 GreenScreen List Translator (GS LT)

Various authoritative bodies and expert organizations evaluate chemicals and generate lists of chemicals with certain hazardous properties. These organizations may be governmental, authorized by governments, or comprised of experts. The simplest approach to CHA is to screen chemicals to see if they are identified as hazardous for specific endpoints based on existing authoritative or screening hazard lists. An *authoritative list* is typically one that is created by, for example, a national regulatory agency, such as the US EPA, or an organization granted authority by regulators, such as the National Toxicology Program which reviews, classifies, and lists carcinogens. Screening lists are typically lists generated by industry (IFRA) or NPO (TEDX) experts or advocacy organizations [53,54]. *Screening lists* may also include chemicals with properties that merit further testing. List screening can be performed using software that is user friendly and for which one does not need toxicological expertise. The GS LT continues to evolve as a comprehensive list screening resource used as the basis for the Pharos Chemical and Material Library and Toxnot [55,56]. GS LT includes rules for making classifications when different lists conflict. For example, authoritative lists are programmed to override screening lists and a more conservative classification by one authoritative body will override a less conservative classification by another. Software companies, such as ChemAdvisor, have developed list searching tools such as the List of Lists or LOLI database that includes up to 5500 regulatory lists from 122 countries around the world [57]. In general, the GreenScreen List Translator is designed to support hazard classifications that lead to the identification and elimination of chemicals of concern while tools like LOLI focus on supporting international regulatory compliance.

The GS LT builds on the idea that high-hazard classifications for priority endpoints can help users to quickly identify a chemical of concern. The GS LT comprises over 850 lists from 36 primary authoritative and screening sources that include national and international regulatory and hazard lists, influential NGO lists of chemicals of concern (screening lists), authoritative scientific bodies, European Risk and Hazard Phrases, and chemical hazard classifications by countries using the Globally Harmonized System of Classification and Labelling System.

Healthy Building Network's Pharos Assessment Tool software was created to automate the GreenScreen List Translator. The automated GS LT screens for chemicals that would achieve a GreenScreen benchmark 1 score, which is the lowest score reserved for the most hazardous chemicals, if a full GreenScreen assessment were performed; therefore, the scope of the GS LT tool is limited to capturing benchmark 1 and possible benchmark 1 chemicals only. Full Green-Screen assessments are required to determine if the final GreenScreen benchmark score is greater than 1.

GreenScreen LT scores fall under one of three following classifications:

- *LT-1: Benchmark 1* – A LT-1 chemical score is based on clear agreement among authoritative lists that it is a chemical of high concern and is likely to be a benchmark 1 chemical using the full GreenScreen method.
- *LT-P1: Possible benchmark 1* – A LT-P1 chemical score translates to possible BM1 and reflects the presence of the chemical on screening A or B lists and some uncertainty about the classification for key endpoints. Further research is needed to determine if the chemical is indeed a GreenScreen benchmark 1.
- *LT-UNK: Unspecified benchmark* – A LT-UNK chemical score indicates that there is insufficient information to provide a benchmark score for the chemical. Typically, only known hazardous chemicals are found on authoritative and screening lists. However, lack of presence on hazard lists can also mean that the hazard of the chemical has not been fully characterized. Therefore, the resulting conclusion using the List Translator is that the benchmark U score is unspecified, pending full GreenScreen review.
- Note that a full GreenScreen assessment is needed to determine if a chemical is a benchmark 2, 3, or 4.

Use of the automated GS LT tool requires little expertise and can generate information quickly and inexpensively (if not freely). The downside of GS LT is that some screening lists are overly conservative and will label an entire class of chemicals as highly hazardous when only one form or member of the class truly demonstrates the hazard. For example, hazards associated with silica depend on the size and form of the particle. But it is difficult to discriminate forms when the CAS number used for all forms of silica links to chemical lists generated based on the most hazardous form. In addition, if a chemical is not on a particular hazard list, it may mean that it does not demonstrate that hazard. But it may also mean that the chemical has not been tested. Only a comprehensive CHA can indicate what is known, and not known, about the hazard properties of a chemical. List screening is good for identifying well-known hazards, but it should not be used to identify inherently safer alternatives.

8.4
How Chemical Hazard Assessment is Used

As described in Section 8.2, chemical hazard assessment provides a structured framework for classifying hazards and communicating *what is known, and not known* about the hazards associated with particular chemicals, including potentially safer alternatives. It provides information to support product design and development. It is also included as a component of several environmentally preferable product certification systems and for procurement. It is also a critical step in an alternatives assessment.

More comprehensive CHAs can help companies identify safer chemical substitutions such as alternative flame retardants or plasticizers. The US EPA Design for the Environment Program evaluated chemicals for applications in a variety of products such as circuit boards, furniture, and thermal paper [58]. Ideally, a CHA provides a science-based solution that identifies and characterizes chemical hazards, promotes the selection of less hazardous chemical ingredients, and avoids unintended consequences of switching to a poorly characterized chemical substitute. CHAs can also be built into standards and software tools for communicating environmentally preferable attributes and for ongoing chemicals management and product development. For multiattribute standards, CHA and inherently safer chemical ingredients are just one attribute of an overall sustainable product framework.

The use of CHA is tied to whatever goals and strategies an organization chooses to adopt. But there are some core elements required. First of all, the identity of chemicals in products must be known in order to apply CHA and to inform decision-making. This may involve inventorying product ingredients or process chemicals. Once ingredients are known, various levels of CHA can be performed as discussed in the prior section. One may choose to simply screen ingredients against RSLs, MRSLs, or to use tools like GreenScreen List Translator. Alternatively, one may choose to do assessments that are progressively more resource and data intensive. Finally, one may go beyond hazard screening or assessment with broader product optimization to address human and environmental impacts that occur over the life cycle of the product, including impacts on natural resources and generation of problematic wastes.

Figure 8.6 lays out a number of tools and programs/standards that use varying levels of the CHA process starting with inventory, screening, assessment, and optimization. While GreenScreen for Safer Chemicals and GS LT were discussed earlier in this chapter, this section provides an overview of additional tools and programs that incorporate CHA. Table 8.5 provides information on associated costs and license periods associated with the use of various CHA-related tools, programs, and certifications.

8.4.1
Chemical Footprint Project

Clean Production Action (CPA) launched the Chemical Footprint Project in 2014. It is a voluntary, survey-based, non-verified approach to provide a basic

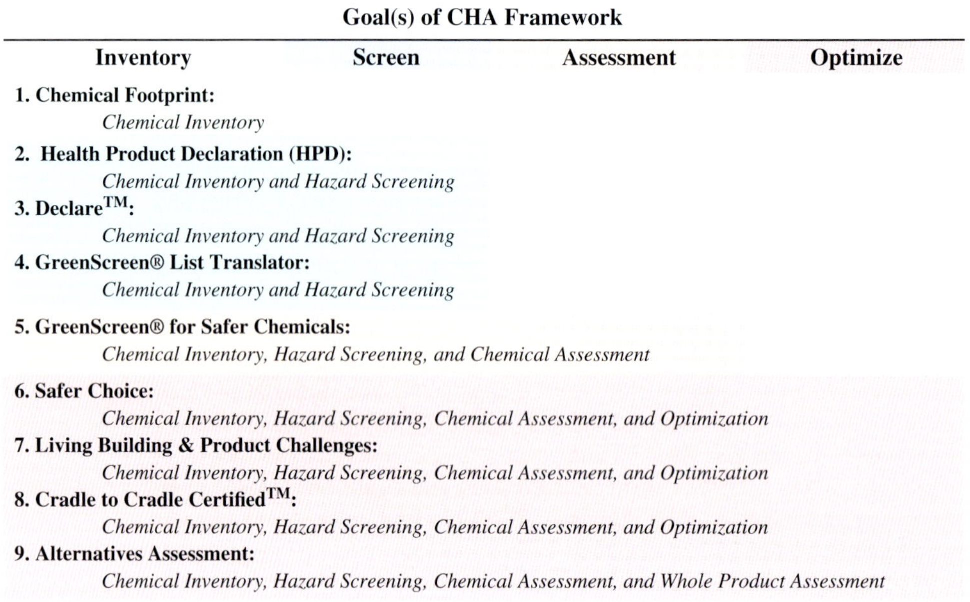

Goal(s) of CHA Framework			
Inventory	**Screen**	**Assessment**	**Optimize**

1. **Chemical Footprint:**
    Chemical Inventory
2. **Health Product Declaration (HPD):**
    Chemical Inventory and Hazard Screening
3. **Declare™:**
    Chemical Inventory and Hazard Screening
4. **GreenScreen® List Translator:**
    Chemical Inventory and Hazard Screening
5. **GreenScreen® for Safer Chemicals:**
    Chemical Inventory, Hazard Screening, and Chemical Assessment
6. **Safer Choice:**
    Chemical Inventory, Hazard Screening, Chemical Assessment, and Optimization
7. **Living Building & Product Challenges:**
    Chemical Inventory, Hazard Screening, Chemical Assessment, and Optimization
8. **Cradle to Cradle Certified™:**
    Chemical Inventory, Hazard Screening, Chemical Assessment, and Optimization
9. **Alternatives Assessment:**
    Chemical Inventory, Hazard Screening, Chemical Assessment, and Whole Product Assessment

Figure 8.6 Tools using CHA in the consumer products sector.

understanding of a company's overall corporate chemical management performance. The Chemical Footprint comprises 20 questions scored to 100 points that assess four key performance categories related to managing chemicals in products and supply chains [59]:

- *Management Strategy* (20 points): This section asks about the scope of corporate chemical policies and their integration into business strategy, accountability, and employees' incentives for safer chemical use, as well as support of public policies for safer chemicals.
- *Chemical Inventory* (30 points): This section asks about the efforts a company has taken to identify chemicals of concern (CoHCs) in its products, the extent of chemical data collected from its suppliers, and its systems for managing chemical data and ensuring supplier compliance with its reporting requirements.
- *Footprint Measurement* (30 points): This section asks about the goals that a company sets to reduce chemicals of high concern, its efforts to establish a baseline chemical footprint and measure progress, and its process for assessing and implementing safer alternatives.
- *Disclosure and Verification* (20 points): This section asks if a company publicly discloses the chemicals in its products beyond regulatory requirements, if it discloses its participation in the Chemical Footprint Project and its answers to the questions, and if its answers have been independently verified by a third party.

Table 8.5 Estimated cost and time of chemical hazard assessment programs.

Program	Preliminary cost estimate	Completion time estimate	Licensing period	Comments
Chemical footprint	N/A	N/A	N/A	Can be performed by manufacturer
Health Product Declaration 2.0 (HPD)	$2000–$6000	1 month	3 years	An HPD can be prepared even if a formulation contains hazardous chemicals. An HPD allows for nondisclosure of proprietary chemicals, but it must disclose all relevant hazards, % compositions, and function
Declare Label	$1500–$3000	1 month	1 year	A Declare Label can still be prepared even if a formulation has one or more hazards/allows for nondisclosure of all relevant hazards but must disclose up to 99% of all chemical names and CAS#s present in formulation
Pharos	$240–$1200	NA	1 year	Building product library, chemical and material library, certifications and standards library; subscription fee is based on number of projects
GreenScreen® for safer chemicals	$895–$1800 per chemical screen	1 month	3 years	A GreenScreen assessment is performed on chemicals in a material or product at a level greater than 100 or 1000 ppm (level set by manufacturer)
US EPA safer choice	$1000–$8000	3–5 months	3 years	No certification fee; lower costs by using ingredients on SCIL or CleanGredients; cost does not include performance testing
Living Building and Living Product Challenges	$15 000–$50 000	6 months	3 years	The Living Building Challenge applies to buildings; The Living

(*continued*)

Table 8.5 (Continued)

Program	Preliminary cost estimate	Completion time estimate	Licensing period	Comments
				Product Challenge applies to individual products. Imperatives include water, energy, materials, equity, health and happiness, beauty, and more
Cradle to Cradle Certified™ Full Certification, Cradle to Cradle Material Health Certificate	$20 000–$50 000 (full certification)	6 months	2 years	A Material Health Certificate is more affordable ($6,000 – $20,000) and faster (3 months) than a full Cradle to Cradle Certification while still earning all of the benefits of a full Cradle to Cradle Certification under LEED v4
Alternatives Assessment	~$170 000	1.5 years	NA	This is an extensive, multiattribute assessment based on the National Research Council, the Interstate Chemicals Clearinghouse and/or the Washington frameworks for Alternatives Assessment

The CFP defines a Chemical of High Concern (CoHC) as a (1) carcinogenic, mutagenic, or toxic to reproduction (CMR); (2) persistent, bioaccumulative, and toxic substance (PBT); (3) any other chemical for which there is scientific evidence of probable serious effects to human health or the environment (e.g., an endocrine disruptor or neurotoxicant); or (4) a chemical whose breakdown products result in a CoHC meeting one or more of the above criteria. This is essentially the definition of a substance of very high concern under REACH, however it is based on the identification of chemicals already on hazard lists. CFP's CoHC list is based on 14 lists of hazardous chemicals developed by governments and other authoritative bodies. The 2016 CoHC list aligns with GreenScreen for Safer Chemicals to identify CoHCs classified as "List Translator-1" chemicals (LT-1's) with two differences. First, CFP uses the European Union's Candidate List of Substances of Very High Concern to identify CoHCs, while GreenScreen uses the European Union's List of Substances of Very High Concern requiring authorization to identify LT-1 chemicals. In addition, CFP does not include EU-REACH Annex XVII CMRs.

The mission of the Chemical Footprint Project is to transform global chemical use by measuring and disclosing information about a business' progress toward incorporating safer chemicals in its operations and products. The CFP enables companies to benchmark their own progress and their progress against other companies as they inventory the chemicals they use and reduce the use of specific chemicals of high concern. A chemical footprint can be performed in-house by a manufacturer as it is tied to management and self-disclosure. It may be useful for companies that do not already have chemical policies and chemicals management systems in place, but it is unclear whether or not the CFP will actually drive greener chemistry. However, it does push companies away from the use of chemicals already defined as chemicals of high concern and may increase a corporation's awareness of its own processes (or lack thereof) to assess and implement safer alternatives.

8.4.2
Health Product Declaration Version 2.0 (HPD)

The Health Product Declaration Collaborative's Health Product Declaration® (HPD) is a program based upon the disclosure of ingredients and their associated human health and environmental hazards [60]. HPDs were launched in 2011, and contain ingredient content and hazard disclosure pertaining to ingredients in screened products. HPDs are prepared using an online tool known as the HPD Builder 2 [61], which follows the Health Product Declaration® Open Standard, version 2.0. To date, more than 2000 products have had an HPD prepared (either following Version 1.0 or 2.0).

HPDs are created following rules in the HPD Standard, Version 2.0 [60] and compare a product's individual chemical ingredients against a large database of "hazard" lists that are populated in the Pharos Project Chemical Screen Library (Pharos), published by both scientific associations and government authoritative bodies. HPDs can be prepared by a consumer products company using the online HPD Builder 2 by registering for access. Alternatively, a consumer products company may contract with third party assessors to prepare an HPD. An example HPD is shown in Figure 8.7.

8.4.3
Red List – Declare Label

The Declare Program was launched in October 2012 by the International Living Future Institute (ILFI). The Declare Label is an ingredient disclosure initiative and is designed to answer three questions: (1) Where does a product come from, (2) What is it made of, and (3) Where does it go at end of life? To date, over 1000 products and materials have had a Declare Label prepared, primarily in the building products sector.

As part of a Declare Assessment, a manufacturer completes an ILFI product declaration form disclosing all intentionally added ingredients plus residuals up

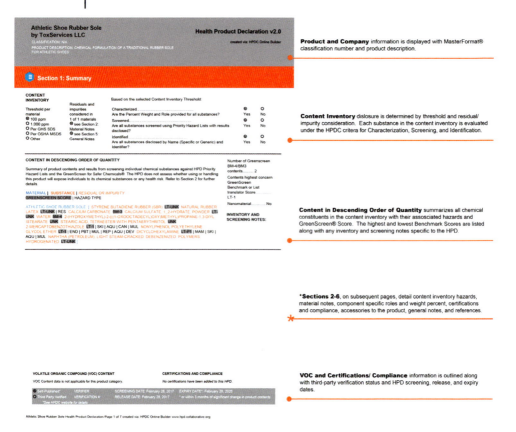

Figure 8.7 Health Product Declaration. (Reproduced with permission from Health Product Design Collaborative.)

to 100 ppm and submits this form to ILFI. ILFI screens ingredients of a building material against ILFI's Red List, a catalog of the world's >800 most toxic chemicals, in order to classify products [62]. Naturally occurring, unintentionally added ingredients, and process chemicals do not need to be reported. To reflect current market limitations, ILFI has instituted a temporary exception that allows manufacturers to keep either one ingredient or up to 1% of the product by weight and volume proprietary. The manufacturer must affirm that any ingredients kept proprietary are either not on the ILFI Red List or qualify under an exception. Figure 8.8 depicts a Declare Label [63].

8.4.4
United States Environmental Protection Agency: Safer Choice Program

Another leadership standard is the US EPA Safer Choice Program. Safer Choice provides an example of how to support transition to inherently safer chemicals throughout the supply chain in particular product sectors in a way that benefits

Figure 8.8 Declare Label. (Reproduced with permission from ILFI.)

all who participate. Generic chemicals that meet the Safer Choice functional class criteria are listed on its Safer Chemicals Ingredients List (SCIL) [64,65]. Trade name-specific chemical ingredients that meet Safer Choice functional class criteria are listed on CleanGredients [66]. CleanGredients is a database of ingredients that meet Safer Choice criteria for use primarily for formulating cleaning products. Manufacturers of chemicals listed on SCIL or CleanGredients are rewarded in the marketplace because these ingredients are purchased by product formulators who are seeking inherently safer ingredients for products, such as cleaning products, in order to qualify for the Safer Choice logo. Manufacturers of products carrying the Safer Choice label are then rewarded with preferred procurement by government and leading retailers, such as Walmart and Target [67,68]. This is a win/win/win for those who participate. There is no penalty to companies who continue to produce chemicals that are inherently hazardous but not restricted by regulations. But there are clear benefits and incentives for those chemical ingredients that are inherently safer when they become one of the over 2500 products that carry the Safer Choice label and experience preferred procurement in the marketplace.

8.4.5
International Living Future Institute's Living Product Challenges

The Living Product Challenge™ (LPC) is a product certification program, advocacy tool, and philosophy created in 2015 by the International Living Future Institute (ILFI) that defines the most advanced measure of sustainability in the built environment possible today and act to rapidly diminish the gap between current limits and the end-game positive solutions [15,69]. The ILFI challenges comprise seven performance categories called Petals: Place, Water, Energy, Health & Happiness, Materials, Equity, and Beauty. ILFI Petals are subdivided into a total of 20 imperatives, each of which focuses on a specific sphere of influence. This compilation of imperatives can be applied to almost every conceivable product.

ToxServices' Full Materials Disclosure™ (ToxFMD™) Program is the approved methodology for the Living Product Challenge's Material Health Petal assessment. The ILFI LPC follows criteria set by ToxServices FMD Program. ToxFMD criteria assesses product content inventory and specific hazard endpoint results relating to human health effects, aquatic toxicity, and environmental fate and toxicity for each CAS# present in a consumer product. Chemicals assessed and scored as GreenScreen benchmark 1 will need to undergo a risk and exposure review that encompasses manufacturing, use/installation, and disposal of the product formulation.

8.4.6
Cradle to Cradle Certified Product Program

The Cradle to Cradle Certified™ Product Program applies to materials, subassemblies, and finished products. Materials and subassemblies can be considered "products" for certification purposes. Certification criteria are the same for all product types. To become certified, an applicant must submit formulation information and related data to a Cradle to Cradle Certified assessor and have the product assessed against the requirements of Version 3.1 of the Cradle to Cradle Certified Product Standard. The Cradle to Cradle Products Innovation Institute (C2CPII) reviews all final certification documents and issues certification.

There are five categories of criteria for Cradle to Cradle Certified certification: material health, material reutilization, renewable energy use, water stewardship, and social responsibility. In order to achieve Cradle to Cradle Certified certification at a certain level, the product and manufacturing processes must meet the criteria at that level in all five categories. Additionally, the C2C PII offers the Cradle to Cradle Material Health Certificate (MHC). The Material Health Certificate is a modified version of the full C2C certification, in which it only covers category 1: Material Health, for the C2C Standard, and does not cover the remaining Categories 2–5.

A Cradle to Cradle hazard screen must be performed on each chemical ingredient present at levels above 0.01% in a homogenous material used in a consumer product. In addition to assessing hazards for 24 different human health, environmental, and climate-related hazard points, assessed products are searched for the presence of chemicals on the Cradle to Cradle Banned List. The Banned Lists contain chemicals and substances that are banned for use in *Cradle to Cradle Certified* products as intentional inputs above 1000 ppm (0.1%) [70].

8.4.7
Chemical Alternatives Assessment

Chemical Alternatives Assessment or more broadly, Alternatives Assessment is a method for identifying alternative chemicals, materials, or products that can provide the same service as a chemical of concern in a product. It is increasingly used in regulations, such as the California Safer Consumer Products Act, to increase the likelihood that chemical restrictions do not result in substitutions that lead to unintended negative consequences. Alternatives assessments involve assessing potential alternatives for multiple attributes, including, at a minimum, cost and availability, performance, hazard, and exposure. Additional modules, including stakeholder engagement, life cycle assessment, and materials management can be included. Each module can be used at increasingly detailed and data intensive levels. How the information is used for decision-making is not prescribed but priorities must be well documented.

Northwest Green Chemistry is currently leading the Washington Boat Paint Alternatives Assessment to evaluate alternative technologies to copper-based antifouling coatings that can be used on recreational boats under 65 ft. [71]. Hard and soft fouling can destroy boat surfaces and increase drag, thus reducing speed and efficiency. Copper is used as an antifouling agent in boat paints for its biocidal properties. But copper is extremely hazardous to salmon because it is toxic and interferes with their ability to smell, to defend themselves, to imprint on a home stream, and to reproduce. Some of the alternatives contain different biocides such as zinc pyrithione, Econea, or SEANINE. Other technologies are called foul release instead of antifouling. These coatings create very hard surfaces that are impenetrable to barnacles and damaging forms of hard fouling. Grasses may attach but they fall off when the boat moves through water. Nonchemical alternatives are also being considered, including ultrasound and low frequency pulse emitters that can be applied via attachments at various locations on the boat hull. NGC has engaged stakeholders throughout the process to ensure relevance and accuracy of information and to understand the needs of the marketplace. CHA is being used to evaluate chemicals in all of the coating formulations. Due to cost limitations, full CHAs are being commissioned only for the active ingredients in the biocides. GreenScreen List Translator is being used to screen for human health hazards and modeling using the freely available EPISUITE that is being used to predict aquatic toxicity, persistence, and bioaccumulation potential. Additional chemical information, such as VOC content, is also being gathered. All of this will be compiled along with information on exposure, cost and availability, and performance to create a selection guide that will inform users. Anticipated release of the report is September, 2017.

8.5
Case Studies Showing How CHA Leads to Safer Consumer Products

8.5.1
Case Study 1. US EPA Safer Choice Product Certification

Safer Choice began in the 1990s as a partnership program of the US Environmental Protection Agency's Design for the Environment (DfE) Program. DfE's original mission was to assess alternatives to US EPA priority chemicals and to recognize companies for making best in class products with safer ingredients. In 2016, DfE changed its name to the Safer Choice Program, although the EPA still uses the DfE Program name for alternatives assessments projects.

The Safer Choice Program (then DfE) invited primarily cleaning product manufacturers to submit their formulations for review and to receive recommendations for improvement. Manufacturers who formulated or reformulated products with inherently low toxicity and best in class chemicals were rewarded with a Safer Choice Partnership Agreement between the manufacturer and Safer Choice allowing them to use the Safer Choice logo on their product labels and other marketing materials. There was no cost to participate in the program and there were benefits such as increased understanding of the health and environmental attributes of chemicals in products, increased knowledge of inherently safer alternatives, and the positive recognition that the logo brought to products in the marketplace that enabled companies to stand out from their competitors as formulators of safer products. As the program caught on, a long waiting list of manufacturers seeking partnerships developed. To adjust to the growing demand, Safer Choice decided to remodel the program into a recognition program where product manufacturers work directly with scientific consulting firms to have their products reviewed, while maintaining Safer Choice oversight. NSF International became the first organization to become an approved third-party profiler for Safer Choice, followed by ToxServices LLC.

In order to obtain Safer Choice certification and to feature the Safer Choice label on certified products, third-party profilers work with product manufacturers under a confidentiality agreement to review products against the Safer Choice Standard and associated ingredient-specific criteria to determine if the overall product is recommended for Safer Choice recognition. Requirements of the Safer Choice Standard include but are not limited to a formulation review of the product and assessment of its constituent ingredients, product performance, packaging sustainability, product pH, ingredient disclosure, and VOC content, in addition to an audit [65]. During the product review, the third-party profiler compiles human and environmental health profiles for each chemical in a formulation (in addition to assessing other requirements, such as ingredient disclosure). Safer Choice then reviews a third-party assessment report to ensure harmonization among third-party profilers and awards recognition, as appropriate.

Outsourcing the responsibility of product reviews to external third-party profilers allowed the Safer Choice Program to expand considerably; however, it also meant that the Safer Choice review was no longer free. While there is no charge for recognition by Safer Choice, manufacturers do need to pay for the independent toxicology assessment services performed by the third-party profilers. These costs are considerably lower relative to other green certification programs and Safer Choice has taken innovative steps to further reduce costs for product manufacturers, through their partnership with GreenBlue for the CleanGredients database and the creation of the Safer Chemicals Ingredients List (SCIL).

Safer Choice designed a two-pronged strategy to identify inherently safer ingredients for use in Safer Choice products. Safer Choice funded the creation of the CleanGredients database to identify commercially available and trade name ingredients that meet the applicable Safer Choice functional class and/or Master Criteria [66]. (*Disclosure*: Development of CleanGredients was led by coauthor Heine). These trade name ingredients are usually mixtures and many times contain proprietary ingredients. For example, a surfactant manufactured by Akzo Nobel whose trade name is listed on CleanGredients may be dissolved in water or another solvent and contain a preservative. All of the intentionally added ingredients in the trade name ingredient (and residuals above 0.01%) must meet the appropriate Safer Choice functional class criteria. Once the ingredient formulation is reviewed by a third-party profiler and is approved for listing by Safer Choice on CleanGredients, the trade name ingredient can be used in products seeking to qualify for Safer Choice without additional assessment.

In addition to CleanGredients, which lists trade name-specific chemical ingredients, Safer Choice also created the Safer Chemical Ingredients List (SCIL) to identify theoretically pure substances that meet functional class and/or Master Criteria [64]. The SCIL is a list of chemical ingredients arranged by functional class that the Safer Choice Program has evaluated and determined to be safer than traditional chemical ingredients used for the same function. There are currently over 800 ingredients on the SCIL. The difference between the SCIL and CleanGredients is that SCIL chemicals are discrete chemicals, whereas CleanGredients ingredients are whole ingredient formulations.

Safer Choice now has a total of nine different ingredient functional class criteria (chelating and sequestering agents, colorants, polymers, preservatives and related chemicals, defoamers, enzymes and enzyme stabilizers, fragrances, oxidants and oxidant stabilizers, processing aids and additives, solvents, surfactants) and a set of Master Criteria for ingredients not covered by a functional class [65].

When formulating with or purchasing ingredients identified on the SCIL, a supplier or formulator will still need to assess any impurities (e.g., residuals such as 1,4-dioxane) and additional intentionally added chemicals in the ingredient (e.g., preservatives, denaturants) against the Safer Choice Criteria, in addition to certain physical properties for some of the functional classes. In contrast, CleanGredients lists trade name ingredients specific to one manufacturer for which every intentionally added ingredient, physical properties, and all known associated impurities and/or residual chemicals have already been reviewed and approved and no further review is necessary.

The use of hazard assessment was fundamental to developing Safer Choice Criteria. Because risk is a function of hazard and exposure, one can compare ingredients used in a similar way and for similar function for a particular product class primarily based on hazard. If the exposure is relatively constant, then reducing hazard will reduce risk. By focusing on the key variable – degree of hazard – the Safer Choice and CleanGredients approach simplifies the complex, often uncertain and elusive exposure calculations.

The CleanGredients database started as a proverbial chicken or the egg scenario. Formulators were not interested in subscribing to the database unless chemical suppliers listed ingredients that qualified for Safer Choice recognized formulations, and chemical suppliers were not interested in subscribing to CleanGredients unless there was a market and audience of formulators interested in purchasing their ingredients. Ultimately, both sides took initial steps to participate and subscriptions to the CleanGredients database grew rapidly. The annual subscription cost for CleanGredients is low. Since one of the challenges associated with environmentally preferable product certification is the cost, having a database of ingredients that are prequalified for use in Safer Choice eligible cleaning products was one way to reduce costs. Over time, the Safer Choice label has become a recognizable logo to consumers and there are now more than 2500 products that carry the logo. Several large retailers, including Walmart and Target, use Safer Choice in two ways as part of their sustainable chemicals policies. First, they give preference to products made by others that carry the Safer Choice logo. Second, they manufacture their own private label products to meet the Safer Choice Criteria. Government procurement by federal government entities and by numerous state government entities also specify Safer Choice products.

The Safer Choice Program demonstrates the effectiveness of a systematic approach to create a marketplace for safer chemicals by setting criteria and supporting tools and information that advance both supply and demand. Safer Choice achieved consensus on the attributes of ingredients used for specific functions that are preferable for human health and the environment. Chemical manufacturers (aka suppliers such as Akzo Nobel, BASF, Novozymes, Rivertop Renewables, and Stepan, to name a few) can market existing ingredients or manufacture new ingredients according to Safer Choice ingredient criteria. Product formulators (e.g., Amway, Clorox, Ecolab, Seventh Generation, Clean Control, Anderson Chemical, Custom Compounders, 1908 Brands, WD-40, and many more) are then able to select preassessed chemical ingredients to formulate products that meet the Safer Choice product criteria. Retailers (e.g., Walmart, Target, Wegmans, Safeway, Amazon, etc.) and government procurement entities specify products that carry the Safer Choice logo. This systemic approach essentially grows a market for greener chemicals and chemical intensive products. It can also contribute to implementation of the new Frank R. Lautenberg Chemical Safety for the twenty first century Act (revised Toxic Substances Control Act). The hundreds of SCIL-listed chemicals can serve as the Agency's starting point for designating low-priority substances.

Safer Choice drives green chemistry by promoting the selection of existing chemicals that are inherently safer but also incentivizes innovation so that chemical manufacturers work toward creating new, safer chemicals. For example, Eastman Chemical achieved success with its nonphthalate plasticizer Eastman168-SG, and was the first plasticizer listed in CleanGredients after meeting Safer Choice criteria [72]. Eastman's Omnia™ solvent (CAS number *[53605-94-0]*) was designed to meet Safer Choice solvent criteria [73].

8.5.2
Case Study 2. Levi Strauss & Co. Screened Chemistry

In 2001, Levi Strauss & Co. (LS&Co.) was one of the first companies in the apparel sector to develop and apply a restricted substances list (RSL) that meets and, in most cases, exceeds, global regulatory requirements [3]. Chemicals on the LS&Co. RSL [74], and on the Zero Discharge of Hazardous Chemical (ZDHCs) manufacturing restricted substances list (MRSL) [75], include the following:

- The 11 priority chemical groups identified by Greenpeace and ZDHC members. These include alkylphenols, phthalates, brominated and chlorinated flame retardants, azo dyes, organotin compounds, perfluorinated chemicals, chlorobenzenes, chlorinated solvents, chlorophenols, short-chain chlorinated paraffins, and heavy metals.
- Substances relevant to the apparel sector that are found on the California Department of Toxic Substances Control (DTSC) list of chemicals of concern, including alkylphenols, aromatic amines, azo dyes, perfluorinated chemicals, formaldehyde, phthalates, and triclosan.

To ensure compliance, LS&Co. ensures that all chemical formulations used in their manufacturing processes are reviewed by a third party to make sure that none of these substances are present. If the formulation or substance contains RSL or MRSL substances, then the chemical supplier is notified.

One of the primary challenges to a company, such as LS&Co., is understanding hazard and risk profiles of chemicals present in raw materials, including chemicals present in supplier proprietary formulations. LS&Co. is able to navigate this roadblock by engaging with approved third party assessors to evaluate supplier formulations (*Disclosure*: ToxServices performs CHA work for LS&Co.). LS&Co. uses GreenScreen and US EPA Safer Choice criteria to evaluate individual chemicals found in supplier formulations. Using a third-party assessor allows suppliers to have their confidential formulations reviewed for chemicals of concern without revealing them to LS&Co. Individual chemicals within a formulation are assigned a numeric score based on their GreenScreen or Safer Choice hazard rating ranging from 50 to −50 points: Green rating (preferred chemical): 35–50; Yellow rating (need improvement): 20–34; and Red rating (phase out): 19 to −50. A formulation-level hazard score is calculated based on a weighted average of ingredient scores [3]. Such a system allows LS&Co. suppliers to formulate with less hazardous alternatives within an ingredient class and also quantifies the hazard profile of the overall formulation to LS&Co.

In addition to simply avoiding chemicals listed on its RSL or the ZDHC MRSL, LS&Co. engages with suppliers to evaluate substances of concern and move to safer alternatives. As an example, in 2014, LS&Co. engaged with Schoeller, a formulator of durable water repellants, to assess a substance of concern in one water repellant formulation and successfully switched to a safer water repellant without sacrificing performance. Schoeller's Nanosphere®, a nanotech-based finishing technology is used to impart water and dirt repellency on denim and nondenim apparel. However, the Nanosphere formulation contained short-chain perfluorinated chemicals (C6) that LS&Co. identified as a class of chemicals to eliminate from its products by December 2015.

Although short-chain PFCs are not currently regulated and do not have a mandatory hazard classification under the European Union's Classification and Labelling (CLP) Regulation, there are data gaps and uncertainty about the safety of such chemicals in part based on experience with longer chain PFCs, such as perfluorooctonoic acid (PFOA) and perfluorooctanesulfonic acid (PFOS) that are harmful to human health and the environment. Therefore, LS&Co. selected to apply a precautionary approach and anticipated that shorter chain PFCs might also exhibit hazardous properties. LS&Co. took six steps toward replacing short-chain PFCs with less hazardous ingredients [76]:

1) LS&Co. identified the chemical of concern by describing the hazard, the function of the substance, and the current conditions to make it work at the desired performance level.
2) Using its RSL process, LS&Co. set requirements to eliminate unsafe chemicals and aligned with the regulatory and legal requirements of the countries in which it operates and sells.
3) LS&Co. engaged with chemical suppliers to discuss chemical sustainability, hazard, risk and exposure, and works with them to find safer alternatives for chemicals of concern.
4) LS&Co. asked suppliers to share what hazard assessment methodology and tools they use to identify safer substitutions.
5) LS&Co. piloted alternatives approved through its chemical screening process to ensure that their performance met customer expectations.
6) After third-party verification, LS&Co. then encouraged the supplier to post a substitution case study on the Substitution Support Portal [77].

The primary focus of this effort was to ensure that the alternative chemical did not turn out to be a "regrettable substitution." At a minimum, the alternative(s) ingredient should not have PBT or carcinogenic hazard properties (which are the critical effect endpoints for long-chain PFCs).

There were a number of business realities that played out in this case. For example, the coating containing the chemical of concern was marketed by Schoeller Textiles. LS&Co. wanted to move away from the chemical but also wanted to maintain its positive relationship with the supplier. LS&Co. decided to work with Schoeller Textiles to find a safer alternative to short-chain PFCs. In response to the global need for a safer alternative to PFCs, Schoeller launched a PFC-free replacement technology called ecorepel®, based on substances that are not classified as hazardous [78]. Schoeller supplied LS&Co. with the following:

- Safety data sheets for all components
- Chemical hazard assessment data generated on ecorepel
- Registration through the European REACH regulation, including access to the REACH dossier

Over an 18-month period, LS&Co. used third-party toxicology experts to work with Schoeller to assess ecorepel using a CHA framework. Using results from successive CHAs, Schoeller Textiles was able to optimize the ecorepel formulation and screen out less desirable ingredients. Results from the final CHA demonstrated that ecorepel's ingredients met LS&Co.'s screened chemistry requirements for both human health and environmental endpoints. Subsequently, the cost and performance of the coating were assessed and found to be equivalent to that of the original product. Based on this level of research and due diligence, LS&Co. was able to switch to a PFC-free water and dirt repellant that performed as well as a PFC-containing formulation, demonstrating the utility of CHA in the safer chemical selection process.

8.5.3
Case Study 3. Development of an Alternative Food Can Liner

Bisphenol A (BPA) is an endocrine disrupting chemical used in the lining of many food and beverage cans. BPA can leach from the lining of cans into food resulting in human exposure. A growing number of scientific studies link small amounts of BPA exposure to health problems, including breast and prostate cancer, asthma, obesity, behavioral changes (including attention deficit disorder), altered development of the brain and immune system, low birth weight, and lowered sperm counts. With growing consumer concern over BPA's health impacts, there has been increased demand for BPA-free products including in food packaging. The power of consumers to move markets was demonstrated by the rapid movement of industry away from the use of bisphenol A (BPA) in baby bottles, sippy cups, sports water bottles, and canned infant formula. Five US cities and counties, 13 states, and ultimately the FDA banned BPA from baby bottles, sippy cups, and infant formula cans [79].

But chemical bans do not always lead to sustainable solutions. In the case of BPA in baby bottles, substitutions were made and led to regrettable substitutions.

For a number of the alternatives, the identity of the chemicals used were not disclosed and, therefore, their hazards could not be assessed. In other cases, BPA in some baby bottles and receipt paper was replaced with other bisphenols such as BPS and BPAF that were found to be lacking in test data and similar in structure to BPA. The US EPA Design for the Environment Program did an alternatives assessment for BPA used in thermal paper [80]. Other bisphenols were identified that were considered alternatives to BPA. Because these chemicals had limited supporting toxicological data, expert chemists and toxicologists at EPA used what is known as a *read-across* approach where existing data for a chemical analog or surrogate with similar structure to the chemical of interest can be used to evaluate the hazard endpoints for the chemical of interest. In this case, the best analog for the related bisphenols was BPA. Therefore, using best available information, the alternatives were predicted to have the same hazard properties as BPA.

An assessment, published March 2011 in *Environmental Health Perspectives* (EHP), of more than 500 commercially available plastic products labeled BPA free, found many to be leaching chemicals that in some cases were more estrogenic than BPA-containing plastics [81]. Thus, BPA free did not mean that the products were free of endocrine activity [82].

The chemical company Valspar recognized the need and the opportunity and began a 5-year journey to develop a new, safer BPA alternative using tetramethyl bisphenol F (TMBPF). Valspar decided to engage the scientific and advocacy communities for feedback and advice on how best to demonstrate the safety of their new polymeric material in a way that would be credible to stakeholders. They wanted to understand what levels of transparency and what types of testing would be necessary. They shared safety data and their model for ensuring safety called *Safety by Design* that was used in the creation of TMBPF. Valspar took an interesting approach by not only seeking to use the best available internal science to assess the safety of their product but also by seeking out some of their likely external critics (Breast Cancer Fund, Environmental Defense Fund) while enlisting trusted, independent scientists to help with the assessment work.

Scientists from Tufts University School of Medicine worked with Valspar to test the new can lining material for estrogenic activity. They looked at both the final material and at the monomer used to create it. They also tested to see if any chemicals would leach into foodstuffs from the final coating polymer. In January 2017, Ana Soto *et al.* published their findings in Environmental Science & Technology in a paper entitled "Evidence of Absence: Estrogenicity Assessment of a New Food-Contact Coating and the Bisphenol Used in its Synthesis."

According to the authors, "TMBPF did not show estrogenic activity in the uterotrophic assay, did not alter puberty in male and female rats or mammary gland development in female rats. Neither TMBPF nor the migrants from the final polymeric coating increased proliferation of estrogen-sensitive MCF7 cells. TMBPF did not show estrogen-agonist or antagonist activity in estrogen receptor-transactivation assay. TMBPF migration was below the 0.2 parts per billion detection limit. Our findings provide compelling evidence for the absence of EA by TMBPF and the polymeric coating derived from it, and that human exposure to TMBPF would be negligible"[83].

Valspar's Safety by Design creates a model for stakeholder engagement and for transparent and credible safety testing of chemicals in food can linings that could be applied to other product sectors as well. It requires transparent testing for endocrine activity, not because the FDA requires it but because it is a source of public concern. The NPO partners praised Valspar for their Safety by Design work and also requested that Valspar share its toxicological assessment results for the other hazard endpoints as well.

8.6
Challenges: Beyond Chemical Hazard Assessment

Chemical hazard assessment is one tool in the toolbox to guide the development of products that are based on principles of green chemistry and engineering. Such products should be safe and sustainable across the supply chain and the full product life cycle. In the current state of practice, there are the following three major challenges associated with CHA:

1) Transparency of chemical and material ingredients used in products and processes
2) Filling data gaps for existing and emerging hazard endpoints
3) Integrating CHA into sustainable product design

8.6.1
Transparency

Transparency of ingredients in products and processes is both a business-to-business and a business-to-consumer challenge. Manufacturers are challenged to know what is in their products beyond what is reported on safety data sheets. Their suppliers are often unwilling to disclose full formulations for fear of losing valuable intellectual property they consider a competitive advantage. In some cases, suppliers are willing to confirm what chemicals (i.e., chemicals on RSLs or MRSLs) are *not* in the products or materials. In other cases, suppliers may be willing to submit full ingredient formulations to trusted third-party assessors for review as in the case of Safer Choice and certification programs like Cradle to Cradle. Assessors often bemoan the challenge and cost of chasing down suppliers to obtain full ingredient disclosure and decisions must be made about who will bear the cost. With respect to consumer disclosure, enforcement of the law in NY requiring ingredient disclosure in cleaning products and proposal of the related bill in California (SB 258) discussed earlier is another signal to manufacturers that consumers are interested in greater ingredient disclosure in order to make informed decisions about products.

8.6.2
Filling Data Gaps for Existing and Emerging Hazards: Predictive Toxicology and Tox21

Standard toxicological testing is time-, resource-, and animal intensive. Historically, chemical hazards were investigated one chemical and one health effect endpoint at a time. As described earlier in this chapter, human health hazards comprise numerous endpoints, such as repeat dose toxicity, developmental toxicity, reproductive toxicity, neurotoxicity, carcinogenicity, sensitization potential, and so on. Few chemicals in consumer products have been tested for all of these health effect endpoints. As a result, a chemical enters the supply chain with multiple data gaps pertaining to one of more hazard endpoints. Predictive toxicology is seen as the path forward. According to the National Research Council of the National Academies of Sciences,

> "Advances in molecular biology, biotechnology, and other fields are paving the way for major improvements in how scientists evaluate the health risks posed by potentially toxic chemicals found at low levels in the environment. Advances in predictive toxicology would make toxicity testing quicker, less expensive, and more directly relevant to human exposures. Predictive toxicology could also reduce the need for animal testing by substituting laboratory tests based on human cells [84]."

Almost 15 years ago, the US National Toxicology Program described a vision for the twenty first century to: "support the evolution of toxicology from a predominantly observational science at the level of disease-specific models to a predominantly predictive science focused upon a broad inclusion of target specific, mechanism-based, biological observations" [85]. These activities are collectively called *Tox21*.

Practically, this means adding or integrating new tools for predictive toxicology to standard methods and regulatory toxicological approaches. The methods are still emerging and in development and professional toxicologists will need to learn new ways to integrate very different lines of evidence into their assessment work. Several emerging publications will address strategies for filling data gaps using predictive toxicology to advance alternatives assessment. The challenge is well-recognized and advancements are occurring. As predictive toxicology advances, assessing standard hazard endpoints may no longer be seen as the goal. Rather, an integrated approach to biological pathways, mechanisms, and adverse outcome pathways will become the way to understand toxicity. When put into practice, we hope that predictive toxicology will make regrettable substitution a distant memory.

8.6.3
Integrating CHA into Green Product Design

CHA is just one tool in the toolbox needed to advance green chemistry and green engineering in product design and development. CHA supports sustainable material flows. It is necessary but not sufficient. A number of aspects must be co-optimized in order for this to occur. The American Chemical Society's Green Chemistry Institute considered the various sets of principles of Green Chemistry and Green Engineering and boiled them down into three broad Design Principles for Sustainable and Green Chemistry and Engineering (*Disclosure*: Developed by authors Constable and Jiménez-González) [86]. These include the following:

1) Maximize resource efficiency
2) Eliminate and minimize hazards and pollution
3) Design systems holistically and use life cycle thinking

Likewise, in 2010, the OECD established Policy Principles for Sustainable Materials Management (*Disclosure*: Developed by coauthor Heine with Marc Major) [87].

1) Preserve natural capital.
2) Design and manage materials, products, and processes for safety and sustainability from a life cycle perspective.
3) Use the full diversity of policy instruments to stimulate and reinforce sustainable economic, environmental, and social outcomes.
4) Engage all parts of society to take active, ethically based responsibility for achieving sustainable outcomes.

These principles have a great deal in common. Leading companies are coming to the same conclusions that sustainable products must be designed to eliminate and minimize hazards and pollution while preserving natural capital and optimizing value recovery and minimizing impacts across the life cycle. This must also be integrated with economic and policy drivers while being inclusive of stakeholders and vulnerable populations. The sooner society can come to consensus on the broad aspects of sustainable materials and products, the sooner determinations can be made on how best to use the tools such as CHA, risk assessment, life cycle assessment, exposure assessment, materials management, and social life cycle assessment. In some ways, the tools have come to drive the vision and not the other way around. Too many arguments focus on how to trade-off life cycle impacts against toxicity. Rather, a product should be optimized across its life cycle for multiple attributes. A product, no matter how safe the ingredients, can still cause risk and harm if it is not managed properly. For example, even plastics made with benign chemicals can end up causing harm as litter and eventually degrade into microplastics in the ocean, lakes, rivers, and so on. Likewise, products derived from sustainable feedstocks can still be made into toxic chemicals such as bio-based benzene [88].

Assessing whole products involves moving beyond assessment of individual chemicals to mixtures and polymers and different forms of chemicals including nanomaterials. It involves assessing products that are not chemical intensive. For example, tanning salmon leather is chemical intensive, but the benefits of using waste salmon skin from sustainably harvested salmon presents a more complicated material assessment question.

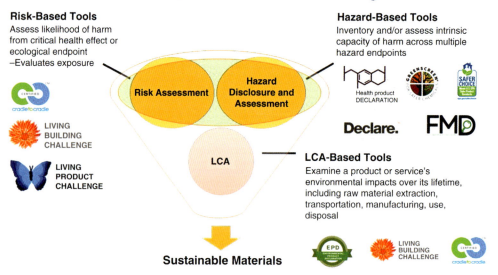

Figure 8.9 Sustainability toolbox.

Figure 8.9 illustrates some of the tools in the Sustainability Toolbox for assessing products and materials used in the built environment.

- Hazard-based tools
  - Tools that inventory and/or assess intrinsic capacity of harm across multiple hazard endpoints.
- Risk-based tools
  - Tools that assess likelihood of harm from critical health effects or ecological endpoint (evaluates hazard and exposure).
- LCA-based tools
  - Tools that assess a product or service's environmental impacts over its lifetime, including raw material extraction, transportation, manufacturing, use, and disposal.

Where and when to use the existing tools, and developing new tools is part of the challenge of moving toward consensus on principles for green and sustainable design and best practices for the young field of alternatives assessment. Much is still to be learned, and product development is the flip side of the alternatives assessment coin.

8.7 Conclusion

CHA is a powerful tool that supports elimination of chemicals of concern and supports the development and identification of inherently safer alternatives that can be used in consumer products and manufacturing processes. It is necessary but not sufficient to shift to a sustainable materials economy. There are a variety of CHA tools and approaches that one can use depending on one's needs, budget, and timeline.

Consumer product companies that flourish exhibit an interesting mix of fidelity to constant reinvention, along with the ability to break the mold and change when needed. Incorporating the practices of sustainability in today's marketplace is not an easy challenge, as investment in research and development is required, along with the use of new assessment tools to identify less hazardous chemicals. Companies that have committed to this approach have begun to see the benefits of an improved position in the growing economy for sustainably made products, community and social recognition, greater employee satisfaction, environmental benefits, and the ever-heavy weighing factor of increased profitability.

Weaving sustainable product design throughout all phases of the consumer products supply chain is smart business sense because resources are conserved while problematic regulatory compliance is avoided through intelligent chemical/material selection. For companies to compete and thrive in the coming years, innovation, not repetition of bad habits, is the key.

References

1 Anastas, P.T. and Warner, J.C. (1998) Green Chemistry: Theory and Practice, Oxford University Press, New York, p. 30.
2 United States Environmental Protection Agency (U.S. EPA) (2014) TSCA Work Plan for Chemical Assessments: 2014 Update. Office of Pollution Prevention and Toxics. Available at https://www.epa.gov/sites/production/files/2015-01/documents/tsca_work_plan_chemicals_2014_update-final.pdf.
3 Strand, R. and Mulvihill, M. (2016) Levi Strauss & Co.: Driving the Adoption of Green Chemistry. Haas School of Business. University of California Berkeley, July 15.
4 Coleman-Lochner, L. (2016) It's so hard to make blue jeans without nasty chemicals. Bloomberg News, November 1. Available at https://www.bloomberg.com/news/articles/2016-11-01/those-nasty-chemicals-in-your-blue-jeans-aren-t-easy-to-replace.
5 State of Washington (2017) Green Chemistry. Frequently Asked Questions. Available at http://www.ecy.wa.gov/GreenChemistry/faq.html.
6 KEMI (2015) Phthalates which are toxic for reproduction and endocrine-disrupting – proposals for a phase-out in Sweden. Report 4/15. Available at https://www.kemi.se/global/rapporter/2015/report-4-15-phatalates.pdf.
7 Office of Environmental Health Hazard Assessment (OEHHA) (2017) The Proposition 65 List. Available at https://oehha.ca.gov/proposition-65/proposition-65-list.
8 Office of Environmental Health Hazard Assessment (OEHHA) (2013) Evidence on the Carcinogenicity of Diisononyl phthalate (DINP). Reproductive and Cancer Hazard Assessment Branch, October. Available at https://oehha.ca.gov/media/downloads/proposition-65/chemicals/dinphid100413.pdf.
9 Greenpeace International (2017) Eleven hazardous chemicals which should be eliminated. Available at http://www.greenpeace.org/international/en/campaigns/detox/fashion/about/eleven-flagship-hazardous-chemicals/.

10 Greenpeace International (2017) Journey toward a toxic-free future. Available at http://www.greenpeace.org/international/en/campaigns/detox/timeline/.

11 ZDHC (2017) Zero Discharge of Hazardous Chemicals (ZDHC) Programme. Available at www.roadmaptozero.com/.

12 Ellen Macarthur Foundation (2017) The Circular Economy 100: The Programme. Available at https://www.ellenmacarthurfoundation.org/ce100/the-programme/enabling-collaboration.

13 Green Chemistry and Commerce Council (GC3) (2017) About GC3. Available at www.greenchemistryandcommerce.org.

14 Cradle to Cradle Products Innovation Institute (C2CPII) (2016) Cradle to Cradle Certified™ Product Standard. Version 3.1. Available at http://s3.amazonaws.com/c2c-website/resources/certification/standard/C2CCertified_ProductStandard_V3.1_160107_final.pdf.

15 International Living Future Institute (ILFI) (2015) Living Product Challenge Standard. Version 1.0. Available at https://living-future.org/lpc/.

16 Scott, L. (2005) Twenty first century leadership. October 23, 2005 Remarks. Available at http://news.walmart.com/executive-viewpoints/twenty-first-century-leadership.

17 National Retail Federation (2017) 2017 top 250 global powers of retailing, January 16. Available at https://nrf.com/news/2017-top-250-global-powers-of-retailing.

18 Green Chemistry and Commerce Council (GC3) (2016) Joint statement on using green chemistry and safer alternatives to advance sustainable products. Available at http://greenchemistryandcommerce.org/documents/RLC-JointStatement.pdf.

19 Green Chemistry and Commerce Council (GC3) (2011) Retailer portal: tools to evaluation chemical ingredients in products. September. Available at http://greenchemistryandcommerce.org/downloads/RetailerPortal.pdf.

20 Green Chemistry and Commerce Council (GC3) (2011) Retailer tools for safer chemistry. Available at http://greenchemistryandcommerce.org/retailer-portal/retailer-tools/.

21 Target (2017) Chemical policy and goals. Available at https://corporate.target.com/article/2017/01/chemical-policy-and-goals.

22 Walmart (2013) Sustainable chemistry policy. Available at http://www.walmartsustainabilityhub.com/app/answers/detail/a_id/310/session/L2F2LzEvdGltZS8xNDYxNTk2Njg1L3NpZC9GUGs0UFZPbQ%3D%3D.

23 Walmart (2016) Walmart 2016 global responsibility report: enhancing sustainability. Available at http://corporate.walmart.com/2016grr/enhancing-sustainability/promoting-product-transparency-and-quality.

24 Target (2016) Target sustainable product index. Available at https://corporate.target.com/_media/TargetCorp/csr/pdf/Target-Sustainable-Product-Index_1.pdf.

25 Safer Chemicals, Healthy Families (2016) Who's minding the store? — A report card on retailer actions to eliminate toxic chemicals. Available at http://saferchemicals.org/newsroom/new-report-shows-big-retailers-cracking-down-on-toxic-chemicals/.

26 Office of Environmental Health Hazard Assessment (OEHHA) (1986) Safe Drinking Water and Toxic Enforcement Act of 1986. Available at https://oehha.ca.gov/proposition-65/law/proposition-65-law-and-regulations.

27 State of Washington (2008) Washington Children's Safe Product Act. Chapter 70.240 RCW. Available at http://apps.leg.wa.gov/RCW/default.aspx?cite=70.240.

28 State of Washington Department of Ecology (Ecology) (2017) The Reporting List of Chemicals of High Concern to Children (CHCC). State of Washington. Available at http://www.ecy.wa.gov/programs/hwtr/RTT/cspa/chcc.html.

29 State of Minnesota (2009) Minnesota Toxic Free Kids Act. Minn. Stat. 116.9401 – 116.9407. Available at http://www.health.state.mn.us/divs/eh/hazardous/topics/toxfreekids/.

30 State of Minnesota (2016) Toxic Free Kids Act. Chemicals of High Concern and Priority Chemicals. September, 2016. Available at http://www.health.state.mn.us/divs/eh/hazardous/topics/toxfreekids/.

31. State of Oregon (2015) Toxic-Free Kids Act. OAR 333-016-2000. Available at https://public.health.oregon.gov/HealthyEnvironments/HealthyNeighborhoods/ToxicSubstances/Pages/Toxic-Free-Rules.aspx.
32. New York State of the State Book (2017) New York State: Ever Upward. Governor Andrew M. Cuomo. Available at https://www.governor.ny.gov/sites/governor.ny.gov/files/atoms/files/2017StateoftheStateBook.pdf.
33. State of California (2017) California Senate Bill 258, the Cleaning Product Right to Know Act of 2017. Available at http://leginfo.legislature.ca.gov/faces/billNavClient.xhtml?bill_id=201720180SB258.
34. HEC Paris and Ecovadis (2017) The 2017 sustainable procurement barometer. Available at www2.ecovadis.com/sustainable-procurement-barometer-2017.
35. United Nations Environment Programme (UNEP) (2017) Sustainable Public Procurement. Available at http://www.unep.fr/scp/procurement/.
36. Seventh Generation (2017) Responsible business. Available at https://www.seventhgeneration.com/transforming-commerce/responsible-business.
37. Method Home (2017) About us. Available at http://methodhome.com/about-us/.
38. Patagonia (2017) Patagonia: the activist company. Available at www.patagonia.com/the-activist-company.html.
39. Kline, M. (2016) Social impact B corporations are on the risk. January 14, 2016. Inc. Available at http://www.inc.com/maureen-kline/the-rise-of-the-benefit-corporation.html.
40. Laws, E.A. (2000) *Sewage Treatment, In Aquatic Pollution: An Introductory Text*, 3rd edn, John Wiley & Sons, Inc., New York, pp. 140–171.
41. Whittaker, M.H. and Heine, L. (2013) Chemicals Alternatives Assessment (CAA): tools for selecting less hazardous chemicals, in *Chemical Alternatives Assessments* (ed. R. Hester), Royal Society of Chemistry, Cambridge, pp. 1–43.
42. United Nations (2015) Globally Harmonized System of Classification and Labelling of Chemicals (GHS) Available at: http://www.unece.org/trans/danger/publi/ghs/ghs_welcome_e.html.
43. United Nations (1992) Chapter 19. Environmentally sound management of toxic chemicals, including prevention of illegal international traffic in toxic and dangerous products. In, Agenda 21. United Nationals Conference on Environment and Development. Rio de Janeiro, Brazil. 3–14 June, 1992. Available at https://sustainabledevelopment.un.org/content/documents/Agenda21.pdf.
44. United National Economic Commission for Europe (UNECE) (2017) GHS Implementation. Available at http://www.unece.org/trans/danger/publi/ghs/implementation_e.html.
45. United States Environmental Protection Agency (U.S. EPA) (2011) U.S. EPA's Design for the Environment Program Alternatives Assessment Criteria for Hazard Evaluation. Version 2. Available at https://www.epa.gov/sites/production/files/2014-01/documents/aa_criteria_v2.pdf.
46. Clean Production Action (CPA) (2016) The GreenScreen® for Safer Chemicals Chemical Hazard Assessment Guidance. Version 1.3 Guidance, March. Available at www.greenscreenchemicals.org/.
47. SciVera (2017) SciVera Lens. Available at www.scivera.com.
48. Chemical Compliance Systems (CCS) (2017) Enhanced GreenSuite®. Available at www.chemply.com.
49. 3E Company (2017) Ariel® WebInsight. Available at http://3ecompany.com/products-services/regulatory-research/ariel-webinsight.
50. Clean Production Action (CPA) (2016) GreenScreen® Version 1.3 Hazard Criteria, March. Available at www.greenscreenchemicals.org/.
51. ToxServices (2015) GreenScreen for safer chemicals assessment of tralopyril (CAS# [122454-29-9]). GS-465. January 4. Available at http://www.newmoa.org/prevention/ic2/projects/assessments/2014-12-06_Tralopyril_GS.pdf.
52. State of Washington Department of Ecology (Ecology) (2017) QCAT: Quick Chemical Assessment Tool. Version 2.0.

Available at http://www.ecy.wa.gov/greenchemistry/qcat.html.

53 International Fragrance Association (2017) Standards Library: Restricted and Prohibited Substances. http://www.ifraorg.org/en-us/standards.

54 TEDX (2017) TEDX list of potential endocrine disrupters. Available at http://www.endocrinedisruption.org/endocrine-disruption/tedx-list-of-potential-endocrine-disruptors/overview.

55 Healthy Building Network (HBN) (2017) Chemical and Material Library. The Pharos Project. Available at pharosproject.net/.

56 Toxnot (2017) Toxnot database. Available at toxno.com/.

57 ChemAdvisor, Inc (2017) LOLI database. Available at https://www.chemadvisor.com/products/loli-database.

58 United States Environmental Protection Agency (2017) Design for the Environment Alternatives Assessments. Available at https://www.epa.gov/saferchoice/design-environment-alternatives-assessments.

59 Clean Production Action (CPA) (2016) Chemical Footprint Project (CFP). Available at www.chemicalfootprint.org/.

60 Health Product Declaration (HPD) Collaborative (2015) Health Product Declaration Standard Version 2.0. Available at http://hpd-collaborative.org/standard-documents/HPD_Open_Standard_Format_V2_0_FORM_FINAL.zip.

61 Health Product Declaration (HPD) Collaborative (2016) HPD Builder 2. Available at builder.hpd-collaborative.org/.

62 International Living Future Institute (ILFI) (2017) ILFI's Red List 3.1 Guide. Available at https://living-future.org/declare/declare-about/red-list/#red-list-cas-guide.

63 International Living Future Institute (ILFI) (2017) What is included in a Declare Label? Available at https://living-future.org/declare/declare-about/.

64 United States Environmental Protection Agency (U.S. EPA) (2017) Safer Chemicals Ingredients List. Available at https://www.epa.gov/saferchoice/safer-ingredients#overview.

65 United States Environmental Protection Agency (U.S. EPA) (2017) Safer Choice Standard and Functional Class Criteria. Available at https://www.epa.gov/saferchoice/standard.

66 GreenBlue (2017) CleanGredients database. Available at www.cleangredients.org/about.

67 Schade, M. (2015) Target takes another significant step to address toxic chemicals. Mind the Store. September 28. Available at http://saferchemicals.org/2015/09/28/target-takes-another-significant-step-to-address-toxic-chemicals/.

68 Coleman-Lochner, L. and Martin, A. (2017) Target tightens grip over chemicals in bid to make goods safer. Bloomberg. January 25. Available at https://www.bloomberg.com/news/articles/2017-01-25/target-asks-suppliers-to-list-ingredients-in-sweeping-overhaul.

69 International Living Future Institute (ILFI) (2017) The Red List. Available at https://living-future.org/declare/about/red-list/#red-list-cas-guide.

70 MBDC (2012) Banned List. Cradle to Cradle Certified™ Product Standard, Version 3.0. Available at http://s3.amazonaws.com/c2c-website/resources/certification/standard/C2CCertified_Banned_Lists_V3_121113.pdf.

71 Northwest Green Chemistry (2017) Boat paint alternatives assessment. Available at www.northwestgreenchemistry.org/boatpaint.

72 GreenBlue (2016) First plasticizer listed in CleanGredients. Available at http://greenblue.org/first-plasticizer-listed-in-cleangredients/.

73 Eastman (2017) Eastman Omnia™ high performance solvent. Available at http://www.eastman.com/Pages/ProductHome.aspx?product=71093918.

74 Levi Strauss & Co (2016) Restricted substances list (RSL). Concerning materials, parts, chemicals, components, packaging and other goods (including Sundries). October 2016. Available at http://levistrauss.com/wp-content/uploads/2016/12/RSL-2016.pdf.

75 Zero Discharge of Hazardous Chemicals Programme (ZDHC) (2015) Manufacturing restricted substances list. Version 1.1. Available at http://www.roadmaptozero.com/programme/

76 Levi Strauss & Co. (LS&Co.) (2016) Case study on phase out of short-chain C6 perfluorinated chemicals (PFCs) from apparel. May. Available at http://www.levistrauss.com/wp-content/uploads/2016/05/160311_Case-Story_Levi-Strauss_May252016final1.pdf. manufacturing-restricted-substances-list-mrsl-conformity-guidance/.

77 Subsport (2017) SUBSPORT substitution support portal. Available at www.subsport.eu.

78 Schoeller (2017) ecorepel® bio. Available at https://www.schoeller-textiles.com/en/technologies/ecorepel.

79 Nudelman, J. and Rasanayagam, S. (2017) Drum roll, please: no estrogenic activity found in tests of new replacement for BPA in food can linings. February 6. Available at http://www.preventionstartshere.org/bpa_replacement/.

80 United States Environmental Protection Agency (U.S. EPA) (2015) Bisphenol A alternatives in thermal paper. Final report. August 15. Available at https://www.epa.gov/saferchoice/publications-bpa-alternatives-thermal-paper-partnership.

81 Yang, C.Z., Yaniger, S.I., Jordan, V.C., Klein, D.J., and Bittner, G.D. (2011) Most plastic products release estrogenic chemicals: a potential health problem that can be solved. *Environmental Health Perspectives*, **119** (7), 989.

82 Bittner, G.D., Yang, C.Z., and Stoner, M.A. (2014) Estrogenic chemicals often leach from BPA-free plastic products that are replacements for BPA-containing polycarbonate products. *Environmental Health*, **13**, 41.

83 Soto, A.M., Schaeberle, C., Maier, M.S., Sonnenschein, C., and Maffini, M.V. (2017) Evidence of absence: estrogenicity assessment of a new food-contact coating and the bisphenol used in its synthesis. *Environmental Science and Technology*, **51** (3), 1718–1726.

84 National Research Council (2007) Toxicity Testing in the 21st Century: A Vision and a Strategy. Available at https://www.nap.edu/catalog/11970/toxicity-testing-in-the-21st-century-a-vision-and-a.

85 National Toxicology Program (NTP) (2004) A National Toxicology Program for the 21st Century Roadmap to Achieve the NTP Vision. Available at https://ntp.niehs.nih.gov/ntp/about_ntp/ntpvision/ntproadmap_508.pdf.

86 American Chemical Society Green Chemistry Institute (2017) Design Principles for Sustainable and Green Chemistry and Engineering. Available at https://www.acs.org/content/acs/en/greenchemistry/what-is-green-chemistry/principles/design-principles-for-green-chemistry-and-engineering.html.

87 Organization for Economic Cooperation and Development (OECD) (2010) OECD Environment Directorate. Policy Principles for Sustainable Materials Management. Available at http://www.oecd.org/env/waste/46111789.pdf.

88 Biobased News (2013) Anellotech announces ability to product large volume product development samples of biomass-derived benzene and toluene. Available at http://news.bio-based.eu/anellotech-announces-ability-to-produce-large-volume-product-development-samples-of-biomass-derived-benzene-and-toluene/.

9
Tying it all together to drive Sustainability in the Chemistry Enterprise

David J.C. Constable and Concepción Jiménez-González

Most people who are not in the field of chemistry may not have a high opinion of chemistry and may have negative impressions about chemicals. While the fundamentals of chemistry are taught at some level in many of the science courses students take during their primary and secondary education, most students never develop more than a cursory understanding of chemistry and later in life, when they hear about chemistry, they usually think of the negative consequences of chemicals in their food, water, clothes, personal care products, and other things they come in contact with every day. Chemistry and chemicals are seen as more or less the same things, and the distinctions chemists sometimes make about the differences between chemistry specialties are lost on the wider population. It would seem that the general public has not developed an appreciation of how dependent society is on chemicals and the global chemistry enterprise that produces chemicals, develops chemical technologies, and the products that are in everyday use that owe their existence to chemistry.

Chemists need to provide believable stories about the benefits of integrating sustainable and green chemistry into the practice of chemistry that will make it more sustainable. It is perhaps ironic that green chemistry is arguably one of the more optimistic and helpful paradigms for engaging the broader public, but many chemists struggle to communicate in meaningful ways about it because of the inherent complexity and the multidisciplinary nature of thinking about how to practice sustainable and green chemistry. On the other hand, many chemists do not prioritize sustainability and sustainable or green chemistry, nor are they interested in incorporating those concepts into the traditional practice of chemistry.

Sustainable development has now been discussed for over 30 years since it was first articulated by the Brundtland Commission in 1987 [1]. It is an idea that has been discussed and debated at length, but it is now embraced by many countries throughout the world and by large international corporations, both of whom have developed a variety of goals and key performance indicators to assess progress toward advancing and balancing societal, environmental, and economic opportunities. The promulgation of the sustainable development goals in 2015 is a good indication that considerable progress has been made in articulating what things are important and measurable when it comes to sustainability or sustainable development.

The global chemistry enterprise plays an incredibly important role in advancing many aspects of sustainable development. Chemistry is an enabling science foundational to many of the things society depends on – food, water, the built environment, transportation, medicine, electronics, and so on – and people's lives today would be very different without the many innovations chemistry has brought to the world. The global chemistry enterprise is also responsible for considerable societal and economic benefits to many. However, the work of the global chemistry enterprise is also directly or indirectly causing or associated with considerable adverse environmental impacts.

Many organizations and individuals have attempted to articulate what sustainability might be for the global chemistry enterprise, or more simply, what sustainable chemistry might look like. One of the earliest attempts to define sustainable chemistry was undertaken by the Organization of Economic Cooperation and Development (OECD) in 1998 at a conference in Venice, Italy [2]. A brief definition of sustainable chemistry from that workshop follows:

"Within the broad framework of Sustainable Development, we should strive to maximize resource efficiency through activities such as energy and non-renewable resource conservation, risk minimization, pollution prevention, minimization of waste at all stages of a product life cycle, and the development of products that are durable and can be re-used and recycled. Sustainable Chemistry strives to accomplish these ends through the design, manufacture and use of efficient and effective, more environmentally benign chemical products and processes."

The OECD has continued its development of ideas about sustainable chemistry since this first foray into sustainable chemistry and a more recent definition from the OECD is as follows:

"Sustainable chemistry is a scientific concept that seeks to improve the efficiency with which natural resources are used to meet human needs for chemical products and services. Sustainable chemistry encompasses the design, manufacture, and use of efficient, effective, safe and more environmentally benign chemical products and processes. Sustainable chemistry is also a process that stimulates innovation across all sectors to design and discover new chemicals, production processes, and product stewardship practices that will provide increased performance and increased value while meeting the goals of protecting and enhancing human health and the environment."

A second international initiative that is seen by many to be part of the movement toward more sustainable chemistry is working to improve chemicals management internationally and is known as the Strategic Approach to International Chemicals Management (SAICM) [3]. SAICM is a policy framework established in 2006 by the International Conference on Chemicals Management (ICCM) to foster the sound management of chemicals throughout their life cycle. The SAICM policy framework supports the 2002 goal set in Johannesburg at the

World Summit on Sustainable Development that by 2020, "chemicals are produced and used in ways that minimize significant adverse impacts on the environment and human health."

More recently, the German government has sponsored the creation of the International Sustainable Chemistry Collaborative Centers (ISC3) [4]. One part of their discussion over the past year or so has been to define sustainable chemistry and they did this in 100 words, as follows:

"Sustainable chemistry contributes to positive, long-term sustainable development. With new approaches, technologies and structures it stimulates technical and social innovations and develops value-creating products and services.

Sustainable chemistry uses approaches, substances, materials and processes to deliver functionalities. Therefore, it applies substitutes, alternative processes, resource recovery and efficiency. Thus, it avoids rebound effects, damage and impairments to human beings, ecosystems and natural resources.

Sustainable chemistry is based on a holistic approach, setting policies and measurable objectives for a continuous process of improvement. Interdisciplinary scientific research, education, consumer awareness, corporate social responsibility and sustainable entrepreneurship serve as an important basis for sustainable development."

In general, a common theme in the OECD, the SAICM, and the recent ISC3 efforts is the idea of "sound chemicals management." In short, this means better management of existing chemicals throughout the supply chain, and there is undoubtedly considerable opportunity to improve how chemicals are managed. It is encouraging to see the evolution in the definitions from 1998 to the present incorporating life cycle thinking, alternative chemicals and processes, and a focus on products and services. However, if one thinks a bit more about sustainability and what people are trying to accomplish in pursuing sustainability, one may draw the conclusion that the current ways in which chemistry is being practiced using existing chemicals and their associated supply chains is far from sustainable. These definitions also fall short in terms of language that may inspire and challenge chemists, or provide guidance on what might be practical, measurable, objective approaches to changing the practice of chemistry. Many chemists implicitly accept the hazards and risks associated with chemistry because it is inseparable from how chemistry is practiced and that is one reason it is so difficult to get chemists to think differently about it.

As the authors of this book have shown, there are a variety of ways to measure how green something is or what is the sustainability performance of chemistry, chemicals, or different kinds of chemical processes. In recent years, as noted in Chapter 2, there has been increased activity in the area of rational molecular design based on a deeper understanding of toxicology and chemistry. There has also been a rise in the development of a variety of "design for X" approaches, systems, or methodologies where X may be a word like manufacturing, recycling, reuse, environment, and so on. Each of these approaches is being developed to help product designers design and develop products that address various concerns associated with a product from the perspective of impacts in its use and end-of-life phases of its life cycle.

Each of the chapters in this book has illustrated that the development of sustainable and green chemistry metrics is mostly a context-specific task, and while there are common themes in metrics approaches, the proposed metrics usually differ in some fashion. One common theme that pervades various green chemistry approaches like those found in the chapters of this book is that the development of metrics requires one to embrace the inherent complexity associated with sustainable and green chemistry. A second theme is that metrics approaches require a multivariate metrics-driven assessment if they are going to be meaningful. While a desire to have a single metric is perhaps understandable, it will invariably be insufficient.

Chapter 3 demonstrated a very detailed and mathematically rigorous approach to assessing synthetic organic routes to a single compound. The use of chemical synthesis trees and radial pentagons enables one to rapidly and visually compare key metrics. The environmental profile enables the chemist to investigate key aspects of the process associated with a route, and includes excess reagents and stoichiometric excess, solvent, catalysts, workup, and purification. This approach is important to highlight that chemistry is in essence a very blunt instrument; that is, a chemist requires in many cases significant excesses of reagents, reactants, catalysts, solvents, and so on to ensure that a reaction goes to completion, while the isolation and workup of the desired compound often requires large amounts of solvent and energy. The original methodology has been expanded from largely mass metrics to include other metrics like energy and several environment-, safety-, and health-related metrics. The approach is straightforward but requires a large amount of data for the environment-, safety-, and health-related metrics that is not always easily obtained, or is completely unavailable.

Chapter 4 made the case for including some form of life cycle thinking and assessment as part of any chemical development effort. Life cycle thinking is central to any determination of whether or not a chemical, or a final product many manufacturing steps away from the chemical process, is more sustainable and green than an existing chemical or product. The use of streamlined life cycle approaches has facilitated the broader use of life cycle in industry but all practitioners of any life cycle assessment methodology still face considerable challenges in obtaining trusted and validated data, especially for chemicals that are new to the market. Life cycle thinking and assessment bring a number of important considerations to the forefront of any sustainability assessment. The first is the idea of systems thinking, the second is the idea of establishing scope and boundaries, and the third is the idea of a functional unit. At a minimum, systems and life cycle thinking should be integrated into how chemistry is taught and practiced to enable chemists to choose chemicals, chemical routes, materials, and so on more wisely and from a sustainability perspective. Ideally, the notion of imparting a function (i.e., color, adhesion, transparency) rather than just creating a new molecule (e.g., in the case of color creating a dye, pigment, or some refractive meta-molecule) should also be taught in the context of a functional unit (i.e., a shirt of a particular color, or the amount of glue/sticky note, etc.).

Chapter 5–7 are examples of how green chemistry metrics may be applied in a chemical processing or manufacturing context. The majority of chemical production by mass and volume is done continuously; a comparatively smaller amount is carried out in some type of batch operation, or in a semicontinuous fashion as when using a continuous stirred reactor in batch chemical operations. Because continuous processing is used extensively in the petrochemical industry, the overall mass efficiency of that industry is in excess of 99% and this mass efficiency has been accompanied by a relentless pursuit of energy efficiency obtained through extensive energy integration across the petrochemical complex. The petrochemical industry produces a majority of the organic chemical building blocks used in the chemical processing industry as well as a majority of the toxic compounds.

Because of the inherent efficiency of continuous processing, there has been effort over the past 20 years to explore the use of mini-, milli-, and microreactors along with other unit operations like separations trains, continuous crystallizers, and so on to replace batch chemical operations. Chapter 5 discussed at length the use of green chemistry metrics and included life cycle inventory/assessment to evaluate microflow continuous processing. Through several case studies spanning the range from the laboratory to larger scale and across different industrial applications, the chapter shows that significant cradle-to-gate environmental life cycle improvements may be obtained when a batch chemical process is converted to a continuous process. The work illustrates several important aspects of metrics; their approach is multivariate, takes a life cycle approach, and it is comparative.

Chapter 6 discussed the design of sustainable batch processing from an economic, social, and environmental perspective. In general, batch chemical operations are much less mass and energy efficient than continuous processes, and there are fewer opportunities to exploit mass and energy integration in a typical multipurpose chemical plant. There also tend to be multiple chemical transformations in batch processing and while the yields may be quite high, the overall mass intensity is also high due to the use of large volumes of solvent.

Although bio-based and renewable processes for chemical production are becoming more prevalent, the world is still a distance from a bio-based, renewable, and sustainable bioeconomy. Metrics approaches to assessing bio-based processes are generally similar to approaches for green chemistry metrics. For example, one approach is the biomass utilization efficiency, which is analogous to atom economy [5]. Bio-based processes are generally perceived as being more sustainable because they are typically water-based, low-temperature, and low-pressure processes, but this ignores broader life cycle impacts and downstream processing steps such as separation and isolation from a dilute aqueous fermentation or reaction liquor that requires significant amounts of solvent and a distillation or concentration step to remove the solvent. Chapter 7 touches on an example of a bio-based process of increasing utility and importance to the chemistry community and that is the production of enzymes for biocatalysis. As noted in Chapter 1, the reliance upon platinum group metals for catalysis has increased over the past 20 years, but there is a need to move away from metals that will never be sustainably sourced. While some researchers have been exploring base metal catalysts, others have been exploring organocatalysis, and others, biocatalysis. It is also noted that it is critically important to assess the relative sustainability of biocatalytic approaches given that some biocatalytic-based synthetic processes are no more sustainable than some organometallic-based catalytic systems and the same is true of organocatalytic and base metal approaches. A comparative life cycle-based assessment of any of these alternative approaches is essential in order for one to adequately assess the sustainability of a new approach.

One area that has always been challenging for the implementation of green chemistry and engineering is in consumer-facing businesses. Consumer products companies are challenged by their customers to produce products that contain "safer" chemicals, where a "safer" chemical is defined as being of low inherent hazard to human health and the environment. Many of these companies produce restricted substance lists that they use to inform suppliers that the chemicals on the lists are not to be used in making a consumer product for that company. The difficulty here is manifold. First, these lists invariably presume that an alternative chemical is available to provide the function that is desired in a product. In almost all cases, this is not the case; alternative chemicals need to be discovered and tested to ensure they meet the functional need, maintain the desired performance, or possess superior performance, they do not have any human health or environmental impacts, and they are at the same price point or cost less. Second, it is very difficult to guarantee compliance with the restricted substance list beyond a first tier supplier; that is, the immediate supplier to the company that is selling the product. Chapter 8 has thoroughly discussed these issues and many others. Suffice it to say, green chemistry and engineering at the end of the supply chain is extremely challenging.

9.1
New Areas of Sustainable and Green Chemistry Metrics Research

There remain a few areas of the global chemistry enterprise where metrics are not being applied as meaningfully as has been described in this book. For example, a rapidly growing area of chemistry research is in the area of materials chemistry. This is admittedly a broad area, but what is in view here are materials like ceramics, mixed polymer–inorganic blends, a variety of structured materials, and metamaterials including nanoparticles and so on. Many of the concepts and approaches behind the metrics described in the chapters of this book could be applied to these areas, but they typically are not. One area of materials research that is seeing a growing interest in assessing its green or sustainability performance is nanomaterials, but the assessments are generally limited to toxicity studies of one kind or another [6].

Another area of green or sustainability assessment that is neglected in nanomaterials research is in the use of critical materials. In life cycle terms, this is a matter of abiotic depletion. A very large amount of nanomaterial research is carried out with gold (Au) and silver (Ag), two elements that are not only rare, as described in Chapter 1, they are associated with significant environmental impacts that result from extracting these elements from ore and purifying the elements to a usable purity. At the other end of the life cycle, after the use phase, it will also be extremely difficult to close the loop on recycling products containing nanoparticles of Au and Ag, if the Au and Ag survive the use phase intact. From a chemistry perspective, the use of these materials may make sense; that is, Au is "well behaved"; it is chemically inert, which means it does not react with many things, it is relatively easy to make the particles within a predictable and narrow size range, and making them can be done reproducibly. From a perspective of recycle and reuse, in combination with depletion of the element from the earth's crust, it is hard to argue that the use of these metals is truly sustainable, so a case could be made for their restricted use in only those applications serving a critical societal need that cannot be met in any other way. However, this is not doable without performing an appropriate assessment of the sustainability of these metals, particularly from a life cycle standpoint accounting for emerging and social challenges.

While it was noted that some green chemistry work has been completed for portions of analytical chemistry, it is an area that is in need of further green chemistry metrics development. Most work to date has focused on metrics that influence the use of solvents in analytical techniques such as high-performance liquid chromatography (HPLC), and there are metrics that drive strategies for avoiding, minimizing, or reusing the large amounts of solvents used. However, this remains a challenging area given the use of mixed aqueous/organic mobile phases that limit the opportunity for recycling. HPLC has also been dependent on a variety of solvents that are hazardous, which further complicates their continued use. Another area of green chemistry metrics development in analytical chemistry that has been considered recently is in the area of sample preparation, where once again, a considerable amount has been done to reduce sample size and the volumes of solvent used, or alternative extraction approaches have been used. An area for metrics development that would address the vulnerability in analytical chemistry from a critical elements perspective is in the use metrics that seek to reduce the use of noble gasses in gas chromatography or in cooling, especially in the case of helium, which is rarely recycled and is lost from the earth's atmosphere into space.

As chemistry increasingly explores the use of enzymes to carry out chemical transformations, there should be an accompanying effort to assess the greenness and sustainability of the technology. As was described in Chapter 7, there has been work to evaluate certain aspects of biotechnology, especially as it is compared with traditional organic synthesis. A natural "win" for biotechnology is in achieving unparalleled chemical selectivity catalytically, especially in areas like asymmetric chiral synthesis. Because of this, the evolution of chemical catalysis toward biocatalysis and away from organometallic catalysis, especially catalysts containing critical elements like ruthenium, palladium, rhodium, and iridium, is likely only a matter of time. It should not be assumed, however, that biocatalysis is always greener or more sustainable. Enzymes are, after all, foreign proteins, and any protein can evoke allergic responses in some people, depending on an individual's unique genetic makeup. Enzymes also have to be evolved to a point where they are useful, and that still takes considerable time, and in some cases, to get sufficient tonnage quantities, significant impacts can be associated with producing the enzyme, depending on the process used to obtain the enzyme. The point here is that the analysis needs to be done, and there may be some areas that are not being considered in current assessments.

Another area in need of further development is in how to objectively assess bio-based and renewable chemicals manufacture. There are several areas of interest for further metrics development. The first has to do with feedstocks and feedstock conversion. It is now possible to use genetically modified organisms to convert methane to chemicals (e.g., methanol, formic acid, etc.), polymers (e.g., polyhydroxyalkanoates, etc.), and fuels (e.g., diesel and or gasoline replacements, depending on the downstream processing). Depending on the source of methane, for example, from wastewater treatment or from hydraulic fracturing, there are legitimate questions about whether or not that particular route is green and or sustainable.

A second area of interest in biomass conversion is with how one takes agricultural biomass, for example, and converts it to different chemicals. Some take the route of separating cellulose and hemicellulose and converting these to their component sugars, then fermenting them to make ethanol, butanol, or a variety of other chemicals through further downstream processing. In most cases, the lignin component of the biomass is used for its fuel value in the process; for example, for making steam. Others take the route of pyrolyzing the biomass, mostly carbon monoxide (CO) and hydrogen (H_2), and using Fischer–Tropsch technology to convert the CO and H_2 to higher molecular weight chemicals as is done in the petrochemical industry. Each of these approaches is currently being commercially implemented, but legitimate questions remain about the overall greenness or sustainability of these approaches.

A third area is a slightly more philosophical question about the types of chemicals that can be produced from bio-based and renewable sources. In 2004, the US Department of Energy (DOE) produced a report on 12 framework molecules that were believed to hold potential as replacements for molecules typically derived from petroleum that are used to make many common organic chemicals like monomers for plastics, or specialty chemicals used in a variety of products. These molecules could be obtained from biomass or biological sources, and would require additional downstream chemical transformations to convert them into a variety of molecules of interest in a broad range of applications. The DOE-identified molecules contained one or more oxygen atoms and they are different compared to the molecules routinely obtained from the petrochemical industry; for example, alkanes, olefinics, aniline, benzene/toluene/xylene (BTX), and so on. In 2016, the US DOE produced its third report on bio-based and renewable molecules and if one compares the lists from each report, there has been a shift in the types of molecules that are now considered to be desirable bio-based target molecules, with a few, that is, isoprene, butadiene, and *p*-xylene, that are currently almost exclusively derived from petroleum. Biotechnology has evolved to the point where it is now possible to produce molecules that are true drop-in replacements, but the question to be considered is whether or not these kinds of molecules with known hazards are truly sustainable, and are they really molecules that should be part of the chemical supply chain?

Finally, to perform the assessments that we suggest in the paragraphs above, a large amount of data are needed, particularly if one is to drive a life cycle mindset. It is necessary to continuously improve the consistency and transparency of the information and the assumptions used in green chemistry metrics to ensure the quality and the validity of the decisions made. The challenge is to have access to life cycle inventory, toxicology, safety, and other relevant information for all chemicals in commerce (~100 000 manufactured at greater than one metric ton per year) to allow analysis by researchers and users on a global basis [7]. Some particular examples include the following [8]:

- Continuous development of reliable, common databases with higher geographic resolution to be used for green chemistry assessment of chemicals, from the standard toxicology data to the more sophisticated life cycle inventory data. Streamlined Life Cycle Inventory and Assessment tools that are easy to use with an appropriate degree of transparency for a given application.
- Updates to industry-average data sets available, such as the data from Plastics Europe.
- Enhanced understanding of the overall environmental impacts of formulated products, especially as consumers become more aware of these.
- The inclusion of quality indicators while communicating metrics results on a routine basis (such as sensitivity and uncertainty analysis).

Finally, the communication of green chemistry metrics for effective use by wider audiences (even within the technical community) is an ongoing challenge that will require continuous attention, as it may be easy to either overcomplicate or oversimplify the results. All of this can be overwhelming, and could drive a company to play catch up with the external environment. A few thoughts here to improve the incorporation of green metrics into decision making, sometimes without people even needing to know the details behind the metrics:

- Start with the end in mind. Different metrics may be more or less appropriate, depending on the system to analyze. Choose your metrics wisely; you may not need the same metrics for comparing a route than for strategy setting.
- Use life cycle thinking and multivariate approach on the assessments. Trade-offs will be very common among metrics and boundaries, life cycle metrics that go beyond a single issue will get you the best opportunity for a robust assessment.

- Small steps count. Routine use of green chemistry metrics into regular practices remains a gap. The degree of sophistication would depend on resources and data availability, so it is OK to start simply and then built into a more complex model.
- Use the metrics to drive decisions. An imperfect metric that drives improvement and action is better than the perfect LCA that produces no results. Estimating an LCA is just the beginning, the real objective is to be able to use the data to identify the most effective development changes and improvements that can be targeted and actionable at the lab or shop level.

With this book we have provided a survey of the current approaches and development for green chemistry metrics, so it is our hope that the application examples presented will enable you and the chemistry community in general the use of robust metrics to continuously improve the environmental profile of the products we all rely on.

References

1 World Commission on Environment and Development (1987) *Our Common Future (The "Brundtland" Report)*, Oxford University Press, Oxford.
2 OECD (1998) Proceedings of the OECD Workshop on Sustainable Chemistry. ENV/JM/MONO(99)19/PART1, Environment Directorate. Available at http://www.oecd.org/env/ehs/risk-management/reports-on-sustainable-chemistry.htm.
3 SAICM (2017) Strategic Approach to International Chemicals Management. Available at http://www.saicm.org/Home/tabid/5410/language/en-US/Default.aspx (last accessed June 4, 2017).
4 International Sustainable Chemistry Collaborative Centers (ISC3) (2017) Available at https://isc3.org/what-is-sustainable-chemistry-2/ (last accessed June 4, 2017).
5 Iffland, K., Sherwood, J., Carus, M., Raschka, A., Farmer, T., Clark, J., Baltus, W., Busch, R., deBie, F., Diels, L., van Haveren, J., Patel, M.K., Potthast, A., Waegeman, H., and Willems, P. (2015) Definition, Calculation and Comparison of the "Biomass Utilization Efficiencies (BUE)" of Various Bio-based Chemicals, Polymers and Fuels. Nova Institute for Ecology and Innovation. Available at http://bio-based.eu/nova-papers/ (accessed June 4, 2017).
6 Skjolding, L.M., Sørensen, S.N., Hartmann, N.B., Hjorth, R., Hansen, S.F., and Baun, A. (2016). A critical review of aquatic ecotoxicity testing of nanoparticles: the quest for disclosing nanoparticle effects. *Angewandte Chemie: International Edition*, **128** (49), 15448–15464.
7 National Research Council (1986) Board on Environmental Studies and Toxicology Assessment of Research Needs for Hazardous Waste Management, Report on Research Needs, Washington, DC.
8 Jimenez-Gonzalez, C. and Overcash, M.R. (2014) The evolution of life cycle assessment in pharmaceutical and chemical applications: a perspective. *Green Chemistry*, **16**, 3392–3400.

Index

a

abiotic resource depletion 79
acetone 81
acid gas scrubbing 13
acidification 79, 109
– potential 108
active pharmaceutical ingredient (API) 105, 196
– artemether (ARM) 171
– batch manufacturing process 175
– environmental life cycle impacts 106
– production process at Sanofi 174
N-acyl pyrazine 57
adipic acid (ADA) 107, 187
– direct microflow synthesis 188
– life cycle assessment (LCA)
– – for continuous flow 190
– – for two-step conventional synthesis 191
– production routes 188
– quantitative results for 192
– two-step conventional synthesis 189, 191
Aggregated Computational Toxicology Resource (ACToR) 35, 37
alkyl boronic acid 59
Alzheimer's disease 113
American Apparel & Footwear Association (AAFA) 238
American Chemical Society (ACS) 157
– Green Chemistry Pharmaceutical Roundtable 106, 201
Ames mutagenicity assay 38
amidation 181
7-aminocephalosporic acid (7-ACA) 109
analytic hierarchy process (AHP) 130
Andraos algorithm 76
anthraquinone oxidation (AO) process 190, 194
antioxidant 167
Apple® products 115

aqueous phase biocatalytic process 222
artemether 171
Artemisia annua 171
artemisinin-based drugs 171, 172
artesunate 171
aspirin 50, 56, 75
– synthesis, reaction network for 75
Association of the British Pharmaceutical Industry (ABPI) and Carbon Trust's tool 115
atom economy (AE) 12, 16, 158, 192, 214
atorvastatin 218
– biocatalytic synthesis 219
auxiliary material 73
azidation 180
– β-azidation reactions 81, 82

b

balanced chemical equations 64
BASF methodology 114
BASF's Eco-efficiency analysis 114
basification 79
batch processing 175, 178. *See also* sustainable batch design
– chemical manufacturing 192
– and continuous microreactor parameters 165
– and continuous operation, cost analysis of 178
– procedures 199
– reaction conditions of 167
– reactor process 170, 171, 187
– *vs.* continuous operation 160
BatchRetroLC 131, 141
benzene chlorination 5, 77, 88
benzene/toluene/xylene (BTX) 288
benzoic acid 75, 88
benzoyl chloride 168
– phase transfer catalysis of 167

Handbook of Green Chemistry Volume 11: Green Metrics, First Edition. Edited by David J. Constable and Concepción Jiménez-González.
© 2018 Wiley-VCH Verlag GmbH & Co. KGaA. Published 2018 by Wiley-VCH Verlag GmbH & Co. KGaA.

– toxicity potentials, biphasic reaction 169
bioaccumulation 43, 79
bio-based chemicals 6, 109
biocatalytic processes 207, 285
– atom economy 214
– biocatalyst/reaction development 217
– biocatalytic route to atorvastatin 218
– Carbon Mass Efficiency 214
– claimed environmental benefits 211
– design/development 210
– E-factor 216
– efficiency 209
– environmental benefits 211
– methodology 223
– potential staged approach to benchmark 214
– process development 221
– process mass intensity 215
– reaction development, evaluation of 215
– route feasibility, evaluation 214
– route selection 216
– sitagliptin 219
– solvent intensity 215
– structure 208
– sustainability 207, 210
– – early stage assessment 213
– – quantitative measuring 212
– synthetic processes 286
– two-stage development paradigm 211
– water intensity 216
biocide-coated medical textiles 106
bioconcentration factor (BCF) 38
biofuels 112
biomass utilization 5
biomaterials 116
bisphenol A (BPA) 233, 241, 269
blister pack carbon footprint 115
Boc protection
– group 61
– phenylalanine 59
bond economy 12
boronic acid 57
boronic ester 73
– cleavage 67
bortezomib 50, 56, 57, 59, 63, 73, 74
Brundtland Report 131
butadiene 187, 289
1-butyl-3-methylimidazolium chloride 167
1-butylsulfonate-3-methylimidazolium 167

c
CA Green Chemistry Act 16
calcination 88
California Act 16
California Office of Environmental Health Hazard Assessment 234
California Safer Consumer Products Act 263
CAPEX cost analysis 188
caprolactam routes 109
carbon efficiency 12, 158, 162
carbon footprint analysis 116
carbon mass efficiency (CME) 214, 218
carbon monoxide 6
carbonyldiimidazole (CDI) 73
carboxylic acid 89
carcinogenicity 81
carcinogenic, mutagenic, or toxic to reproduction (CMR) 258
cascading reactions, into a microreactor flow network 179
case study
– aspirin 74
– – Case I 84
– – Case II 85
– – Case III 85
– – Case IV 85
– – Case V 86
– – Case VI 86
– – environmental and safety–hazard impact 78
– – input energy 84
– – material efficiency 76
– – reaction network 74
– bortezomib 56
– – material efficiency 64
– – millennium pharmaceuticals' process 59
– – pharma-sintez process 62
– – synthesis strategy, for future optimization 72
catalytic transformation 2
catechol 88
C-efficiency 192
cell death 56
cereal crop preservation 114
chain optimization 103
chemical additives 114
chemical alternatives assessment 18
chemical efficiency 11
chemical engineers 11
Chemical Footprint Project 256, 259
– chemical inventory 256
– disclosure and verification 256
– footprint measurement 256
– management strategy 256
chemical hazard assessment (CHA) 6, 9, 231-233, 244, 271
– case studies 264

– – Levi Strauss & Co. (LS&Co.) 267
– challenges 271
– – filling data gaps 272
– – into green product design 272
– – transparency 271
– characteristics 232
– chemical alternatives assessment 263
– chemical footprint project 255
– comprehensive/abbreviated forms 247
– consumer product
– – companies 232
– – design/development considerations 235
– – sector 242, 256
– Cradle to Cradle Certified™ Product Program 262
– declare label 261
– defined 243
– endocrine disrupting chemical used in lining 269
– estimated cost and time 257
– GC3 joint statement on green chemistry/safer alternatives 238
– GHS hazard classifications 245
– globally harmonized system of classification and labelling 244
– GreenScreen for safer chemicals 248, 250
– GreenScreen list translator (GS LT) 253
– hazard endpoints 248
– health product declaration 260
– Health Product Declaration Version 2.0 (HPD) 259
– Living Product Challenge™ (LPC) 262
– nonprofit organization (NPO) 235
– quick chemical assessment tool (QCAT) 252
– red list declare label 259
– retailer initiatives 237, 239
– state initiatives 240
– substitution and regrettable 233
– United States Environmental Protection Agency 260, 264
– uses 255
– worldwide GHS implementation 246
chemical intensification 180
Chemical of High Concern (CoHC) 258
chemical reactions 10, 12, 81
chemical risk assessment 45
chemical selectivity 10
chemicals legislation 6
Chemicals Management Software 238
chemical transformations 2, 12
chemists, integrating sustainable/green chemistry 281
chemoenzymatic conversions 208
chemoselectivity 10

Children's Safe Products Act (CSPA) 240
chiral compounds 10
chloridehydroxylation 180
chlorine 89
CleanGredients database 266
Clean Production Action (CPA) 255
closed-paths (CP) 144
CML impact scores 173
Coca-Cola Company 96
contamination 159
continuous end-to-end pharmaceutical manufacturing 201
continuous flow 187
– microflow process 167, 195
– millireactor-based process 174
– processing 187, 200
– reactors 73, 190
continuous process 175
– advantages of 160
– production 162
– at small scale 159
– synthesis process 170
continuous tablet production 201
conventional batch API manufacturing process (AP) 177
convergent approaches 57
convergent plans 54
convergent strategy 64
costing metrics 209
Cradle to Cradle Products Innovation Institute (C2CPII) 237, 262
cradle-to-gate life cycle 109
crude batch/continuous operation, ecological profile comparison of 175
cryogenic system 167
crystallization 177
cumene hydroperoxide 75
cumene oxidation 88
cumulative energy demand (CED) 108, 166, 168, 172
– demand 166
– impact 166
– for microreactor processing 167
current continuous processing (CP1) 175
4-cyanophenylboronic acid 197
cyclohexane 187
cyclohexene 187
– production 191

d

database algorithms 49
data gap (DG) 7, 248
data transparency 101
DEAM 98

Declare Assessment 259
Department of Energy (DOE) 288
Department of Toxic Substances Control (DTSC) 267
Design for the Environment (DfE) Program 246
Design Institute for Physical Property Data (DIPPR) database 84
DfE AA method 251
dichloromethane 181
di(2-ethylhexyl)phthalate (DEHP) 234
2,6-difluorobenzyl alcohol 186
2,6-difluorobenzyl azide 180, 181, 183
dihydroartemisinin (DHA) 171
DINP's Prop 65 234
1,3-dipolar Huisgen cycloaddition 181
dipolarophile 185
drug discovery 157

e

eco-efficiency 178, 179
– alternative 115
– assessment 114
EcoIndicator 99 method 95
Eco-Invent method 98
ecological niche 3
economic analysis 195
economic benefits 200
economic competitiveness 162
economic evaluation 170
economic improvements 162
Eco-Profiles Web site 109
ecorepel® 269
ecosystem 3
ecotoxicity 7
– hazard data 8
E-factors 12, 56, 77, 162, 223
effective mass yield 13
EI99 scores 107
electricity demand 167
Ellen MacArthur Foundation 236
(E)-methyl-3-methoxyacrylate (EMMA) 181, 182
enantiomers 10
endocrine disruption 81
energy consumption 87, 158, 167, 168, 175
energy demand 169
energy efficiency 3, 157
energy input consumption 56
engineering metrics 6
enterprise resource planning (ERP) 138
enthalpic contribution 86
enthalpy 84
environmental burdens 178

environmental chemistry 1
environmental efficiency 178
environmental factor (E-factor) 162
environmental, health, and safety (EHS) 35
Environmental Health Perspectives (EHP) 270
environmental indicator 149
Environmental Product Declarations (EPD) 97
environmental protection 179
– technology 31
Environment, Safer Choice program 16
enzymatic pathways 44
enzyme modification 209
EPISuite^{TM} 37
– toolbox 35
Escherichia coli 142, 221
ethanol 171
etherification 172
European Union's Classification and Labelling (CLP) 268
eutrophication potential 3
explosivity 7
– strength 79
– vapor 79

f

fermentation 142
– operation 144
– processes 207, 211
final products profiles 152
flammability 7
FLASC^{TM} tool 105, 109
flash point 7, 81
flowsheet decomposition 144
fossil-based plastics 109
fossil fuels 96
Frank Lautenberg Chemical Safety Act of 2016 6, 16
fresh water
– aquatic ecotoxicity potential 169
– sediment ecotoxicity potential 169
furfural 6

g

GABI 98
gaseous hydrochloric acid 88
gastrointestinal tract (GI) 41
Gauging Reaction Effectiveness for the Environmental Sustainability of Chemistries with a multi-Objective Process Evaluator (GREENSCOPE) 34
– sustainability indicators 35
generic convergent process, consisting steps 55
generic linear process steps 54

GlaxoSmithKline (GSK) 105, 110, 116
– methodology 106
– Solvent Selection Guide 110
Globally Harmonized System (GHS) of Classification and Labeling of Chemicals 244
global warming 7, 79, 96, 109, 115, 140, 198
global warming potential (GWP) 107, 108, 168, 185, 199
glucose dehydrogenase (GDH) 218
– recycle reaction 218
green chemistry 1, 2, 4, 6, 9, 11, 14, 17, 29, 32, 49, 90, 157-159, 174, 194, 200, 231
– ACS-GCI Roundtable 160
– to assess microflow processing metrics 163
– chemicals 194
– – manufacturing processes 209
– metrics 162, 214, 217, 285, 289
– – evolution of 11
– safer chemical design, life cycle thinking 30
Green Chemistry and Commerce Council (GC3) 236
Green Chemistry Institute (GCI) 157
green engineering 29, 157-159, 174, 194, 200
– safer chemical design
– – life cycle thinking 30
greener chemistry 12
greenhouse gas (GHG) 109
– emissions 199
green metrics 158, 200
– analysis 49
– for direct microflow route and two-step conventional routes 193
greenness 44, 53
– measuring with LCI/A applications 103
– – chemical route comparison 106
– – footprinting 115
– – material assessment 109
– – probing case studies 103
– – product LCAs 112
Greenpeace's Detox Campaign 235
GreenScreen 7, 246, 248
– assessments 248, 251
– list translator 255, 263
– – scores 254
– for safer chemicals 249

h

halohydrin dehalogenase (HHDH) 218
hazard 9
– assessment 7, 246
– data 7
– reduction 46

– solvents 42
– substances 157
– waste 3
Hazard Identification and Ranking System (HIRA) 134
2-(H-benzotriazole-1-yl)-1,1,3,3-tetramethyluronium tetrafluoroborate (TBTU) 61, 73
Health Product Declaration® (HPD) 259
Healthy Building Network's Pharos Assessment Tool 254
high-performance liquid chromatography (HPLC) 287
high-temperature flow chemistry protocol 187
high-throughput screening (HTS) 37
Hock oxidation-rearrangement 77
H_2O_2 production 190, 191
H_2O_2 synthesis 194
hotspots identification 32, 96
Huisgen cycloaddition 181, 182
– dipolarophile 185
human toxicity potential (HTP) 108, 168, 169, 185
hydrogenation catalyst 177
hydrogen gas generation (HG) 79
hydrogen peroxide 88, 187, 190
hydroquinone 88

i

ideal material efficiency conditions 53
impact sensitivity 79
incineration 112
Indicator Sensitivity Analysis (ISA) 142
– ISA 95 132
indigo production 114
ingestion toxicity 79
inhalation toxicity 79
Inherently Safer Design (ISD) 135
Inherent Safety Index (ISI) 134
in'situ product removal (ISPR) 218
Integrated Inherent Safety Index (IISI) 135
Integrated Risk Information System (IRIS) 35
intensification, process alternatives for 174
intensified chemistry 187
International Agency for Research on Cancer (IARC) 253
International Conference on Chemicals Management (ICCM) 282
International Living Future Institute (ILFI) 237, 259, 262
International Standards Organization (ISO) 32, 132
– 14000 series (14040, 14044) 97

– standards 101
International Sustainable Chemistry Collaborative Centers (ISC3) 283
Interstate Chemicals Clearinghouse (IC2) 241
ionic liquids 168, 169
iridium 2, 3
isobutyl boronic acid 62, 63
isoprene 289
isosteres 42

k

kernel 74
ketoreductase (KRED) 218
key input material (KIM) 179, 196
key output material (KOM) 175
– recovery 178
Kolbe–Schmitt synthesis 108

l

LCIA categories, reduction 183
Leadership in Energy and Environmental Design (LEED) 249
Lennox-Gastaut syndrome 180
Lewis acid 60
life cycle approach 95
life cycle assessment (LCA) 32, 107, 132, 158, 162, 163, 171, 209, 232
– assessments of formulated products 113
– assess microflow processing 163
– complete picture 191
– components of 33
– for continuous flow 190
– data 115
– evolution of 96
– goal/scope definition 33
– methodology
– – goal and scope 98
– – impact assessment 99
– – interpretation 99
– – inventory analysis 98
– pharmaceutical product, cradle-to-gate 113
– simplified LCA (SLCA) 163, 168
– streamlined 102
– study for two-reactor network process designs 184
– treatment options 112
– for two-step conventional synthesis 191
life cycle costing (LCC) 195
– analysis 159
– applications with continuous microflow processing, snapshot on 196
– chemical production processes 198
– composition of 195

life cycle evaluation 95
life cycle inventory and assessment (LCI/A) 3, 33, 34, 95, 101, 103, 184, 187
– application 104
– batch/continuous multistep processing 185
– cradle-to-gate 106
– data 95
– databases 99
– impact categories 9
– limitations 100
– as measure of greenness 103
– methodologies 101
– score 110
life cycle practice, historical evolution 97
lipophilic compounds 43
liquid-liquid extraction 190
List Translator-1 chemicals (LT-1's) 258
lithium diisopropylamide (LDA) 60
lithium hexamethyldisilazane (LiHMDS) 61

m

macro-economic factors 100
macroscale reactors 161
manufacturing restricted substances list (MRSL) 267
marine aquatic ecotoxicity potential 169
marine sediment ecotoxicity potential 169
market competitiveness 127
mass efficiency 16, 194
mass intensity (MI) 15, 16, 158, 162
mass productivity 16, 158, 162
material efficiency 53, 115
– analysis for synthesis plans 50
– metrics 52
Material Health Certificate (MHC) 262
material performances 72
material recovery parameter (MRP) 52
material selection 110
material value-added (MVA) 144
maximal State-Task Network (mSTN) 128
metal oxides 88
methane 191
methanol 171
methyl 1-(2,6-difluorobenzyl)-1H-1,2,3-triazole-4-carboxylate 184, 186, 187
microchemical process 197
micro-economic factors 100
microflow
– processing 199
– production 200
– reactors 161
– synthesis in interdigital mixer 168
– technology 200

micromixer 168
microprocess engineering 197
microreaction technology-based processes 167, 199
microreactors 107, 166, 167, 171
– flow networks 186
microstructured reactors 197
Millennium Pharmaceuticals' process 64, 65, 67, 68
– bortezomib process 59-61, 62
– – global efficiency metrics for 69
– – intrinsic efficiency metrics 67
– global metrics analysis for 71
– linear process 69
miniaturization 161
miniflow technology 161
Minnesota Department of Health (MDH) 240
Minnesota Pollution Control Agency (MPCA) 240
mixed-integer linear programming (MILP) 128
mixed-integer nonlinear programming (MINLP) 128
Mixed-Logic Dynamic Optimization 129
molecular initiating events (MIE) 37
multifacetted process optimization *versus* process intensification 174
multiobjective genetic algorithm (MOGA)
– fuzzy logic concepts 130
multiparameter variations 179
multistage synthetic route 12
multistep microreactor flow network 181
multivariate approach to assessment 18

n
naphthalene 75
National Fire Protection Association (NFPA) 163
– rating 181
National Retail Federation 238
N-Boc-L-phenylalanine 61
net present value (NPV) 130, 196
nitrobenzene 180
nitrogen gas 88
nitrous oxide 88
N,N'-carbonyldiimidazole (CDI) 63
N,O-bis(trimethylsilyl)-acetamide (BSA) 63
nondominating sorting genetic algorithm (NSGA-II) 130
nonenzymatic process 44
nonrenewable resource 4
– conservation 282
– depletion 126
normal-boundary intersection (NBI) 130

nosocomial infections 106
Novozymes 112
NOx emissions 108
N scaling factors 51

o
occupational exposure limit 79
1-octadecyl-3-methylimidazolium bromide 167
Office of Environmental Health Hazard Assessment (OEHHA) 234
Office of Research and Development (ORD) 38
operation time factor (OTF) 144
optimal batch plant topology 149
optimal plant scheduling 150
optimization
– process alternatives for 174
– strategies 49
OptimRetHeat 131
Oregon Toxic-Free Kids Act 241
organic solvents 16, 178
Organization for Economic Cooperation and Development (OECD) 35, 282, 283
ortho-phthalate plasticizers 233
– diisononyl phthalate (DINP) 234
oxidant 88
oxidative process 89
1-oxybenzotriazole unit 73
oxygen balance 79
o-xylene 75
ozone-depleting 4, 7, 79
– potential 3, 108

p
Pareto nondominant solution 130
– ε-constraint method 131
path flow decomposition approach 129
peptide coupling reagents 73
peptide-like coupling reactions 73
perfluorooctanesulfonic acid (PFOS) 268
perfluorooctanoic acid (PFOA) 268
persistent, bioaccumulative, and toxic (PBT) 232, 258
petrochemical process 12, 107
PFC-free replacement technology 269
pH adjustment 13
pharmacodynamics 44
Pharma-Sintez bortezomib process 63, 64, 66, 69, 72
– convergent bortezomib route 65
– global efficiency metrics for 69
– global metrics analysis for 71
– – individual steps 70

– intrinsic efficiency metrics for 67
Pharma-Sintez route 72
Pharos Project Chemical Screen Library (Pharos) 259
phase transfer catalysis 168-170
phase transition 84, 86
phenol 56, 75, 88, 89, 168
– from glucose, biotechnological route 89
– plans, synthesis tree representation 80
– production 88
– syntheses
– – BI(waste) results 81
– – energy consumption data for 86
– – global E-factor breakdown for 80
– – global material efficiency metrics for 80
– – plans 76
– toxicity potentials, biphasic reaction of 169
phenylalanine 57
L-phenylalanine 73
phenyl benzoate synthesis 167, 168
– cost evaluation 167
phosphates reduce water hardness 244
photochemical ozone creation potential (POCP) 109
photo oxidant creation potential 108
phthalate plasticizers 244
pinanediol boronic ester 62
plasmodium falciparum 171
pollution prevention 96
polyvinyl chloride (PVC) 234
– plastics 232
pot economy 13
pregablin 105
preliminary cost analysis 197
Prezista 201
Pr_2NEt 73
process characteristics
– of considered batch scenarios 176
– of continuous process 177
process-design intensification 161
process intensification tools 194
process mass intensity (PMI) 15, 158, 162
process optimization opportunities 50
process safety 160
process sensitivity analysis 146
process simulation 188
– and CAPEX cost study 188
Product Category Rules (PCR) 97, 110
product diversity 159
production cost breakdown 196
product quality 159, 160
Program for Assisting the Replacement of Industrial Solvents (PARIS)

– user-friendly graphical user interface 42
Prop 65-listed substances
– into drinking water sources 240
proteasome 56
purification 142
– protocols 73
2-pyrazine carboxylic acid 62

q

quantitative and other structure activity relationship (Q/SAR) models 251
quantitative structure-activity relationship (QSAR) 18, 35
– methodologies 38
– modeling 38, 249
Quick Chemical Assessment Tool (QCAT) 252

r

raw material (RM) 148
raw materials consumption 194
reaction mass efficiency (RME) 14, 49, 50, 158
– RME_{kernel} values 72
reaction network strategies
– to synthesize bortezomib 59
– to synthesize pyrazine 58
– variety of 58
REACTION spreadsheets 51, 56
reaction yield 51
ReCiPe 132
recycling, role of 171
Redlich–Kwong equation 86
Registration, Evaluation, Authorization and Restriction of Chemicals (REACH) 16
– legislation 7
regrettable substitution 234, 268
renewable bio-based feedstocks 207
renewable catalyst 212
renewable energy 194
restricted substances list (RSL) 267
Retailer Regulation 237-239
reverse Lipinski 41
risk reduction strategy 41
rufinamide process 180, 181
– routes, comparison of 182

s

safer chemical design 29
– alternatives assessment (AA) 45
– attributes and measurement metrics 39
– chemicals attributes
– – good character 36
– – potency/efficacy, maximize uses 40

– – tools for characterizing 37
– economic efficiency 40
– hazardous functional groups, minimize the incorporation 42
– life cycle assessment (LCA) 31, 32
– life cycle, stages 33
– life cycle thinking 30, 31
– limited bioavailability 41
– limited environmental mobility 41
– limited persistence/bioaccumulation 43
– pH 44
– protocol for approaching 45
– quick transformation to innocuous products 44
– risk assessment 45
– seven virtues and four turpitudes of 36
– strive to reduce/eliminate the use 40
– toxicity 41, 42
– training 46
Safer Chemical Ingredients List (SCIL) 261, 265
safer chemicals, challenge 29
Safer Choice
– on CleanGredients 265
– criteria 261, 266
– logo 267
– program 264-266
safety–hazard constraints 89
safety–hazard impact 56
safety–hazard index (SHI) 74, 79
Sanofi Company 174
scaling factors 51, 53, 56
SciVeraLENS 246
SEEBalance® 113
selectivity 187, 199
sertraline 105
shikimic acid 89
SimaPro 98
simulated annealing (SA) 130
single innovation drivers 171
single-step analyses in batch and flow 181
sitagliptin
– antidiabetic compound 219
– biocatalytic synthesis 220
skin dose 79
skin sensitization 232
small scale reactors 161
smog formation 79
social indicators 134
societal values 100
sodium borohydride 171
sodium hydroxide 75
sodium tripolyphosphate (STPP) 244
solvent free methodologies 90

solvent load 200
– improvements 187
solvent rate 162
solvent selection 110
solvent swap
– to recover product 222
space-time yield 208
state-of-the-art batch process 136
State-Task Network (STN) 147
– representation, product recipe 148
Strategic Approach to International Chemicals Management (SAICM) 282, 283
streamlined life cycle assessment 102
sulfonic acid 89
sulfur dioxide 181
SuperPro Design
– insulin process, flowsheet of 143
sustainability 17, 30, 162
– assessment 287
– batch design (see sustainable batch design (SBD))
– of biocatalytic processes 207
– chemical process evaluation-metrics 34
– chemistry 282, 283
– development 281, 282
– metrics 31
– molecular design 31
– procurement 241
– safer chemical design, life cycle thinking 30
– toolbox 274
sustainable batch design (SBD) 7, 125
– batch design 125
– – assessment 131
– – challenges 126
– – characteristics 125
– – design and retrofit 127
– – process sustainability 126
– – State of the Art 126
– framework (SBD-FRAME) 126
– – assessment 147
– – batch indicators for insulin production 145
– – business planning 138
– – case studies 142
– – compound specification 137
– – design of batch process 147
– – economic assessment 136, 138
– – environmental assessment 135, 139
– – fermentation 144
– – flow diagram 137
– – integration task characterization 148
– – ISA 95 framework 132, 133
– – mass indicators for insulin production 145
– – methodologies 141

– – primary recovery 145
– – process recipe 138
– – purification 146
– – retrofit 142
– – retrofitting 136
– – scheduling 138
– – series of reactions 145
– – social assessment 140
– – waste recovery/disposal 138
SustainPro tool 131, 142
symmetric fuzzy linear programming (SFLP) 130

t
Tait equation 86
tegretol 113
terrestrial ecotoxicity potential 108, 169
tetrahymena pyriformis IGC$_{50}$ 38
tetralone 105
tetramethyl bisphenol F (TMBPF) 270
The Society of Environmental Toxicology and Chemistry 97
thionyl chloride 181
three-reaction network process designs 184
TidalVision USA 242
titanium silicates 88
toluene 5, 75
toluene oxidation 76, 88
Tool for the Reduction and Assessment of Chemical and other environmental Impacts (TRACI) 35
total cradle-to-gate energy consumption 106
toxicity 29, 41
– hazard profile 37
Toxicity Estimation Software Tool (TEST) 35, 38
toxicity pathways 44
toxicity potentials 168
toxicological sciences 30
toxicology tests 35
toxicophores 42, 45
Toxic Substances Control Act (TSCA) 6
ToxServices' Full Materials Disclosure^{TM} (ToxFMD^{TM}) Program 262
trade-offs 96, 100
transformation 179
– equipment flowsheet 144

trifluoroacetic acid (TFA) 61
Trp-LE'-MET-proinsulin 142
true greenness of a process 95
two-liquid phase biocatalysis 223

u
UMBERTO 98
United Kingdom's National Health System's "Green House Gas 115
α,β-unsaturated ketones 81, 82
US EPA Safer Choice Program 260
US National Academy of Science 18

v
Valspar's Safety 271
vanadium pentoxide 82
vancomycin hydrochloride 106
vector magnitude ratio (VMR) 83
– scores 83
Ventolin 116
very persistent or very bioaccumulative in the environment (vPvB) 232

w
Washington Department of Ecology (WA DOE) 252
waste disposal 199
waste materials 81
Waste Reduction Algorithm (WAR) 35, 132
wastewater treatment plants 13
water intensity (WI) 217
water reuse optimization 128
World Summit on Sustainable Development 283

x
xylene 5
– p-xylene 289

z
zeolites 88
zero discharge of toxic chemicals (ZDHC) program 236
zinc chloride 60
Z-isomeric compound 175
Zoloft$^®$ 105